# Studies in Systems, Decision and Control

Volume 639

The series "Studies in Systems, Decision and Control" (SSDC) covers both new developments and advances, as well as the state of the art, in the various areas of broadly perceived systems, decision making and control–quickly, up to date and with a high quality. The intent is to cover the theory, applications, and perspectives on the state of the art and future developments relevant to systems, decision making, control, complex processes and related areas, as embedded in the fields of engineering, computer science, physics, economics, social and life sciences, as well as the paradigms and methodologies behind them. The series contains monographs, textbooks, lecture notes and edited volumes in systems, decision making and control spanning the areas of Cyber-Physical Systems, Autonomous Systems, Sensor Networks, Control Systems, Energy Systems, Automotive Systems, Biological Systems, Vehicular Networking and Connected Vehicles, Aerospace Systems, Automation, Manufacturing, Smart Grids, Nonlinear Systems, Power Systems, Robotics, Social Systems, Economic Systems and other. Of particular value to both the contributors and the readership are the short publication timeframe and the world-wide distribution and exposure which enable both a wide and rapid dissemination of research output.

Indexed by SCOPUS, DBLP, WTI Frankfurt eG, zbMATH, SCImago.

All books published in the series are submitted for consideration in Web of Science.

Alla G. Kravets · Alexander A. Bolshakov
Editors

# Human-Centric Cyber-Society

Springer

*Editors*
Alla G. Kravets
CAD&RD
Volgograd State Technical University
Volgograd, Russia

Alexander A. Bolshakov
Peter the Great St. Petersburg Polytechnic University
St. Petersburg, Russia

ISSN 2198-4182 ISSN 2198-4190 (electronic)
Studies in Systems, Decision and Control
ISBN 978-3-032-06786-9 ISBN 978-3-032-06787-6 (eBook)
https://doi.org/10.1007/978-3-032-06787-6



This Springer imprint is published by the registered company Springer Nature Switzerland AG
The registered company address is: Gewerbestrasse 11, 6330 Cham, Switzerland

# Preface

The book *Human-Centric Cyber-Society* embarks upon an extensive examination of emergent phenomena and breakthroughs reshaping contemporary society through a multi-disciplinary lens, focusing particularly on cybernetic engineering, ecological sustainability, personalized medicine, and immersive digital technologies. Structured into four distinct sections, the book engages critically with foundational questions concerning the evolution of human-centric societal constructs, sustainable ecological paradigms, precision-focused healthcare modalities, and experiential innovations in interactive media.

Initially, we probe into organizational configurations geared toward fostering adaptability in industrial ecosystems. Central themes encompass formulating principles of human-centric governance in flexible enterprises, implementing digital configuration controls for optimized systemic performance, employing Haken's paradigm to evaluate regional socioeconomic indices, investigating the interplay of demand/cost functions in oligopolistic markets, devising a genetic-algorithm-informed approach to airline crew scheduling, and constructing ontological models for managing cargo port operational risks.

Subsequently, the discourse shifts to environmental sustainability, wherein methodological advances are expounded upon. These include deploying multivariate statistical tools for forecasting cereal crop yields, modeling fluid-dynamic interplays in hydrological contexts, applying machine learning-assisted techniques to deduce fluvial flow characteristics, simulating particulate transport in atmospheric boundary layers, and quantifying secondary pollutants resulting from varied automotive emissions and traffic flows.

Next, attention turns to transformative biomedical informatics. Machine learning-aided algorithms for identifying breast-cancer susceptibility, image-processing-based evaluation of breast thermogram asymmetries, ISIDA-framework-derived classifications of cerebral lesions, server-side implementations for remote depression screening, and robust neural-network architectures capable of overcoming noisy datasets in sign-language recognition constitute some salient highlights.

Ultimately, the treatise concludes with an excursion into immersive extended-reality domains. Novel level-generation protocols for solo-play three-dimensional games, scenario-persona hybridizations in souls-like narratives, utilization of behavioral cues extracted from gameplay for mental health diagnostics, and fully realized virtual-reality recreations of investigatory procedures anchored in high-resolution 3D reconstructions illustrate the breadth of inquiry.

By synthesizing theoretical rigor with empirical validation, this compendium seeks to equip researchers, practitioners, and enthusiasts alike with invaluable resources pertinent to navigating the multifaceted landscapes of twenty-first-century technological advancement.

Volgograd, Russia
St. Petersburg, Russia
July 2025

Alla G. Kravets
Alexander A. Bolshakov

# Contents

# Human-Centric Society

# Formation of Human-Centricity in the Management of Flexible Industries and Organizations

**Natalya Aristova, Valentin Chadeev, and Olga Yakimova**

**Abstract** In the conditions of digitalization, the influence of the adverse human factor in the production system is minimized, and a person gets the opportunity to engage in creative activities rather than routine ones and work in more comfortable conditions, that is, in parallel with technocentricity, human-centricity is formed in industries. IPU RAS has developed a methodology for modeling the process of technological preparation of flexible production of various products (including automatic machines) in various production conditions, taking into account the probabilities of high-quality technological operations. It is noted that flexible production can be realized only in conditions of digitalization. The proposed methods and algorithms can flexibly adjust to different production conditions and help the technologist decide on the possibility and expediency of putting the product into production. As a further development of this methodology, the relevance of the scientific problem of creating a unified formalized approach to studying the process of organizing human creative work in flexible industries is substantiated. It is shown that one of the components of the formalized description is the probability of the targeted use of working time, which is proposed to be determined using the results of population surveys.

**Keywords** Self-reproduction in industry · Automaton · Human-centered · Technocentric · Flexible production · Machine learning · Probability of targeted use of working time

N. Aristova (✉) · V. Chadeev
Institute of Control Sciences (RAS), 65 Profsouznaya Str, Moscow 117997, Russia
e-mail: avtprom@ipu.ru

V. Chadeev
e-mail: chavama@ipu.ru

O. Yakimova
HSE University, 11 Myasnitskaya Str, Moscow 101000, Russia
e-mail: oyakimova@hse.ru



A. G. Kravets and A. A. Bolshakov (eds.), *Human-Centric Cyber-Society*, Studies in Systems, Decision and Control 639, https://doi.org/10.1007/978-3-032-06787-6_1

## 1 Introduction

In the second decade of the XXI century, there was talk all over the world about the beginning of the Fourth Industrial Revolution or Industry 4.0, which allows the integration into a single system all the main achievements of automation, informatization, algorithmization, modeling, data collection, and transmission, as well as developing new technologies [1]. In the digital age, the priority task of a discrete production process is to ensure its customer orientation, that is, flexibility, characterized by [2, 3]:

- the ability to produce different types of products on the same equipment,
- minimal human involvement in the production process,
- reconfiguration in a short time for the release of new products,
- production of products that are maximally tailored to the requirements of specific customers,
- the possibility of making changes to the finished product design at the client's request.

In the era of digitalization, the IPU RAS has pioneered a groundbreaking methodology for modeling the process of technological preparation of flexible production [4–8]. This innovative approach encompasses the production of all types of products in various production conditions, commencing from the stage of technological preparation of flexible discrete production.

A contemporary discrete technological process can comprise tens, hundreds, or even thousands of individual technological operations (TO) that can be executed by automata (robots, CNC machines, etc.) or a person. All types of maintenance used in the manufacturing of products, during maintenance (readjustment, loading/unloading of blanks, etc.) and maintenance of automatic machines are taken into account. Specific automata perform THAT qualitatively with a certain probability. Hence, in the context of preparing flexible automated production, it is crucial at the very initial stage to scrutinize the feasibility and efficiency of manufacturing a high-quality product using specific automata (automation efficiency). The minimum time and minimum cost of manufacturing a product are regarded as the primary criteria for production efficiency. The cost is quantified in terms of time spent [4, 5].

The estimates obtained as a result of modeling contribute to reducing the execution time of the technological preparation of production, which is especially important for flexible automated production facilities that are regularly reconfigured to produce a new product and adapt to customer requirements [6].

## 2 Methodology of Modeling the Process of Technological Preparation of Flexible Production

The methodology is based on the developed formal description of hierarchical discrete production, which does not depend on the types of maintenance, manufactured products, or automata used. The probability of high-quality maintenance by automatic machines is taken into account.

The formalized representation of discrete production includes [4]:

- *description of technological operations* using the time of completion of the TO ($b$) and the probability of high-quality performance of the TO ($p$),
- *description of automata* using parameters: the cost of the machine (C), which can be recalculated through the time spent on its manufacture; the lifetime of the machine (T); the coefficient of the unit of working time of the machine:

$$\lambda = C/T, \tag{1}$$

- *description of the procedure for manufacturing products and maintenance of automatic machines*: the number of TOs required for the manufacture of the product and the number of executions of each type of TO,
- *a model of the maintenance automation* process presented in the form of a matrix $A = \|\alpha_{ij}\|$:

$$A = \begin{pmatrix} \alpha_{10} & \alpha_{11} & \alpha_{12} & \cdots & \alpha_{1j} & \cdots & \alpha_{1f} \\ \alpha_{20} & \alpha_{21} & \alpha_{22} & \cdots & \alpha_{2j} & \cdots & \alpha_{2f} \\ \dots & \dots & \dots & \dots & \dots & \dots & \dots \\ \alpha_{i0} & \alpha_{i1} & \alpha_{i2} & \cdots & \alpha_{ij} & \cdots & \alpha_{if} \\ \dots & \dots & \dots & \dots & \dots & \dots & \dots \\ \alpha_{m0} & \alpha_{m1} & \alpha_{m2} & \cdots & \alpha_{mj} & \cdots & \alpha_{mf} \end{pmatrix} \tag{2}$$

where $\alpha_{ij}$ are the parameters of the automation process that determine which types of tasks are performed by one another automaton and vary within $0 \le \alpha_{ij} \le 1$. If a person performs all operations of the same type, then $\alpha_{ij} = 0$; if an automaton, then $\alpha_{ij} = 1$. The matrix contains the number of rows corresponding to the number of different types of TO ($m$) and the number of columns according to the number of performers ($f$) of these TOs (a person or a $j$–type automaton).

The task is to distribute the work between machines (as well as people) by the leading indicators:

- minimum time and minimum average production time of the product,
- the minimum cost and the minimum average cost of manufacturing the product.

In the formalized description of discrete hierarchical production, assumptions are made:

- we will consider the execution time of a particular type by a person and an automaton to be a constant (not random) value,
- the order of execution is not tanken into account.

Within the framework of the methodology, mathematical models have been developed that allow us to obtain estimates of the following production indicators [4–8]:

- the cost and the average cost of manufacturing the product,
- time and the average production time of the product,
- the cost of the serviced/maintenance machine to perform maintenance,
- the time spent by the serviced/maintenance machine to perform maintenance,
- the coefficient of the unit of working time of the machine and its average value.

Here is a practical example of using the developed methodology to obtain an estimate of the minimum average cost of manufacturing a node, including the execution of three TOs: 1, 2, and 3. The first two TOs are performed once, and the third three times: $n1 = n2 = 1, n3 = 3$. Four types of automata R1, R2, R3, and R4 can be used to perform these maintenance tasks. Table 1 shows the parameters of maintenance and automata, demonstrating the real-world application of our methodology.

*The model for estimating the cost of manufacturing a product by a k automaton system* has the form:

$$C_k(A) = \sum_{j=0}^{f} \lambda_j \sum_{i=1}^{m} \alpha_{ijk} b_{ij} n_{ik}, \tag{3}$$

where $b_{ij}$ is the execution time of the *i*-th type by an automaton of the *j*-th type,

$\alpha_{ijk}$ is a parameter of the automation process of the i–th type by a *j*-type automaton in the manufacture of a *k*-type product,

$n_{ik}$ is the number of identical items of the i-th type required for the manufacture of one copy of the *k*-th type product,

$\lambda_j$ is the coefficient of the unit of working time of the *j*-type automaton,

$f$ is the number of different types of automata in the system,

$m$ is the number of types of maintenance necessary for the manufacture of the entire product system.

*The model for estimating the probability of high-quality manufacturing of a product k* has the form:

**Table 1** Parameters of node execution by automata

| | R1, $\lambda 1 = 0.21$ | | R2, $\lambda 2 = 0.20$ | | R3, $\lambda 3 = 0.22$ | | R4, $\lambda 4 = 0.21$ | |
|---|---|---|---|---|---|---|---|---|
| TO | *b* | *p* | *b* | *p* | *b* | *p* | *b* | *p* |
| 1 | 4 | 0.999 | 4 | 0.999 | 5 | 0.997 | 6 | 0.999 |
| 2 | 5 | 0.998 | 6 | 0.998 | 5 | 0.997 | 5 | 0.999 |
| 3 | 10 | 0.999 | 10 | 0.99 | 9 | 0.999 | 9.5 | 0,997 |

$$P_k(A) = \prod_{j=0}^{f} \prod_{i=1}^{m} p_{ij}^{\alpha_{ijk} n_{ik}} \tag{4}$$

where $p_{ij}$ is the probability of high–quality TO of the *i*-type by a *j*-type automaton.

*The model for estimating* the average cost of manufacturing a product k has the form:

$$\Phi_k(A) = C_k(A)/P_k(A) \tag{5}$$

An important property of each of these models is the possibility of converting them to a matrix form, where the rows correspond to different types of TO performed, and the columns correspond to different types of R1…Rf automata.

For example, let's write down a matrix representation of the model for estimating the average cost of manufacturing a product by a system of automatic machines:

| TO | | R1 | R2 | … | R*f* |
|---|---|---|---|---|---|
| 1 | $\Phi_k(A)$ | $\frac{\alpha_{10} b_{10} \lambda_0 n_1}{p_{10}{}^{\wedge}\alpha_{10} n_1} +$ | $+\frac{\alpha_{11} b_{11} \lambda_1 n_1}{p_{11}{}^{\wedge}\alpha_{11} n_1} +$ | … | $+\frac{\alpha_{1f} b_{1f} \lambda_f n_1}{p_{1f}{}^{\wedge}\alpha_{1f} n_1} +$ |
| 2 | | $+\frac{\alpha_{20} b_{20} \lambda_0 n_2}{p_{20}{}^{\wedge}\alpha_{20} n_2} +$ | $+\frac{\alpha_{21} b_{21} \lambda_1 n_2}{p_{21}{}^{\wedge}\alpha_{21} n_2} +$ | … | $+\frac{\alpha_{2f} b_{2f} \lambda_f n_2}{p_{2f}{}^{\wedge}\alpha_{2f} n_2} +$ |
| … | | … | … | … | … |
| *m* | | $+\frac{\alpha_{m0} b_{m0} \lambda_0 n_m}{p_{m0}{}^{\wedge}\alpha_{m0} n_m} +$ | $\frac{\alpha_{m1} b_{m1} \lambda_1 n_m}{p_{m1}{}^{\wedge}\alpha_{m1} n_m} +$ | … | $+\frac{\alpha_{mf} b_{mf} \lambda_f n_m}{p_{mf}{}^{\wedge}\alpha_{mf} n_m}$ |

Let's build a matrix representation of the model for estimating the average cost of manufacturing a product by a system of automata for the example under consideration:

| 1 | 2 | 3 | 4 | 5 | 6 | 7 |
|---|---|---|---|---|---|---|
| TO | R1, λ1=0,21 | R2, λ2=0,20 | R3, λ3=0,22 | R4, λ4=0,21 | R | $\Phi_k(A)$ |
| 1 | c11=0,841 | c12=0,800 | c13=1,103 | c14=1,261 | R2 | 7,81 |
| 2 | c21=1,052 | c22=1,202 | c23=1,103 | c24=1,051 | R4 | |
| 3 | c31=6,319 | c32=6,183 | c33=5,958 | c34=6,039 | R3 | |

The algorithm for working with the matrix representation of the model. In the first stage, we enter in columns 2, 3, 4, and 5 the values calculated according to (6) of the average cost of performing each of the three TOs by machines R1, R2, R3, and R4, respectively. In each row of the matrix representation, we look for the minimum average cost of performing each TOs. The minimum values are highlighted in color. In column 6, we write out the machines that perform these TOs with the minimum average cost. In the last column, we get the sum of the minimum cost values for performing TOs by the R2, R4, and R3 automata.

Thus, thanks to the matrix representation of the model for estimating the average cost of manufacturing a product and the proposed algorithm, performers performing TOs with a minimum average cost were identified promptly, without going through all possible options, and an estimate of the average cost of manufacturing a given node was obtained.

The developed methods and algorithms are implemented in the form of an original software package, which is characterized by the ability to flexibly adjust to a given type of discrete production, to the range of products, the fleet of automatic machines, and the TOs performed by them. The software package is flexibly configured for various production conditions; it provides the technologist with a decision support system on the possibility and expediency of putting the product into production. The software package is implemented in the Delphi object-oriented environment and is designed for integration with other technological production systems.

## 3 Organization of Human Creative Work in the Context of Digitalization

Introducing digital technologies in production is not about replacing humans but enhancing their capabilities and competitiveness [9]. Some of the functions traditionally performed by humans are now handled by automation and digitalization tools. However, this shift doesn't exclude humans from the production circuit. Instead, it empowers them to perform intellectual-level functions [10]. This transition presents two aspects of human employment. Firstly, there are process operators who manage all automation and digitalization tools and systems for flexible production. These operators are not being replaced, but their roles are evolving. They need to be trained, constantly improve their skills and monitor their activities. Industrial simulators are actively developing in this direction, making it possible to assess the personnel training level [11–13].

Secondly, a person can leave production directly but be engaged in design, programming, logistics, etc., and perform intellectual work. For such categories of employees, the availability of personal computers, various gadgets, cloud technologies, data collection, and processing technologies, and information security tools opens up the opportunity to work not only in the office but to perform some functions remotely or according to a flexible schedule.

To date, flexible working hours are used in the IT sector, marketing, design, engineering organizations, and even in individual industries (at the level of management personnel). In 2022, 2.7% of our country's population worked remotely in Russia. A high level of education characterizes the vast majority of remote workers (74%), and they are highly qualified specialists (https://rosstat.gov.ru).

Among the advantages of a flexible work schedule, employees note significant time savings on the way to the office, the ability to allocate personal and work time productively, increased efficiency, and less fatigue. Employers also benefit, noting savings on utility bills and the need to organize workplaces. However, there is a mutual concern about the productive use of working hours. This raises the question of how to control employees who work remotely effectively. Despite this concern, the benefits of flexible work schedules are clear and should be a source of optimism for the future of work.

For these purposes, personnel management develops various qualitative and quantitative indicators for evaluating personnel engaged in intellectual activity [14]. The complexity of the task is related to the probabilistic nature of forecasts of employee behavior, taking into account the complex dynamic, physical, and mental processes of the individual. In this regard, a unified methodology is needed to effectively and reliably assess the effectiveness of work in the intellectual field [15–18].

Given the rapid changes in the nature of work, the scientific problem of creating a unified formalized approach for studying the process of organizing work schedules, including remote ones, of employees from the intellectual sphere of activity, is becoming increasingly relevant. This approach should consider the probabilistic indicator of the target use of working time, providing a comprehensive and reliable assessment of work effectiveness. The urgency of this issue cannot be overstated, as it is crucial for the future of work.

To solve this problem, it is proposed that the methodology basics be used to model the technological preparation of flexible production. This is possible because the methodology describes automata and the person involved in the production process.

A formalized description of an intellectual task performed by a person (an industrial operator or an office worker) will generally include the time of completion of the task ($b$) and the probability of targeted use of working time ($p$).

## 4 Estimation of the Probabilistic Indicator of the Target Use of Working Time

To determine the probabilistic indicator of the target use of working time by an employee, it is proposed to use machine learning approaches and databases formed based on monitoring surveys of the population (blocks of variables related to intellectual labor and employment):

- RLMS-HSE is the only non-governmental monitoring of the socio-economic situation and health status of the population of the Russian Federation in Russia, carried out by the Higher School of Economics (conducted annually from 1994 to the present). The database contains more than 500 thousand records; that is, more than 15 thousand respondents participate in surveys annually,
- ESS—European Social Research, an academic cross-country project involving more than 30 European countries (held every two years from 2001 to the present).

To ensure the utmost accuracy and reliability, meticulous methods and algorithms will be created. These will be specifically designed to identify and extract data that characterizes a person's ability to responsibly approach work remotely. Subsequently, a variety of machine-learning approaches will be employed to build models. These models will determine the probability of targeted use of working time by a person, taking into account specified personal, motivational, and other parameters. The correctness of the trained model will be rigorously checked based on the results of surveys conducted since 2021, which include sections about working according to a remote or flexible schedule. The use of different approaches to machine learning will enable us to select the model that most accurately solves the task.

Thorough and meticulous preparatory work was undertaken to delve into the depths of this voluminous database. To illustrate the proposed idea, a comprehensive test study was meticulously performed on a small fragment of a database of 69 records provided by a recruitment agency that selected designers for the project organization.

The recruitment agency's task was to select specialists based on resumes posted by applicants on the HH portal, conduct interviews, and provide the employer with recommendations on candidates capable of working in a project organization, including in a flexible format. The database, rich in relevant fields, includes results (0/1)—whether the employee was hired; gender—male/female; education—higher/incomplete higher education; age; work experience; number of organizations in which the applicant worked previously; number of skills indicated in the questionnaire; proficiency in foreign languages; availability of additional education or advanced training courses; the number of specified personal characteristics; indication of the desire to develop professionally; desired salary.

The sample provided is compiled in such a way that it contains approximately the same number of employees hired and rejected by the employer. At the same time, all employees were ready for a flexible work schedule in case of employment.

To process the existing data set, logistic regression was used—a classification algorithm in machine learning to predict the probability of occurrence of some event by comparing it with a logistic curve. Regression gives an answer in the form of the probability of a binary event (1 or 0)—hiring or rejection.

In the first stage, a correlation matrix was built, which showed the maximum dependence of the positive result of hiring an employee on the parameters: the availability of additional education or advanced training courses, the number of specified personal characteristics, an indication of the desire to develop professionally, build a career. The following parameters correlate in this sample: length of service and age,

**Table 2** The study of human parameters based on the constructed model

| №. | Education | Work experience | Number of organizations, units | Availability of advanced training | Number of personal characteristics | Desire to develop | Probability of being hired |
|---|---|---|---|---|---|---|---|
| 1 | 1 | 2 | 1 | 0 | 4 | 1 | 0.43 |
| 2 | 0 | 1 | 1 | 1 | 5 | 0 | 0.66 |
| 3 | 1 | 15 | 2 | 1 | 3 | 0 | 0.91 |
| 4 | 1 | 10 | 4 | 1 | 7 | 0 | 0.97 |
| 5 | 0 | 21 | 5 | 0 | 3 | 1 | 0.91 |

age, and number of organizations. Therefore, the age parameter was excluded from further analysis.

Next, a logit model was built, which demonstrated the high accuracy and reliability of the following parameters: education, work experience; the number of organizations in which a person worked previously; the availability of additional education or advanced training courses; the number of specified personal characteristics; the desire to develop. The available database was divided in a one-to-three ratio into a training (two-thirds) and a test (one-third) sample. The accuracy of the logistic regression classifier on the test set was an impressive 0.94%. All coefficients of the model ($\beta i$) have a positive sign except the coefficient $\beta 3$, which refers to the number of organizations in which the person worked. This is quite logical—employers do not like job seekers who often change their place of work. The most informative were the same parameters as in the correlation matrix (Table 2).

A Python program was written to implement logistic regression. The table shows the program's results, showing the probability that the designer will accept an applicant with a given set of parameters into a design organization with the option of remote work.

The results obtained cannot claim to accurately describe human behavior in remote work. However, they allow us to analyze questionnaires and predict the likelihood that the applicant will satisfy the request of the organization for which the logit model was trained. In other words, the proposed approach helps to decide whether to invite a person for an interview with a specific employer.

## 5 Conclusion

In the Fourth Digital Revolution context, digitalizing flexible discrete industrial production is among the most urgent tasks. The methods and algorithms developed in this area allow online modeling of complex production situations, evaluating production performance, and helping operational personnel make decisions. The proposed mathematical models can be part of digital product twins and discrete technological processes [19–21].

Shortly, this area will be complemented by mathematical and sociological models and methods that formalize the creative activities of operational personnel. Production managers will receive tools for flexible management of automatic machines and human resources to strengthen labor discipline, loyalty to the organization, and employee motivation. The transition to human-centricity will open up new potential for flexible industries in increasing efficiency and competitiveness.

## References

1. Romero, D., Mattsson, S., Fast-Berglund, Å., Wuest, T., Gorecky, D., Stahre, J.: Digitalizing occupational health, safety and productivity for the operator 4.0. In: Proceedings of the IFIP Advances in Information and Communication Technology, Seoul, Korea, 23–26 August 2018, vol. 536, pp. 473–481. Springer: New York, NY, USA (2018)
2. Cohen, Y., Naseraldin, H., Chaudhuri, A., Pilati, F.: Assembly systems in industry 4.0 era: a road map to understand assembly 4.0. Int. J. Adv. Manufact. Technol. **105**, 4037–4054
3. Rossit, D.A., Tohme, F., Frutos, M.: An industry 4.0 approach to assembly line resequencing. Int. J. Adv. Manufact. Technol. **105**, 3619–3630 (2019)
4. Chadeev, V.M., Aristova, N.I.: Simulation of discrete manufacturer with an end product feedback. IFAC Proceed. **46**(9), 1252–1255 (2013)
5. Aristova, N.I.: Intelligence in industrial automation. Autom. Remote. Control. **77**(6), 1071–1076 (2016)
6. Chadeev, V.M., Aristova, N.I.: Automation of cluster large-scale production systems. Adv. Syst. Sci. Appl. **20**(4), 105–112 (2020)
7. Aristova, N.I., Chadeev, V.M.: A methodology of automation cost assessment under given defective products probability. Autom. Remote. Control. **79**(5), 940–946 (2018)
8. Aristova, N.I., Chadeev, V.M.: A methodology for estimating the benefits of mass production automation. Autom. Remote. Control. **79**(2), 366–371 (2018)
9. Bhandurge, G.M., Bhide, M.S.: Industry 5.0: The Convergence of AI and HI (Human Intelligence). URL: https://www.researchsquare.com/article/rs-693806/v1 (2021)
10. Longo, F., Padovano, A., Umbrello, S.: Value-oriented and ethical technology engineering in industry 5.0: A human-centric perspective for the design of the factory of the future. Appl. Sci. **10**, 4182 (2020)
11. Manca, D., et al.: Procedure for automated assessment of industrial operators. Chem. Eng. **36**, 391–396 (2014)
12. Marcano, L. et al.: Using the concept of data enclosing tunnel as an online feedback tool for simulator training. In: Procceding 59th Conference Simulation and Modelling, Oslo, Norway. Linköping Electronic Conference Proceedings, vol. 153(19), pp. 132–139 (2018)
13. Manca, D., et al.: Performance indicators for training assessment of control-room operators. Chem. Eng. Trans. **26**, 285–290 (2012)
14. Beer, H., Micheli, P., Besharov, M.: Meaning, mission, and measurement: how organizational performance measurement shapes perceptions of work as worthy. Acad. Manag. J. **65**(6), 2014–2033 (2022)
15. Gilbreth, L.M.: The Psychology of Management: The Function of the Mind in Determining, p. 360. Nabu Press, Teaching and Installing Methods of Least Waste (2011)
16. Kondo, A.: Does promoting elderly employment hurt young Japanese workers? Soc. Sci. Japan J. **20**(2), 279–285 (2017)

17. Wang, Y., Kim, S., Rafferty, A, Sanders, K.: Employees' perception of HR management practices: a critical review and further directions. Int. J. Hum. Resour. Manag. **31**(1), 128–173 (2020)
18. Yan, J., Luo, J., Jia, J., Zhong, J.: High-commitment organization and employees' job performance: the roles of the strength of the HRM system and taking charge. Int. Labor Force J. **40**(7) (2019)
19. Grieves, M.: Digital twin: developing a 21st century product model. In book: Voices of Practitioner Scholars in Management: The History and Impact of the Doctor of Management Programs at Case Western Reserve University, pp. 7–17. Orange Frazer Press (2020)
20. Grieves, M., Vickers, J.: Digital twin: mitigating unpredictable, undesirable emergent behavior in complex systems. In: Kahlen, F.-J., Flumerfelt, S., Alves, A. (eds.) Trans-Disciplinary Perspectives on System Complexity, pp. 85–114. Springer, Switzerland (2017)
21. Grieves, M.: Intelligent digital twins and the development and management of complex systems. Digit. Twin **2**(8) (2022)

# Development of Technology for Organizational System Digital Configuration Control

**Anna Khanova, Irina Bondareva, Evgenia Nesterova, Eugeny Bondarev, and Anna Muntyanova**

**Abstract** The chapter discusses the developed technology for managing the digital configuration of an organizational system, including cyber-physical systems, implemented as part of digital transformation. The technology consists of building a business model for the joint use of a digital balanced scorecard and a data envelopment analysis method. The digital balanced scorecard consists of a digital transformation strategic map and its complementary key success factors and key performance indicators. Within the framework of the described technology, it is proposed to use a two-stage scheme for the method of analyzing the operating environment, and its formal description is given. The possibility of modeling the configurations of the digital transformation strategy of an organizational system based on the use of a digital twin of the organizational system, implemented in the form of a simulation model, is described. A method for increasing the productivity of organizational systems through their digital transformation is proposed. The chapter proposes to select the configuration of the digital transformation of an organizational system by assessing each individual strategy based on real company data, as well as using forecast data obtained as a result of experiments with a simulation model. An algorithm for selecting the best configuration for the OS digital transformation strategy has been developed and described. The structural diagram of the organizational system digital transformation, including the double cycle of digital transformation, is described step by step.

**Keywords** Organizational system · Balanced scorecard · Operating environment analysis · Simulation model · System configuration · Digital transformation strategy · Key performance indicator

---

A. Khanova (✉) · I. Bondareva · E. Nesterova · E. Bondarev · A. Muntyanova
Astrakhan State Technical University, 16 Tatishcheva St, Astrakhan 414056, Russia



A. G. Kravets and A. A. Bolshakov (eds.), *Human-Centric Cyber-Society*, Studies in Systems, Decision and Control 639, https://doi.org/10.1007/978-3-032-06787-6_2

# 1 Introduction

The digital transformation of organizational systems is primarily understood as increasing productivity (accelerating business processes) (Fig. 1). To increase the productivity of organizational systems (OS), it is necessary to simultaneously improve efficiency and effectiveness [1]. If the OS is effective, but at the same time its activities are not effective, then the OS as a whole is not able to achieve its goals, and that is why it cannot implement its strategies and increase productivity.

The digital transformation of an enterprise is not just a set of activities, a disparate set of digital solutions, but, above all, a fundamental improvement of all business processes of the organization, requiring a full-fledged strategy based on new business models [2, 3]. 42% of Russian companies consider the lack of a unified strategy to be one of the key obstacles to digital transformation.

An organization may have formulated several different digital transformation strategies, and the task of choosing the most productive one becomes important. The problem is that the implementation of each strategy requires a significant investment of resources, including time, but only one strategy can be implemented in practice at the current time. At the same time, the exact effect of introducing certain digital solutions is not known in advance.

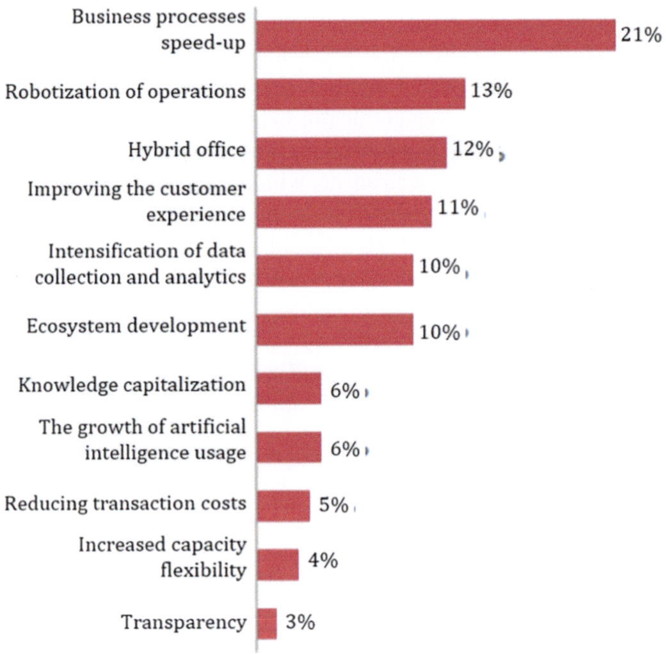


**Fig. 1** Digital transformation trends

The desire to evaluate a digital transformation strategy leads to the need to use computer models that simulate the behavior and current state of a real organizational system in a virtual environment, using real-time data from sensors and enterprise information systems to predict and optimize performance [4–7].

The use of simulation models will make it possible to simulate various configurations of the organizational system based on a variety of digital transformation strategies, consisting of the introduction of technologies, activities, changes in business process schemes, and select the best one. It is becoming relevant to develop business models, methods and algorithms as add-ons to simulation models to compare one or another strategy for the digital transformation of an organizational system.

## 2 Technology for Control of Organizational System Digital Configuration

The building blocks of the vision that make up the Digital Transformation Strategic Map (DTSM) have different systemic relationships [8]. Achieving a complex goal at a higher level of goal setting depends on achieving the control levels goals that develop the composite goals of the levels of levels. The digital transformation strategies under consideration are conducted and prioritized for implementation using the developed Digital Balanced Scorecard (DBSC), which is based on Key Performance Indicators (KPIs) and Success Factors (KSFs), each of which is responsible for one of the designated four DBSC perspectives: Business Management business management, employee/client value, digital business processes, digital transformation. KSF allows you to obtain qualitative assessments of overview characteristics, and KPI—quantitative assessments in accordance with the proposed perspectives. By comparing the obtained data (KPI and KSF values) and the data of the constructed DTSM, the hierarchical management of the DBSC can be studied. Based on real success, DBSC manages each organization of the organizational system and defines specifically for this organization KSF business management, KSF customer and employee values, KSF digital business processes, as well as KSF digital transformations.

DBSC allows for each OS configuration to assess the success and effectiveness of the digital transformation strategies under consideration. The data environment analysis (DEA) method measures the performance of an organization. These two performance factors can be considered simultaneously by combining them in an integrated BSC-DEA model [9, 10]. There are studies that have already combined the BSC and BSC methods, most of which either used the BSC methods to assess effectiveness and did not consider all the relationships between the four dimensions of the BSC [11, 12]. Using a multi-step DEA approach, the relationships and interdependencies of existing BSC perspectives can be considered, that is, in addition to calculating overall effectiveness, the performance of each BSC perspective can be calculated and the source of inefficiency can be identified.

The process of assessing effectiveness using the integrated BSC-ASF model in this case is a three-stage process. At the first stage, within each digital BSC perspective, it is proposed to allocate one input $KPI_{IN} = < KPI_{BMIN}, KPI_{CWIN}, KPI_{BPIN}, KPI_{DTIN} >$, and several output $KPI_{OUT} = < KPI_{BMOUT}, KPI_{CWOUT}, KPI_{BPOUT}, KPI_{DTOUT} >$ (see Fig. 4, stage 1). In almost all perspectives, the input indicator represents certain OS costs for various items. KPIs must be quantifiable for all strategies considered. Table 1 shows sample variables distributed across perspectives corresponding to the DBSC and DEA models. [13–15].

In the DEA model, related to the "Business Management" perspective, outputs are associated with the financial results of the operating system, while inputs are associated with various costs to ensure its functioning. The outputs considered in the DEA model of the Customer/Employee Value perspective, on the contrary, are related to the main organization goals in terms of retaining and attracting customers, as well as ensuring a positive reputation of the enterprise among stakeholders. The outputs ща DEA model Digital Business Process perspective are focused on ensuring the effective uninterrupted operation of the organization in the context of the introduction of digitalization in various processes for their subsequent improvement. In the Digital Transformation perspective, the output variables are aimed at monitoring activities carried out with the aim of consistently increasing digital capacity.

The second iteration evaluates the effectiveness of each digital BSC perspective is assessed in comparison with the prospects of other strategies (see Fig. 2, stage 2). Let there be $k$ ($k = 1,2,\ldots,l$) OS configurations of the OS digital transformation strategy that need to be assessed and compared. Let's consider $n$ $KPI_{OUTi}$ ($i = 1,2,\ldots,n$) and $m$ $KPI_{INj}$ ($j = 1,2,\ldots,m$) of some digital BSC perspective. $KPI_{OUTik}$ and $KPI_{INjk}$ correspond to the $k$-th configuration of the digital transformation strategy. We will assume that $u_{out}$ and $v_{in}$ are the weighting coefficients of the corresponding KPIs. For the DEA model, the solution is reduced to the problem of linear programming of the target configuration of the digital transformation strategy $c$ ($c = 1,2,\ldots,l$) [10]:

**Table 1** Input and output variables of DBSC-DEA models

| Prospects | Inputs, $KPI_{IN}$ | Outputs, $KPI_{OUT}$ |
|---|---|---|
| Business management | Costs of operating the organization | Profit; client profitability |
| Customer/employee value | Costs of ensuring customer/employee loyalty | % of attracted clients; % of clients satisfied with the quality of services; % of clients who made a repeat transaction; enterprise reputation |
| Digital business process | Constant | Shipment error rate; average time to process a client request; % of services performed on time |
| Digital transformation | Costs of digital transformation | Amount spent on new equipment purchases; % of corporate data digitized; % of employees who have undergone advanced training |

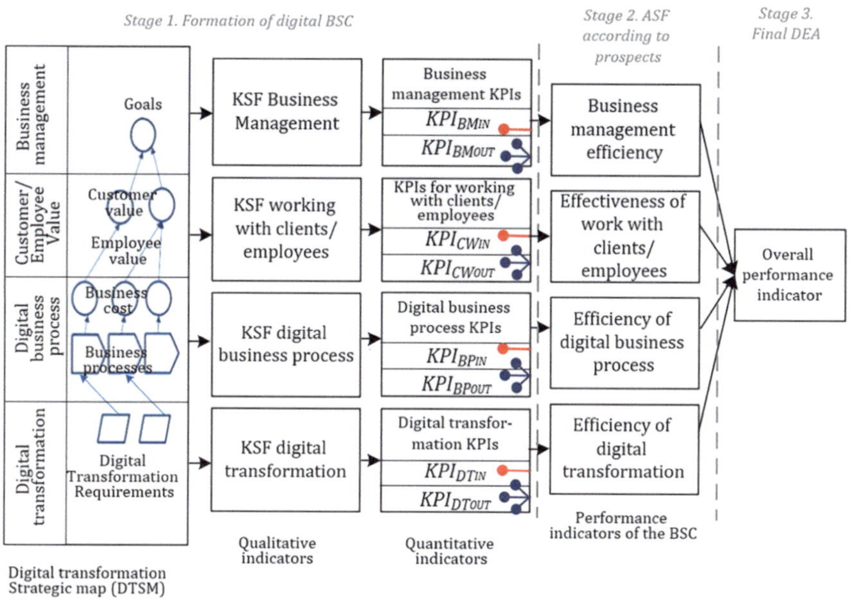


**Fig. 2** Business model for integrated use of DBSC and DEA

$$\frac{\sum_{i=1}^{n} u_{out_i} KPI_{OUT_{ic}}}{\sum_{j=1}^{n} u_{in_j} KPI_{IN_{jke}}} \to \mathbf{max} \tag{1}$$

with restrictions:

$$\frac{\sum_{i=1}^{n} u_{out_i} KPI_{OUT_{ik}}}{\sum_{j=1}^{m} v_{in_j} KPI_{IN_{jk}}} \leq 1,\ k = 1, 2, \ldots, l;$$

$$u_{out_i} \geq 0,\ i = 1, 2, \ldots, n;\ v_{in_j} \geq 0,\ j = 1, 2, \ldots, m.$$

At the end of the second iteration, we obtain four performance indicators using the DEA method, one for each perspective of the digital BSC. Considered DEA models correspond to the developed BSC perspectives and have a similar structure: each is an output-oriented DEA model with restrictions on output weights. The third stage of the evaluation process again uses the DEA approach to aggregate the performance indicators of the four digital BSC perspectives into an overall performance indicator (see Fig. 2, stage 3). This iteration also involves ranking R configurations of the digital transformation strategy of the organizational system according to the aggregate DEA indicator.

A methodology has been developed for selecting the best configuration for the OS digital transformation strategy, the algorithm of which is shown in Fig. 3.

Various OS configurations were selected as input data for the algorithm to implement the strategy. Each configuration is applied as a situation to a simulation model

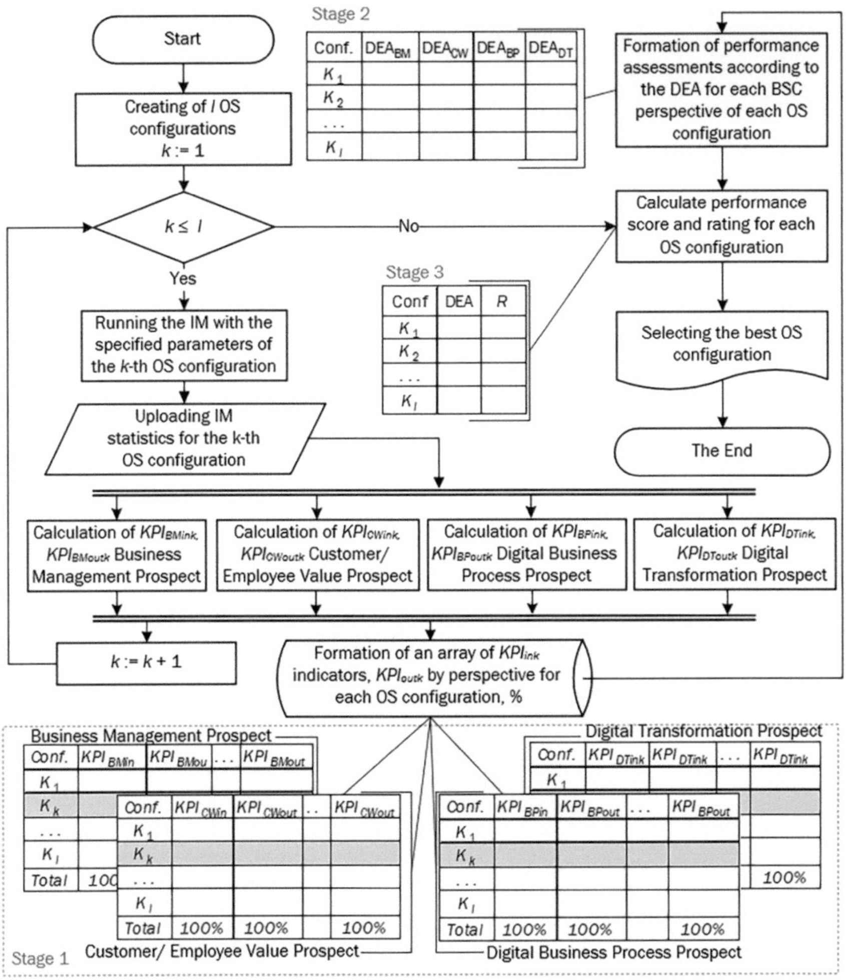


**Fig. 3** Algorithm for methodology for selecting the best OS digital transformation strategy configuration

of the OS, a series of experiments is carried out, and then statistical characteristics are downloaded based on the results of the runs. Based on the obtained statistics, $KPI_{IN}$ and $KPI_{OUT}$ are calculated for each perspective of the digital BSC. Next, the tables of digital BSC indicators according to prospects will be filled out in percentage values for each *k*-th configuration of the OS digital transformation strategy (see Fig. 3, stage 1). Using formula (1), aggregated DEA indicators are calculated for each perspective of the digital BSC (see Fig. 3, stage 2) and the overall performance indicator (see Fig. 3, stage 3). As a result of applying the methodology in accordance with the algorithm, the best configuration of the OS digital transformation strategy is produced.

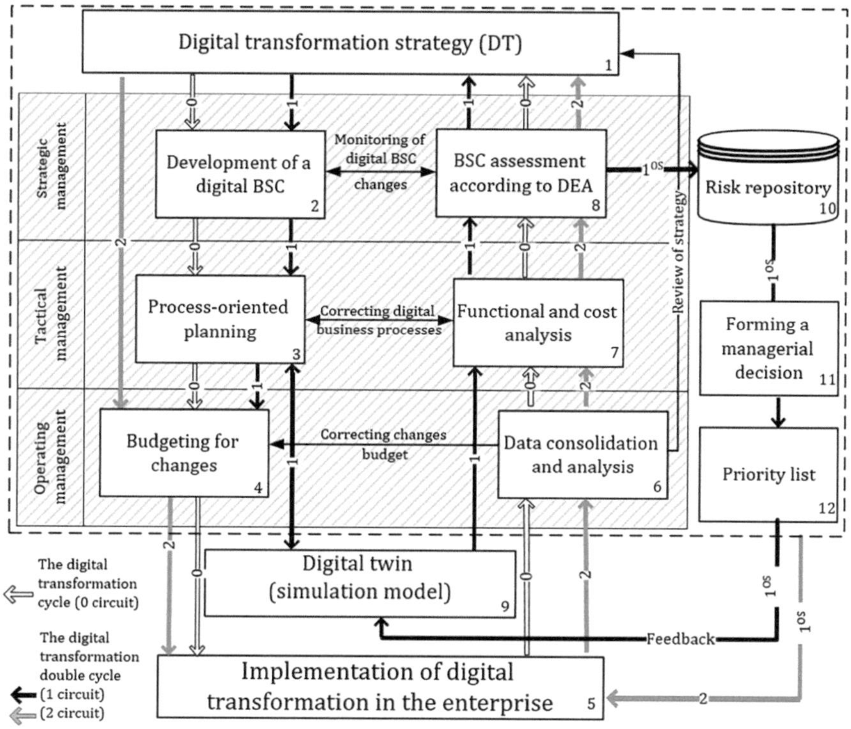


**Fig. 4** Block diagram of OS digital transformation

Using an integrated BSC-DEA business model can improve the process of assessing the effectiveness of various strategy configurations.

Changing the digital state of any organizational system involves a large number of time-consuming transformations carried out in accordance with a set of management decisions (MD). The consistent implementation of MD is aimed simultaneously at achieving the strategic goals outlined in the DBSC, as well as at increasing the capabilities and prospects for digital transformation of the operating system. Such changes in most cases are associated with the emergence of various types of uncertainties, fraught with the emergence of risks of failure to achieve strategic goals [16, 17]. To reduce the level of uncertainty and minimize possible critical risks, a clearly structured sequence of adopted MDs is extremely important.

Quantitative values of indicators for assessing the achievability of goals make it possible to arrange in descending order the sequence of necessary changes. Business management objectives are linked to DTSM, which includes the digital capabilities of the enterprise. The goals that are of greater importance to the operating system are implemented first. In accordance with this principle, a priority list of MD is formed in the form of a roadmap for the digital transformation of the OS [18, 19].

The technological implementation of digital transformation of the operating system is shown schematically in Fig. 4, where, in accordance with DBSC, it is necessary to carry out sequential actions corresponding to modules from the first to the eighth in the diagram, inclusive (circuit 0). Consistent implementation of the activities of models 2–4, direct implementation of the developed strategy (module 5) and analysis of the obtained statistical data after its implementation (modules 6–8) make it possible to monitor the digital transformation of the operating system [20]. The inability to use this methodology to assess and take into account the risks arising when choosing a particular digital transformation strategy dictated the need to supplement the formulated methodology.

Thus, it was necessary to take into account the ability to simulate various scenarios of the most probable states of the OS, as well as to generate MD in automatic mode. In this regard, the digital transformation of the OS is justified by modifying the usual approach based on the use of a digital twin of the OS. This feature is shown schematically in Fig. 4 by the presence of a double circuit, which allows one to assess the prospects for the potential digital transformation of the OS. The main double-loop technology is simulation modeling, which allows, through experiments, to dynamically manage the evolution processes of the OS digital transformation structure, taking into account emerging risks and the influence of various internal and external factors.

The double cycle of digital transformation (Fig. 4) assumes the following sequence of implementation of modules according to the direction of circuit 1: selection of a digital transformation strategy (module 1), development of DBSC (module 2), implementation of process-oriented planning (module 3), construction of a digital twin (simulation models) OS (module 9), implementation of functional cost analysis (module 7), DBSC analysis (module 8) [21]. The risks of automating approvals and confirmations of changes for customers and suppliers pose a threat to the achievement of the corresponding goals of the DTSM and indicators of the digital BSC, are identified and placed in the repository of identified risks (Module 10).

As feedback after the DBSC analysis, a request is made to the risk repository (module 10), on the basis of which MDs are formed (module 11), arranged according to execution priorities (module 12), ranked in descending order according to the quantitative values of the BSC goals. The implementation of modules 1–3 and 7–8 is carried out based on the results obtained from runs of the simulation model, and not from the actual operation of the enterprise. This approach using simulation modeling allows you to quickly, taking into account the compression of the time scale, monitor the digital transformation strategy, adapt the directions of changes, and assess the risks of not achieving the goals of KSF and KPIs, taking into account stochastic factors of the external environment [11]. Based on the list of execution priorities, MDs are initially implemented on a digital twin (simulation model of the organizational system under consideration), which allows generating options for implementing a digital transformation strategy, taking into account the required profitability. Feedback (modules 10–12) in the structure of managing the digital transformation of an enterprise allows you to influence the dynamics of strategy implementation, which makes the organizational system more resistant to stochastic changes in parameters.

The second circuit of the double management cycle (Fig. 4) includes the sequential implementation of the following modules: choosing a digital transformation strategy (module 1), budgeting for changes (module 4), implementing digital transformation (module 5), consolidating and analyzing data (module 6), performing a functional cost analysis (module 7), DBSC analysis (module 8) and returning to the digital transformation strategy (module 1). Thus, a digital transformation cycle has been formulated taking into account the implementation of a priority list of management decisions (changes) into the functioning of the enterprise, the order of which is established in module 12.

## 3 Conclusion

There are differences between the traditional DEA method and the proposed one (two-stage). First, the traditional approach evaluates the inputs and outputs of the system, while the two-stage approach also considers intermediate outputs across perspectives. Secondly, in DEA, the efficiency indicator is the weight ratio of initial costs to final results. In the two-stage approach, the overall performance measure is the sum of the input and output weights of the different strategies. Third, in the proposed DEA, performance can be determined for each organizational configuration and strategy, identifying weaknesses and strengths, while in the traditional DEA, only an overall performance indicator can be determined. Using the integrated BSC-DEA model, it is possible to examine the relationships between the four perspectives of the digital BSC, in addition to calculating overall effectiveness, calculate the effectiveness of each BSC perspective and identify the source of inefficiency.

## References

1. Išoraite, M.: Evaluating efficiency and effectiveness in transport organizations. Transport **20**(6), 240–247 (2005)
2. Iden, J., Bygstad, B.: Sociotechnical micro-foundations for digital transformation. Euro. J. Inform. Syst. 1–16
3. Gökalp, E., Martinez, V.: Digital transformation maturity assessment: development of the digital transformation capability maturity model. Int. J. Prod. Res. **60**(20), 6282–6302 (2021)
4. Zhang, J., Yang, X., Wang, L.: How does digital transformation affect enterprise diversification? Appl. Econ. 1–16
5. Tang, D.: What is digital transformation? EDPACS **64**(1), 9–13 (2021)
6. McCarthy, P., Sammon, D., Alhassan, I.: Digital transformation leadership characteristics: a literature analysis. J. Decis. Syst. **32**(1), 79–109 (2021)
7. Protalinsky, O., Shcherbatov, I., Khanova, A.: Simulation of power assets management process. Stud. Syst. Decis. Contr. **199**, 488–501 (2019)
8. Yamamoto, S.: A strategic map for digital transformation. Procedia Comput. Sci. **176**, 1374–1381 (2020)

9. Protalinsky, O.M., Khanova, A.A., Bondareva, I.O., Nesterova, E.T.: Control of the organizational system configuration in the context of digital transformation. In: *International Conference on Information, Control, and Communication Technologies (ICCT)*, Astrakhan, Russian Federation, pp. 1–5 (2022)
10. Amirbeygi, A.F., Seyyed, E.H., Khorshidvand, B.: Evaluation of green supply chain performance using balanced scorecard and data envelopment analysis. J. Industr. Eng. Manag. Stud. **9**(2), 64–85 (2022)
11. Basso, A., Funari, S.: A three-system approach that integrates DEA, BSC, and AHP for museum evaluation. Decis. Econ. Financ. **43**, 413–441
12. Juiz, C., Colomo-Palacios, R., Gómez, B.: Cascading ISO/IEC 38500 based Balanced Score Cards to improve board accountability. Procedia Comput. Sci. **138**, 417–424 (2018)
13. Kritzinger, W., Karner, M., Traar, G., Henjes, J., Sihn, W.: Digital Twin in manufacturing: a categorical literature review and classification. IFAC-PapersOnLine. **51**(11), 1016–1022 (2018)
14. Green, R.H., Doyle, J.R.: Implementing data envelopment analysis: primal or dual? INFOR: Inform. Syst. Operat. Res. **35**(1), 66–75 (1997)
15. Wöber, K.W.: Data envelopment analysis. J. Travel Tour. Mark. **21**(4), 91–108 (2007)
16. Sengupta, J.K.: Dynamic data envelopment analysis. Int. J. Syst. Sci. **27**(3), 277–284 (1996)
17. Dyson, R.: Performance measurement and data envelopment analysis: rankings are rank! OR Insight **13**(4), 3–8 (2000)
18. Bondareva, I., Khanova, A.A.: Multi-level management of organizational systems on the basis of risk cascading, logical-probabilistic modeling and simulation. In: Studies in Systems, Decision and Control, vol. 2022, pp. 416. Springer, Cham (2022)
19. Protalinskiy, O., Andryushin, A., Shcherbatov, I., Khanova, A., Urazaliev, N.: Strategic decision support in the process of manufacturing systems management. In: Eleventh international conference "Management of large-scale system development" (MLSD, Moscow, 2018), pp. 1–4 (2018)
20. Bondareva, I.O., Shendo, M.V., Luneva, T.V., Khanova, A.A.: Logical-Probabilistic and Simulation Modeling as a Toolkit for Complex Analysis and Risk Management of a Cargo Port E3S Web Conference 224 02027 (2020)
21. Bondareva, I., Khanova, A., Khanova, Y.: Configuring systems based on petri nets, logic-probabilistic, and simulation models. In: Kravets, A.G., Bolshakov, A.A., Shcherbakov, M. (eds.) Cyber-Physical Systems: Modelling and Intelligent Control. Studies in Systems, Decision and Control, vol. 338. Springer, Cham (2021)

# Haken's Model for Assessing the Socio-Economic Condition of the Region

**J. A. Vishnevskaya, A. Sh. Dzhalolov, S. I. Suyatinov, and T. I. Buldakova**

**Abstract** The chapter considers the task of assessing the socio-economic state of the region, analyzes the main purpose of the assessment and provides a scheme of the region management system. The main stages of the decision-making process in relation to the state of the region are given, the peculiarities of the assessment requirements are mentioned. The most frequently used methods for assessing the state of the region are mentioned, the most promising method of assessment—intellectual methods—is highlighted. Haken's model is chosen as an intellectual method for experimental evaluation as a pattern recognition problem. The main stages and key features of the method are considered. The important criteria for assessing the socio-economic condition of the region using Haken's model are noted. The description of the process of data preparation for analysis is given, as well as the main characteristics of the regions and fragments of the formation of the values of the reference vectors' features. Examples of assessments of the state of regions in the format of graphs for two test vectors are given. Explanations of the recognition results for each of the vectors are given, and conclusions are drawn about the applicability of the Haken model for assessing the socio-economic state of the region, the key feature of the study given in the chapter is highlighted, taking into account technological parameters in the assessment.

**Keywords** Socio-economic state · Region · Haken model · Pattern recognition · Reference vectors

J. A. Vishnevskaya (✉) · A. Sh. Dzhalolov · S. I. Suyatinov · T. I. Buldakova
Bauman Moscow State Technical University Russia, Moscow, Russia
e-mail: vishne-yulya@yandex.ru

S. I. Suyatinov
e-mail: ssi@bmstu.ru

T. I. Buldakova
e-mail: buldakova@bmstu.ru



A. G. Kravets and A. A. Bolshakov (eds.), *Human-Centric Cyber-Society*, Studies in Systems, Decision and Control 639, https://doi.org/10.1007/978-3-032-06787-6_3

# 1 Introduction

The socio-economic state in the region is influenced by various processes (environmental, personnel, economic, regulatory and legal, etc.). In addition, when managing regions, the information has a territorially distributed nature, which is especially noticeable in the social sphere, economy, in the aftermath of emergency events, agriculture, etc. As a result, the tasks of regional management are characterized by high labor intensity, multidimensionality, multicriterionality and, as a consequence, the presence of patterns of socio-economic state of the region that cannot be quantitatively analyzed. Therefore, the key task in the process of making strategic decisions on regional development is the development of methods of forecasting and analysis of socio-economic information at the regional level. Moreover, it can be argued that the quality of decisions made on the basis of a large amount of unstructured data largely determines the effectiveness of management of socio-economic state of the regions [1–4].

Thus, the assessment of the socio-economic state of the region requires a systematic approach to solving the problem and the use of intelligent data processing technologies [5–9].

The aim of the chapter is to apply Haken's intelligent model to recognize the socio-economic situation in the region.

# 2 Approaches to Assessing the Socio-Economic Condition of Regions

The relevance of the analysis of socio-economic indicators is due to the fact that its results play an important role in determining the current state and potential for future development of regions.

The tasks of regional management related to the assessment of socio-economic condition of regions can have different objectives, for example: effective distribution of financial and material resources between regions; timely identification of problem situations; ranking and assessment of regions' competitiveness. In addition, knowledge of the current assessment of the socio-economic state of the region is necessary for forecasting its development, including in the conditions of possible man-made and natural emergencies.

Depending on the specific formulation of the task of regional management, different approaches and methods of decision-making are used. This is due to the fact that the region is a complex organizational system with a large number of intersecting flows of control actions, both within the region and coming from the authorities of different levels. In addition, it is also necessary to take into account the impact of the external environment (the influence of neighboring regions, ecology, economic and political processes in the world and much more). At the same time, the main goal of

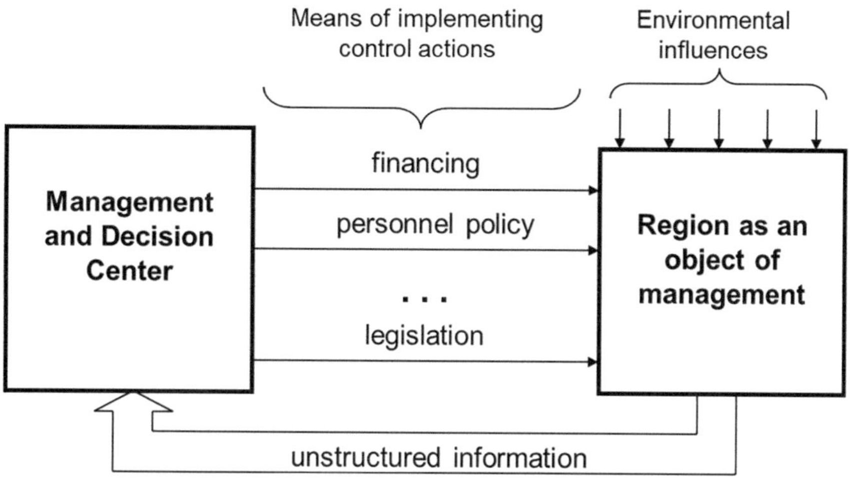


**Fig. 1** Regional management system

regional management remains the improvement of socio-economic condition of the region (Fig. 1).

Several stages of the decision-making process of regional management can be distinguished (Fig. 2). One of the main stages is the stage of assessment of the socio-economic state of the region, since the results of this assessment are the basis for solving most of the regional management tasks.

When assessing the socio-economic state of a region, it is necessary to take into account many requirements, including the coverage of regional processes affecting the socio-economic development of the region; a finite set of indicators characterizing each process; availability of indicators for assessment; the possibility of interpreting the obtained assessment of the socio-economic state of the region [10–12].

Currently, there is no single universal approach to assessing the level of socio-economic development of the region. Different approaches and methods are proposed to assess the socio-economic state of regions, differing in the set of evaluated indicators and the purposes of using the obtained estimates [13–15].

Quantitative assessments based on macroeconomic indicators are most often used.

Depending on the task to be solved, the obtained private indicators are then subjected to integration and generalization using appropriate methods and tools. Integral indicators of socio-economic development assessment allow analyzing the trends in socio-economic development of the region. The advantage of integral assessment is the systematization of various indicators and consideration of the effectiveness of various regional processes.

Rating assessments are also used to evaluate the investment attractiveness of the region. Often the rating assessment is performed with the help of an expert approach.

A promising direction for obtaining an objective assessment of the socio-economic state of regions is intelligent technologies that allow identifying hidden patterns in a

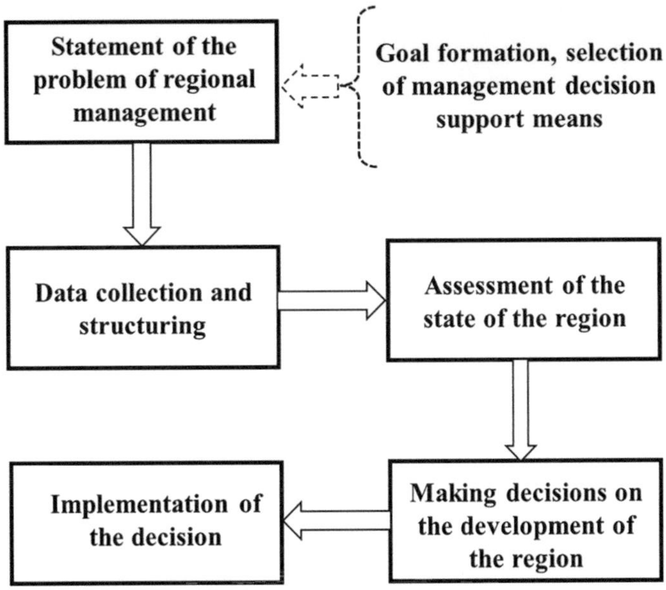


**Fig. 2** Decision-making process

set of unstructured data [16–18]. For example, [19] analyzes the competitiveness of 78 regions using a modified fuzzy method ELECTRE II. To assess the socio-economic situation, the authors applied a multi-criteria decision-making method based on the fuzzy logic approach. The obtained results make it possible to identify the most risky factors of regional competitiveness and distinguish regions with different levels of risk.

## 3 Application of Haken's Model for Pattern Recognition

In general terms, the assessment of the state of a region can be attributed to the problem of pattern recognition, which consists in determining the class to which an object belongs with given (e.g., obtained as a result of observation) values of attributes [20, 21]. In the task under consideration, the objects are regions, and the images are their states, which are evaluated by a set of indicators (attributes).

It should be noted that in the existing studies, the assessment of the state of a region is carried out, as a rule, using one type of indicators, which leads to a decrease in the adequacy of the assessment. However, the more indicators are taken into account and their spectrum is more diverse, the more objective it is possible to determine the state of the region.

A promising method of pattern recognition under conditions of incomplete data is the approach based on Haken's synergetic model [22, 23]. The model is based on the potential function $V$:

$$V = -\frac{1}{2}\sum_{k=1_k}^{L}\left(v_k^+ q\right)^2 + \frac{1}{4}B\sum_{k\neq k'}\left(v_k^+ q\right)^2\left(v_{k'}^+ q\right)^2 + \frac{1}{4}C\left(q^+ q\right)^2.$$

Here $L$ is the number of reference vectors; $v_k^+$ are conjugate vectors defined by the orthogonality relations $v_k^+ v_{k'} = \delta_{kk'}$; $v_k$ are the reference vectors; $q$ is the current vector to be estimated; $B$, $C$ and $\lambda_k$ are constants, parameters of the recognition dynamics.

The first term in the potential function takes the current vector $q$ away from the unstable equilibrium point of the reference landscape given the attention parameters $\lambda_k$. The second term forms the boundaries between different benchmarks, and the third term limits the growth of the values of the elements of the current vector $q$ during recognition.

In this context, recognition amounts to minimizing the value of the potential function by varying $q$ according to special dynamics due to the structure of the potential function. The minimum into which $q$ falls represents the class into which the current vector is classified.

In the software modeling of the recognition process, an opportunity is provided to make a transition of the dependence of the potential $V$ on the current vector $q$ and the set of benchmarks $v_k$ to the dependence on the predicted state $v_k^+ q$ and the parameters of the recognition dynamics. Thus, in the training phase, the dependence between the benchmarks $v_k$ and the predicted states $v_k^+ q$ is formed, and in the recognition process, the dynamics is realized based on the initial values of these predicted states.

It is considered that the current vector $q$ is closest to the benchmark $v_k$ whose predicted state will be maximal, i.e.

$\left|\left(v_k^+ q\right)\right| = \max_k\left\{\left|\left(v_k^+ q\right)\right|\right\}, k = 1, \ldots, L.$

For this purpose, we need to take into account that the potential $V$ is a function of both the current vector $q$ and the set of patterns $v_k^+$. During the recognition process, the values of the patterns are set and fixed, and the potential function is minimized with respect to $q$. During the learning process, the $v_k^+$ vectors are modified by gradient dynamics, and the shape of the potential is created by superposition of all training samples. To realize such a method, the last summand of the potential function is modified:

$$V = -\frac{1}{2}\sum_{k=1_k}^{M}\left(v_k^+ q\right)^2 + \frac{1}{4}B\sum_{k\neq k'}\left(v_k^+ q\right)^2\left(v_{k'}^+ q\right)^2 + \frac{1}{4}C\sum_{kk'}\left(v_k^+ q\right)^2\left(v_{k'}^+ q\right)^2.$$

The parameters of recognition dynamics should be set depending on the problem to be solved. To assess the socio-economic state of the region, the values of the attention parameter $\lambda$ should be balanced and meet the condition: $\lambda = C > 0$, since there is no need to prioritize one of the benchmarks. Parameter $B = 1$ reflects the degree of influence of the internal structure of the system and the level of its self-organization. The value of parameter B determines the level of cooperation or competition between the elements of the system. Parameter $C = 1$ characterizes the degree of external influence on the system, for example, external flow of energy or information. The value of parameter $C$ reflects the external support or destabilizing influence [23, 24].

## 4 Preparation of Initial Data

In the process of conducting research on Haken's model, the mathematical components of the algorithm, which allow for optimal realization of its operation, were determined experimentally. It is worth noting the simplicity of training and variability of the choice of mathematical methods in aspects of the algorithm realization. This property opens wide opportunities for research of the model and formation of a universal algorithm depending on the requirements to the problem to be solved [25].

The information about the regions was analyzed on the basis of four groups of indicators: social, economic, financial and technological.

35 indicators characterizing the socio-economic state of the regions were used as initial information. Among them:

- 13 indicators reflecting social processes in the region (including quality of life, availability of services, degree of social stability in the region);
- 7 economic indicators assessing the economic development of the region, its potential for attracting investment and business development;
- 11 financial indicators assessing the financial condition of the region, its ability to manage the budget and ensure investment in development; technological indicators characterizing the level of technological development of the region and its readiness for digital transformation (the use of renewable energy sources in the region, the share of people employed in research and development, the level of digital literacy of the population, the development of mobile communications and the Internet).

The selected indicators allow us to conduct a more objective and reasonable analysis of the situation in the region and identify priorities for the development and improvement of various spheres of its activities.

The current vector $q$ is the state of the region at the current moment, which should be attributed to one of four classes: excellent state, good state, satisfactory state, unsatisfactory state. Each class corresponds to its own benchmark. Accordingly, there are 4 reference vectors of socio-economic state of the region.

The reference vectors take values from $-1$ to 1 with a step of 0.25 according to expert assessments. The value «$-1$» means extremely negative state of the indicator, and the value «1» is the most favorable state of the indicator.

Tables 1, 2, 3 and 4 present examples of attributes of different groups of indicators for the reference vectors.

**Table 1** Fragment of regional social indicators and reference vectors

| Social indicators | Reference vector $v_1$ | Reference vector $v_2$ | Reference vector $v_3$ | Reference vector $v_4$ |
|---|---|---|---|---|
| High level of employment of the population | 1 | 0.25 | −0.25 | −1 |
| High share of labour migrants in the regional economy | −1 | −0.25 | 0.25 | 1 |
| High share of medical workers per 1000 inhabitants of the region | 1 | 0.25 | −0.25 | −1 |

**Table 2** Fragment of regional economic indicators and reference vectors

| Economic indicators | Reference vector $v_1$ | Reference vector $v_2$ | Reference vector $v_3$ | Reference vector $v_4$ |
|---|---|---|---|---|
| High number of employees of small and medium-sized enterprises in the regional economy | 1 | 1 | −0.5 | −1 |
| High share of small and medium-sized enterprises in the gross regional product of the constituent entity of the Russian Federation | 1 | 1 | −0.5 | −1 |
| High level of tourism development | 1 | 0.25 | −0.25 | −1 |

**Table 3** Fragment of regional financial indicators and benchmark vectors

| Financial performance | Reference vector $v_1$ | Reference vector $v_2$ | Reference vector $v_3$ | Reference vector $v_4$ |
|---|---|---|---|---|
| High share of the budget allocated to infrastructure projects | 1 | 0.25 | 0 | −0.25 |
| Compliance of planned revenue indicators, laid down in the first edition of the legal act approving the regional budget, with the actual indicators of the region's budget execution | 1 | 0.25 | 0.25 | −1 |
| Compliance of planned expenditure indicators, laid down in the first edition of the legal act approving the regional budget, with the actual indicators of the regional budget execution | 1 | −0.25 | −0.75 | −1 |

**Table 4** Fragment of regions' technological indicators and reference vectors

| Technological indicators | Reference vector $v_1$ | Reference vector $v_2$ | Reference vector $v_3$ | Reference vector $v_4$ |
|---|---|---|---|---|
| Utilization of renewable energy sources | 1 | 0.75 | 0.25 | −1 |
| High proportion of people employed in research and development | 1 | 0.75 | 0.5 | 0.25 |
| High level of digital literacy of the population | 1 | 1 | 0.5 | −1 |
| High level of mobile and Internet coverage | 1 | 1 | 0.5 | −1 |

When assessing the socio-economic state of the region using the Haken model, 2 variants are considered, where the current vectors $q_1$ and $q_2$ are formed on the basis of expert assessments of 35 different indicators, taking into account the ratings of the regions for 2022.

## 5 Discussion of the Results Obtained

When forming the current vectors, one region is selected from the first 10 leading regions, and the other is on the border between regions with average rating and lagging regions.

In the first case we have $q_1 = [1; 1; 1; -1; -1; 1; 1; 1; -1; 1; 1; 1; -1; 1; 1; 1; 1; 1; 1; -0.25; 1; 0.25; 1; 1; 1; 1; 1; 1; 1; 1; 1; 1; 1; 1; 1]$.

In the second case we have $q_2 = [1; -0.25; -1; 1; -1; 0.5; -1; 1; -0.25; -1; -1; -1; -0.25; 1; 1; -0.25; -1; -1; -1; -0.25; 0.25; -1; -1; -1; 1; 1; 1; 1; 1; -0.25; -0.25; 1; 0.75; -1; 1]$.

Figure 3 shows the results of assessing the socio-economic state of region 1.

The figure shows that one of the four curves tends to one and reflects the evaluation result, while the other curves tend to zero.

Figure 4 shows the results of assessing the socio-economic state of region 2.

As a result of the study, the ratings of the regions are confirmed: the first region has an excellent socio-economic condition, and the second region has a satisfactory condition.

The modeling results have shown that G. Haken's intellectual model can be successfully applied to assess the state of a region using information reflecting

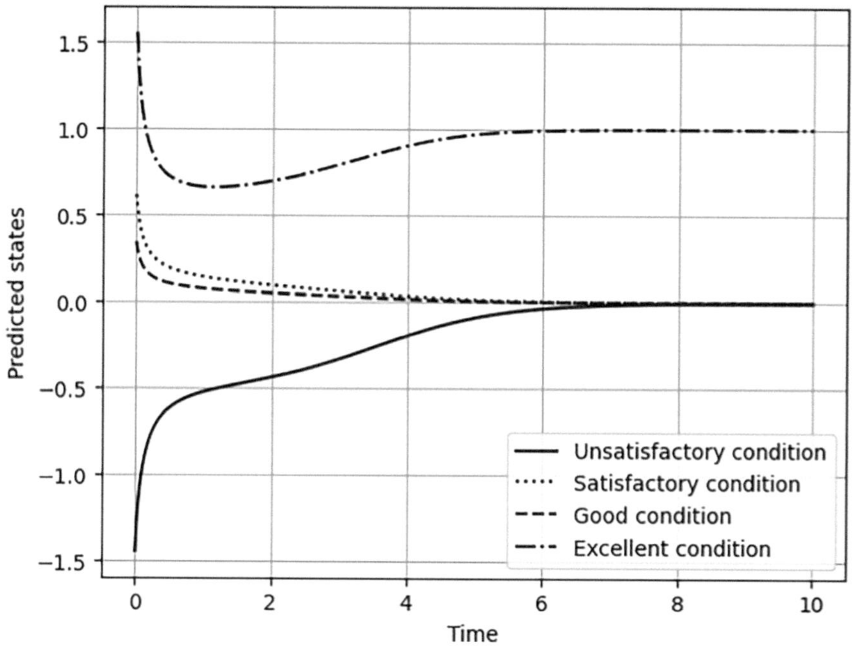


**Fig. 3** Estimation of the state of region 1 using Haken's model

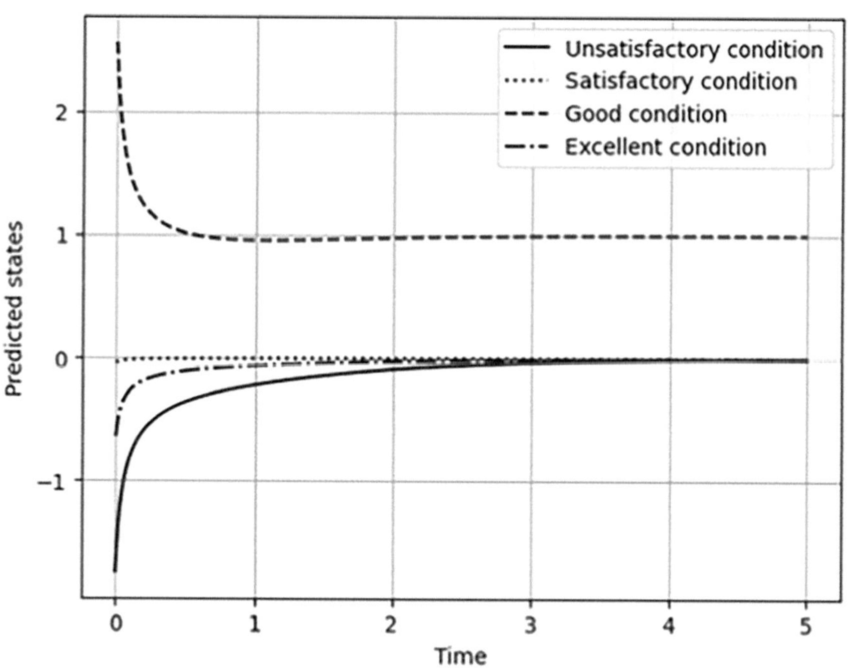


**Fig. 4** Estimation of the state of region 2 using Haken's model

various regional processes. The model allows taking into account a different number of indicators that reflect different regional processes.

## 6 Conclusion

The chapter shows that the assessment of socio-economic state of the region is performed on a set of unstructured data and refers to the problem of pattern recognition.

The trained model of G. Haken is applied to assess the state of the region based on the information reflecting various regional processes.

The peculiarity of this study is the consideration of technological indicators, which are usually not taken into account when rating regions. Therefore, this analysis can serve as a basis for the development of regional development strategies aimed at improving the socio-economic situation and improving the quality of life of the population.

## References

1. Proletarsky, A., Berezkin, D., Popov, A., Terekhov, V., Skvortsova, M.: Decision support system to prevent crisis situations in the socio-political sphere. In: Kravets, A., Bolshakov, A., Shcherbakov, M. (eds.) Cyber-Physical Systems: Industry 4.0 Challenges. Studies in Systems, Decision and Control, vol. 260. Springer, Cham (2020). https://doi.org/10.1007/978-3-030-32648-7_24
2. Dzhalolov, A.S., Buldakova, T.I., Proletarsky, A.: Socio-economic decision support module by unstructured data. In: Proceedings of the 2020 IEEE Conference of Russian Young Researchers in Electrical and Electronic Engineering (EIConRus), St. Petersburg and Moscow, Russia, pp. 1931–1934 (2020). https://doi.org/10.1109/EIConRus49466.2020.9039086
3. Veshneva, I., Chernyshova, G., Bolshakov, A.: Regional competitiveness research based on digital models using Kolmogorov-Chapman equations. In: Kravets, A.G., Bolshakov, A.A., Shcherbakov, M. (eds.) Society 5.0: Cyberspace for Advanced Human-Centered Society. Studies in Systems, Decision and Control, vol. 333. Springer, Cham (2021). https://doi.org/10.1007/978-3-030-63563-3_12
4. Anokhin, A., Burov, S., Parygin, D., Rent, V., Sadovnikova, N., Finogeev, A.: Development of scenarios for modeling the behavior of people in an urban environment. In: Kravets, A.G., Bolshakov, A.A., Shcherbakov, M. (eds.) Society 5.0: Cyberspace for Advanced Human-Centered Society. Studies in Systems, Decision and Control, vol. 333. Cham, Springer (2021). https://doi.org/10.1007/978-3-030-63563-3_9
5. Buldakova, T.I., Suyatinov, S.I.: Biological principles of integration information at big data processing. In: International Russian Automation Conference (RusAutoCon), Sochi, Russia, pp. 1–6 (2019). https://doi.org/10.1109/RUSAUTOCON.2019.8867710
6. Buldakova, T.I., Dzhalolov, A.Sh.: Analysis of data processes and choices of data-processing and security technologies in situation centers. Sci. Techn. Inform. Process. **39**(2), 127–132 (2012)
7. Bolshakov, A.A., Veshneva, I.V., Lushin, D.: Mathematical model of integration of cyber-physical systems for solving problems of increasing the competitiveness of the regions of

the Russian federation. In: Society 5.0: Cyberspace for Advanced Human-Centered Society. Studies in Systems, Decision and Control, vol. 333, pp. 129–139. Cham, Springer Verlag (2021). https://doi.org/10.1007/978-3-030-63563-3_11
8. Guo, S., Bai, Y., Bocus, M.J., Fan, R.: Digital transformation for intelligent road condition assessment. In: Kahraman, C., Haktanır, E. (eds.) Intelligent Systems in Digital Transformation. Lecture Notes in Networks and Systems, vol. 549. Springer, Cham (2023). https://doi.org/10.1007/978-3-031-16598-6_22
9. Liu, D., Peng, L., Zhao, Z.: A review of intelligent methods of health assessment technology. Intell. Robot. **3**(3), 355–373 (2023). https://doi.org/10.20517/ir.2023.16
10. Chernyshova, G., Veshneva, I., Piunova, E.: Analysis of Russian regions innovative competitiveness dynamics using fuzzy methods. In: Kravets, A.G., Bolshakov, A.A., Shcherbakov, M. (eds.) Society 5.0: Human-Centered Society Challenges and Solutions, vol. 416, pp. 115–128. Cham, Springer (2022). https://doi.org/10.1007/978-3-030-95112-2_10
11. Shanabhoga, M.B., Krishnamurthy, B., Suresha, S.V.: Assessment of vulnerability to climate change among the districts of Hyderabad Karnataka region. Int. J. Agricult. Sci. **11**(16), 8932–8935 (2019). https://doi.org/10.6084/m9.figshare.13365152.v1
12. Ma, L.: Management of employees' cooperative of the state universities and colleges in region. Int. J. Multidisciplin. Res. Anal. **07**(01), 289–322 (2024). https://doi.org/10.47191/ijmra/v7-i01-38
13. Raynor, E., Derner, J., Hartman, M., et al.: Secondary production of the central rangeland region of the United States. Ecol. Appl. (2024). https://doi.org/10.1002/eap.2978
14. Antonio-Estrada, C., Cruz-Domínguez, E., Martínez-López, M., Matías-Pérez, D., García-Montalvo, I.A.: Oreochromis niloticus is a blue economy alternative for the Papaloapan region of the state of Oaxaca, Mexico. Front. Sustain. Food Syst. **8** (2024). https://doi.org/10.3389/fsufs.2024.1322838
15. Li, B., Zhang, J.: Subsidies in an economy with endogenous cycles over investment and innovation regimes. Macroecon. Dyn. **18**(6), 1351–1382 (2013). https://doi.org/10.1017/S1365100512000995
16. Zadeh, L., Aliev, R.: Fuzzy logic theory and applications: Part I and Part II. World Sci. 612 (2018). https://doi.org/10.1142/10936
17. Golosovskiy, M., Bogomolov, A., Balandov, M.: Algorithm for configuring sugeno-type fuzzy inference systems based on the nearest neighbor method for use in cyber-physical systems. In: Kravets, A.G., Bolshakov, A.A., Shcherbakov, M. (eds.) Cyber-physical systems: intelligent models and algorithms: studies in systems, decision and control, vol. 417, pp. 83–97. Cham: Springer (2022). https://doi.org/10.1007/978-3-030-95116-0_7
18. Suyatinov, S.I., Khudyakov, A.M., Uvarova, M.S.: A regularization-based method of identification of information objects. Automat. Document. Math. Linguist. **56**(6), 324–330 (2022). https://doi.org/10.3103/S0005105522060097
19. Chernyshova, G., Veshneva, I., Firsova, A.A., Makarova, E.L., Makarova, E.A.: Methodology for assessing the risks of regional competitiveness based on the Kolmogorov-Chapman equations. Mathematics **11**(19), 4206 (2023). https://doi.org/10.3390/math11194206
20. Slavin, O.A.: Using special text points in the recognition of documents. In: Kravets A., Bolshakov, A., Shcherbakov, M. (eds.) Cyber-Physical Systems: Advances in Design & Modelling. Studies in Systems, Decision and Control, vol. 259. Springer, Cham (2020). https://doi.org/10.1007/978-3-030-32579-4_4
21. Arlazarov, V.V., Slavin, O.A., Uskov, A.V., Janiszewski, I.M.: Modelling the flow of character recognition results in video stream. Bull. South Ural State Univ. Ser. Math. Modell. Program. Comput. Softw. (Bulletin SUSU MMCS) **11**(2), 14–28 (2018). https://doi.org/10.14529/mmp180202
22. Wagner, T., Schramm, U., Boebel, F.G.: Synergetic learning for unsupervised texture classification tasks. Physica D **80**, 140–150 (1995)
23. Shen, Z., Meng, J., Wei, Y.: A new method of object recognition based on deep synergetic neural network. DEStech Trans. Mater. Sci. Eng. (2017). https://doi.org/10.12783/dtmse/amme2016/6900

24. Haken, H.: Synergetic computers and cognition–a top-down approach to neural nets, Springer, Berlin. Heidelberg (2004). https://doi.org/10.1007/978-3-662-10182-7
25. Suyatinov, S.I., Buldakova, T.I., Vishnevskaya, Y.A.: Synergistic model of situational awareness of the human operator. In: Kravets, A.G., Bolshakov, A.A., Shcherbakov, M. (eds.) Society 5.0: Human-Centered Society Challenges and Solutions. Studies in Systems, Decision and Control, vol. 416, pp. 331–340. Springer, Cham (2022). https://doi.org/10.1007/978-3-030-95112-2_27

# Influence of Demand and Cost Functions on Conjectured Variations in Oligopoly Game

**Mikhail Geraskin**

**Abstract** For the case of general demand and cost functions, a game-theoretic problem of the quantity competition analysis in an oligopoly is considered. A cyber-physical system is investigated based on conjectural variations, i.e., assumptions of players about the impact of their actions on the action of the counterparty. A comparative analysis of the conjectural variations for various types of inverse demand function (linear and power) and cost functions (linear, power, quadratic) is carried out. The following main results are obtained. In the linear-quadratic model, the influence of the player's sales volume increase on the decrease in the environment's sales volume is less radical than in the linear model. In the power model, a greater absolute value of the environment's sales volume decrease is observed relative to the linear model. The quadratic-power model is similar in characteristics to the power model with a negative scale effect.

**Keywords** Oligopoly · Conjectural variation · Demand function · Cost function · Stackelberg leadership · Reflexive game

## 1 Introduction

An oligopoly game model is related to the aggregative games [1]. The aggregative games mean that a payoff of the game for a player is a function of the total actions of other players, i.e., the game environment. Therefore, the player formulates assumptions about the strategies of the environment, which the classics defined by the term "conjectural variations" [2, 3]. The game taking into account such assumptions was later defined as a reflexive game [4]. This game is interpreted as a cyber-physical system, because it implements the interaction between the thought process of the firm's management and the technical process of varying the volumes of production and sales of products or services. For example, in the telecommunications sector,

M. Geraskin (✉)
Samara National Research University, Samara, Russia
e-mail: innovation@ssau.ru



A. G. Kravets and A. A. Bolshakov (eds.), *Human-Centric Cyber-Society*, Studies in Systems, Decision and Control 639, https://doi.org/10.1007/978-3-032-06787-6_4

the management's decision to increase or decrease the volume of services is implemented through a change in the capacity of radar stations. The process of reflection expresses the players' reasoning regarding the best response of the environment to variations in the volumes of production and sales. The formal definition of the conjectural variation is the following: an increase or decrease in the sales volume of the environmental firm that the player (i.e., the firm being analyzed) expects as a result of a single increase in its sales volume.

Researchers were based on these variations when analyzing the Stackelberg leadership in two directions: firstly, they analyzed the increase in the number of reflexive players, leading to the emergence of many Stackelberg leaders in the game [5], and secondly, they studied the deepening of reflextion, which determines the emergence of leaders at higher levels, that is, a formation of the multi-level leadership [6].

In the latter case, the hierarchy of players' assumptions determines their stratification by levels of the leadership into zero-level leaders (regarded as followers), who do not take into account the actions of the environment and are players with zero conjectural variation. Then, the first-level leaders may arise, who believe that the environment includes zero-level leaders. In the general case, r-level leaders emerge, who consider other players to be (r − 1)-level leaders.

This hierarchy defines the player's reflexion rank r as a number in the described sequence of mental types, which made it possible to derive a formula for the sum of the leader's conjectural variations for an arbitrary rank in the case of convex-concave cost functions. The formula essentially depends on the parameters of the inverse demand function and the players' cost functions, which allows us to analyze the influence of different types of these functions on the nature of the players' assumptions.

The conjectural variation characterizes the optimal response of the player to the counterparty's strategy. Positive variation means the expected increase in the counterparty's action as a result of the player's increase in action, and it occurs rarely. This reaction of the player's environment is not optimal, because it leads to an increase in the total sales volume and then to a decrease in the equilibrium price, which in turn causes a decrease in the payoffs of all players. Therefore, positive variation applies only to players with positive returns to scale. For such players, a decrease in the equilibrium price does not lead to a decrease in the payoffs. In the most common cases of constant or negative returns to scale, the conjectural variation is negative. In this case, the player assumes that the environment intends to reduce the sales volume in order to prevent a decrease in the equilibrium price. In the game with more than two players, the sum of conjectural variations, i.e., the generalized parameter of all agents' response, determines the equilibrium. The greater the number of players, the less the rare cases of positive returns to scale influence on the sum of variations. Therefore, the sum of the conjectural variations is usually negative.

As a rule, the consumer demand was modeled either by a linear function [5–13] with a variable elasticity used to reflect a smooth decrease in price, or by a power function [5, 14–17] with a constant elasticity, which expresses a sharp decrease in price with an increase in the aggregated offers. The player's costs can grow in direct proportion to the player's action according to a linear function [8, 10–12, 14, 16], but they can show a sharp increase with an increase in the player's action, which was described by a power law [6, 15] or quadratic [5, 7, 9, 13, 17] functions. In addition, the costs can demonstrate slow growth, which was modeled by a power function with elasticity less than one. In addition, when describing the modeling methods, it should be noted that the system of interactions between the firm's management and the technical equipment forms man–machine networks, which belong to the cyber-physical systems [18–20].

In summary, the chapter is devoted to modeling the cyber-physical system of the oligopoly game in the context of the influence of the consumer demand function and the players' costs function on the total conjectural variation.

## 2 Description of Oligopoly Game Model

The model for choosing the player's action in an oligopoly game has the following form:

$$\max_{Q_i \geq 0} \prod_i (Q, Q_i) = P(Q)Q_i - C_i(Q_i), i \in N = \{1, ..., n\} \quad (1)$$

$$Q = \sum_{i \in N} Q_i \quad (2)$$

where $Q_i$, $\prod_i$ are the action and the utility function of the player $i$; $Q$ is the aggregate of actions; $N$ is the set of players; $n$ is the number of players; $P(Q)$ is the inverse demand function, $P_Q^{/} < 0$; $C_i(Q_i)$ is the cost function of the player $i$, $C_{Qi}^{/} \geq 0$; $P$ is the action price (i.e., the price aggregate).

The Nash equilibrium in the game $\Gamma = \langle N, \{Q_i, i \in N\}, \{\Pi_i, i \in N\}\rangle$ with a certain known vector of conjectural variations is determined by solving the reaction equations system of the following type:

$$\frac{\partial \prod_i (Q_i^*, \rho_{ij})}{\partial Q_i} = 0, i, j \in N \quad (3)$$

where $\rho_{ij} = Q_{jQ_i}^{/}$ is the conjectural variation of the $i$-th player, that is, the change in the output of the $j$-th player assumed by the player $i$ in response to a unit increase in the output of the $i$-th player; $Q_i^*$ is the equilibrium value.

System (3) with utility function (1) is transformed as follows:

$$P(Q) + (1 + S_i^r)Q_i P_Q^{/} - C_{i_{Q_i}}^{/} = 0, i \in N, S_i^r = \sum_{j \in N \setminus i} \rho_{ij}^r \quad (4)$$

where $S_i^r$ is the sum of the conjectural variations of the $i$-th player at the r-th reflexion rank.

The recurrent formula for the conjectural variations sum at an arbitrary reflexion rank has the following form [6]:

$$S_i^r = \left( \frac{1}{\sum_{j \in N \setminus i} \frac{1}{u_j - S_j^{r-1} + 1}} - 1 \right)^{-1} \quad (5)$$

$$u_i = -1 + \frac{P^{/}_{Q_i} + \left(1 + S_i^{r-1}\right) Q_i P^{//}_{QQ_i} - C^{//}_{iQ_iQ_i}}{\left| P^{/}_{Q} \right|} \quad (6)$$

Note that formula (5) is presented for the conjectural variations, which are independent of the players' actions, that is, for $\rho^{/}_{ijQ_i} = 0$, and in [6] a more general case $\rho^{/}_{ijQ_i} \neq 0$ was described. In addition, in [6] the parameter has the form

$$u_i = -2 - \frac{C^{//}_{iQ_iQ_i}}{b} \quad (7)$$

because it was derived for a linear demand function, then $P^{/}_{Q} = P^{/}_{Q_i} = -b, P^{//}_{QQ_i} = 0$. For the parameter $u_i$ we introduce the term "nonlinearity coefficient", because, in contrast to the linear form (4) at $u_i = -2$, in the case of nonlinear demand and cost functions this parameter has different values.

Thus, from (4) it follows that the calculation of the game equilibrium directly depends on the ability to find the conjectural variations, which, in turn, is predetermined by the features of the functions $P(Q)$ and $C_i\left(Q_i\right)$. Therefore, further we will consider various options for combinations of these functions and methods for calculating variations.

## 3 Influence of Demand and Cost Models on Nonlinearity Coefficient

In the following transformations, we will omit the player index i and use the symbol $q = Q_i \forall i \in N$ for the player's sales volume.

In oligopoly studies, the most common inverse demand functions are:

- the linear demand function [5, 6, 9–15]

$$P_1(Q) = a - bQ, a > 0, b > 0 a >> b; \quad (8)$$

- the power demand function [5, 16–19]

$$P_2(Q) = AQ^{\alpha}, A > 0, A > 0, |\alpha| < 1, \quad (9)$$

where *a, b, A*, α are the constant coefficients; $P_k(Q), k = 1, 2$ is the *k*-type demand function.

The cost functions of oligopolists are modeled in the following forms:

- the linear cost function [10, 12–14, 16, 18]

$$C_1(q) = B_0 + B_1 q, B_0 \geq 0, B_1 > 0 \tag{10}$$

- the power cost function [6, 17]

$$C_2(q) = B_0 + B_1 q^{\beta}, B_0 \geq 0, B_1 > 0, \beta \in (0, 2), \tag{11}$$

- the quadratic cost function [5, 7–9, 11, 15, 19]

$$C_3(q) = B_0 + B_1 q + \frac{B_2}{2} q^2, B_0 \geq 0, B_1, B_2 > 0, \tag{12}$$

where $B_0, B_1, B_2$, β are the constants; $C_m(q), m = 1, 2, 3$ is the cost function of type *m*.

To calculate (6), in cases (8–12), we derive the formulas $P^{/}_{Q}, P^{/}_{q}, P^{//}_{Qq}, C^{//}_{qq}$:

$$P^{/}_{1_Q} = P^{/}_{1_q} = -b, P^{//}_{1_{Qq}} = 0 \tag{13a}$$

$$P^{/}_{2_Q} = P^{/}_{2_q} = A\alpha Q^{\alpha-1}, P^{//}_{2_{Qq}} = A\alpha(\alpha - 1)Q^{\alpha-2} \tag{13b}$$

$$C^{//}_{1qq} = 0, C^{//}_{2qq} = B_1\beta(\beta - 1)q^{\beta-2}, C^{//}_{3qq} = B_2 \tag{13c}$$

As a result, for functions (8–12), the parameters $u_{km}, k = 1, 2, m = 1, 2, 3$ have the form:

$$u_{11} = -2, u_{21} = -2 + \left(1 + S^{r-1}\right)(1 - \alpha)\frac{q}{Q} \tag{14a}$$

$$\begin{aligned} u_{12} &= -2 - \frac{B_1}{b}\beta(\beta - 1)q^{\beta-2}, \\ u_{22} &= -2 + \left(1 + S^{r-1}\right)(1 - \alpha)\frac{q}{Q} - \frac{B_1}{A|\alpha|Q^{\alpha-1}}\beta(\beta - 1)q^{\beta-2} \end{aligned} \tag{14b}$$

$$u_{13} = -2 - \frac{B_2}{b}, u_{23} = -2 + \left(1 + S^{r-1}\right)(1 - \alpha)\frac{q}{Q} - \frac{B_2}{A|\alpha|Q^{\alpha-1}} \tag{14c}$$

Note that in (14) the parameter $B_1$ corresponds only to the case of power function (11).

## 4 Comparative Analysis of Conjectural Variations

Obviously, the reference case in set (14) is the linear demand and cost functions game, for which $u_{11} = -2$. According to (5), for $S_j^{r-1} = 0$, we obtain:

$$S_i^r = \frac{1-n}{n} \in (-1, 0], S_i^r(n = 3) = -0, 67$$

Henceforth, we will study the influence of the demand and cost functions on the conjectural variations at $r = 1$ (i.e., we will accept $S^{r-1} = 0$), therefore, in the designation of the conjectural variations sum, we will omit this index, and the player's index, that is, we denote $S_{km} = S_i \forall i \in N$. We will consider the number of players $n = 3$, when $S_{11} = -0, 67$. In case 13, obviously, $u_{13} < -2, S_{13} > -0, 67$; therefore, we consider nontrivial cases.

First, we estimate the influence of nonlinear demand function (9) with linear costs (10) on S, that is, we analyze the difference $u_{21}$ from $-2$ and $S_{21}$ from $-0.67$ for different values of the demand elasticity $|\alpha|$ and the ratio $\frac{q}{Q}$. For Q = 100, we consider an increase in the share $\frac{q}{Q}$ from 0 to 1 (Figs. 1 and 2). Conclusions: 1) the nonlinearity coefficient $u_{21} > -2$ and increases with increasing in $|\alpha|$ and $\frac{q}{Q}$; 2) the sum of variations $S_{21} < -0, 67$ and decreases with increasing in $|\alpha|$ and $\frac{q}{Q}$, going beyond the range $(-1, 0]$.

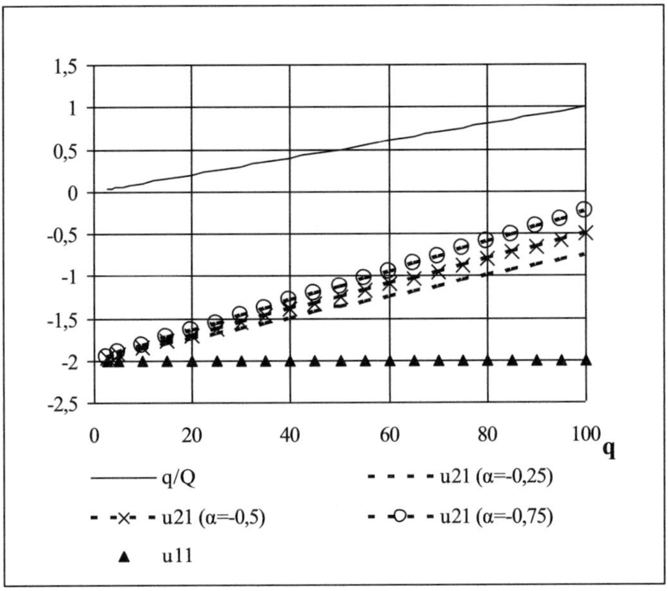


**Fig. 1** Nonlinearity coefficient for power demand function and linear costs

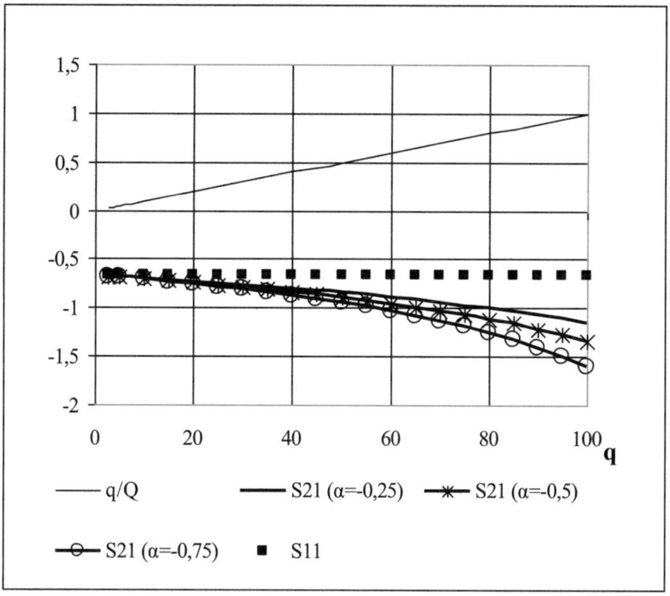


**Fig. 2** Sum of variations for power demand function and linear costs

Consequently, the greater the elasticity of the power function of demand and the player's share in the supply aggregate, the greater the absolute value of the player's expected reduction in the volume of sales of counterparties due to the increase in the player's sales volume.

Secondly, in the case of linear demand function (8), we analyze the influence of power-law cost function (11) on S, that is, we study the difference $u_{12}$ from $-2$ and $S_{12}$ from $-0.67$ for different values of the cost elasticity $\beta \in (0, 2)$ and the ratio $\frac{b}{B_1}$, which we will consider increasing from 0 to 1 (Figs. 3 and 4).

Conclusions:

(1) the nonlinearity coefficient $u_{21}\begin{cases} < -2, \beta < 1, \\ > -2, \beta > 1, \end{cases}$, $\lim\limits_{q\to\infty, B_1\to\infty} u_{21} = -2$ and this coefficient decreases with increasing β, and, with increasing q and $\frac{b}{B_1}$, it decreases for $\beta < 1$ and increases for $\beta > 1$;

(2) the sum of variations $S_{21}\begin{cases} < -0,67, \beta < 1, \\ > -0,67, \beta > 1, \end{cases}$ $\lim\limits_{q\to\infty, B_1\to\infty} S_{21} = -0,67$, in addition, this sum increases with increasing β, and, with increasing q and $\frac{b}{B_1}$, it increases for $\beta < 1$ and decreases for $\beta > 1$.

To generalize, with the linear type of demand and power-law cost functions, the player's assumption about the environment' response significantly depends on the concavity or convexity of these functions. As the exponent β increases, including the transition from positive to negative economies of scale, the asymmetry of players in

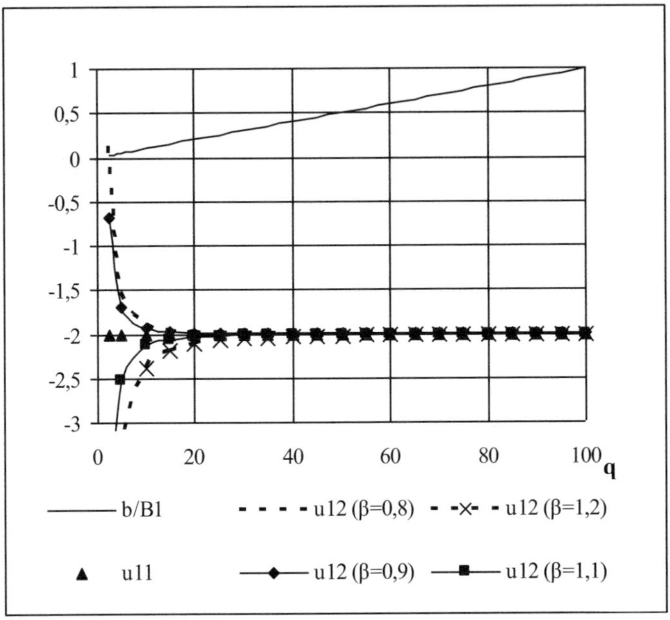


**Fig. 3** Nonlinearity coefficient for linear demand function and power-law costs

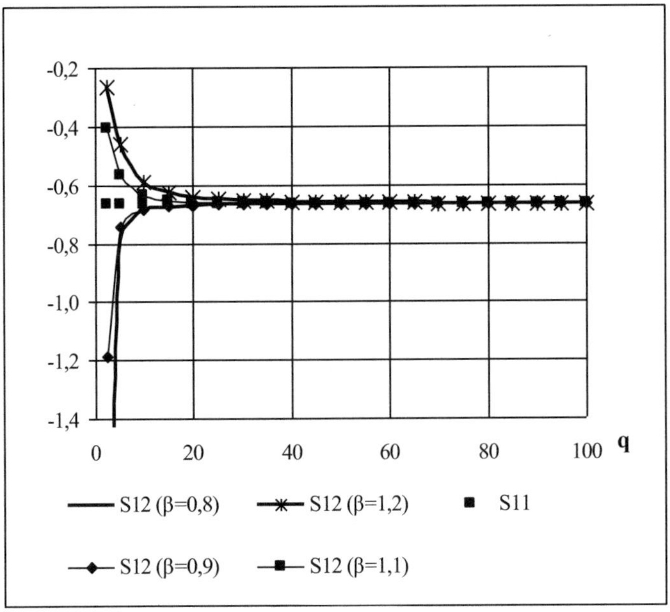


**Fig. 4** Sum of variations for linear demand function and power-law costs

terms of costs increases, and, therefore, the gaming position of the player with a large value of β worsens. Hence, as β increases, the module of the environment's negative reaction expected by the player in response to the increase in the player's action decreases. In other words, the player with higher costs has potentially less impact on the game equilibrium. However, as the coefficient $B_1$ decreases, the influence of the exponent β gradually levels out, and the conjectural variation approaches the level inherent in linear costs. For large values $B_1$ and small values of q, the difference between concavity and convexity plays the most important role. In the case of a positive scale effect (i.e., $\beta < 1$), an increase in $B_1$ causes an increase in the module of $S_{12}$, and in the case of a negative scale effect (i.e., $\beta > 1$), it causes a decrease in this indicator. Thus, the positive scale effect compensates for the increase in the coefficient $B_1$, and the negative one enhances this influence.

Thirdly, in the case of power-law demand function (9), we analyze the influence of power-law cost function (11) on S, that is, we study the difference of $u_{22}$ from $-2$ and $S_{22}$ from $-0.67$ for different values of the cost elasticity $\beta \in (0, 2)$, the demand elasticity $|\alpha| < 1$, the constant Q = 100, variable ratios $\frac{q}{Q} \in [0, 1]$ and $\frac{B_1}{A} \in [0; 0, 2]$ (Figs. 5 and 6). Using numerical experiments, it was established that in the studied ranges of parameters it is possible to approximately replace $\frac{q^{\beta-2}}{Q^{\alpha-1}} = \frac{\beta}{10}\left(\frac{q}{Q}\right)^{\frac{|\beta-2|}{\alpha-1}}$, therefore we present formula (14b) in the form: $u_{22} = -2 + (1-\alpha)\frac{q}{Q} - \frac{B_1}{A|\alpha|}\beta(\beta-1)\frac{\beta}{10}\left(\frac{q}{Q}\right)^{\frac{|\beta-2|}{\alpha-1}}$. Conclusions:

(1) nonlinearity coefficient $u_{22}\begin{cases} > -2, \beta < 1, \\ < \vee > -2, \beta > 1, \end{cases}$ and this coefficient
(a) with growth $\frac{q}{Q}$ changes non-monotonically (first decreases and then increases) for $\beta < 1$ and monotonically increases for $\beta > 1$;
(b) with growth $|\alpha|$ it increases for $\beta < 1$ and decreases for $\beta > 1$;
(c) with growth $\frac{B_1}{A}$ it increases for $\beta < 1$ and decreases for $\beta > 1$;
(2) the sum of variations $S_{22} < \vee > -0, 67$, and this sum
(a) with growth $\frac{q}{Q}$ changes non-monotonically (initially increases and then decreases), has the second kind discontinuities for $\beta < 1$ and decreases monotonically for $\beta > 1$;
(b) with growth $|\alpha|$ decreases (except for intervals with discontinuities) for $\beta < 1$ and increases for $\beta > 1$;
(c) with growth $\frac{B_1}{A}$ decreases (except for intervals with discontinuities) for $\beta < 1$ and increases for $\beta > 1$.

In general, with the power-law demand and costs functions, in the case of the convex cost functions (i.e., $\beta > 1$), the player's assumption about the environment's response clearly increases in absolute value with an increase in its market share, and decreases in absolute value with increasing the parameter $B_1$ and elasticity of demand. In the case of a positive scale effect (i.e., $\beta < 1$), the dependence of the player's assumption about the environment's response on the market share is non-monotonic.

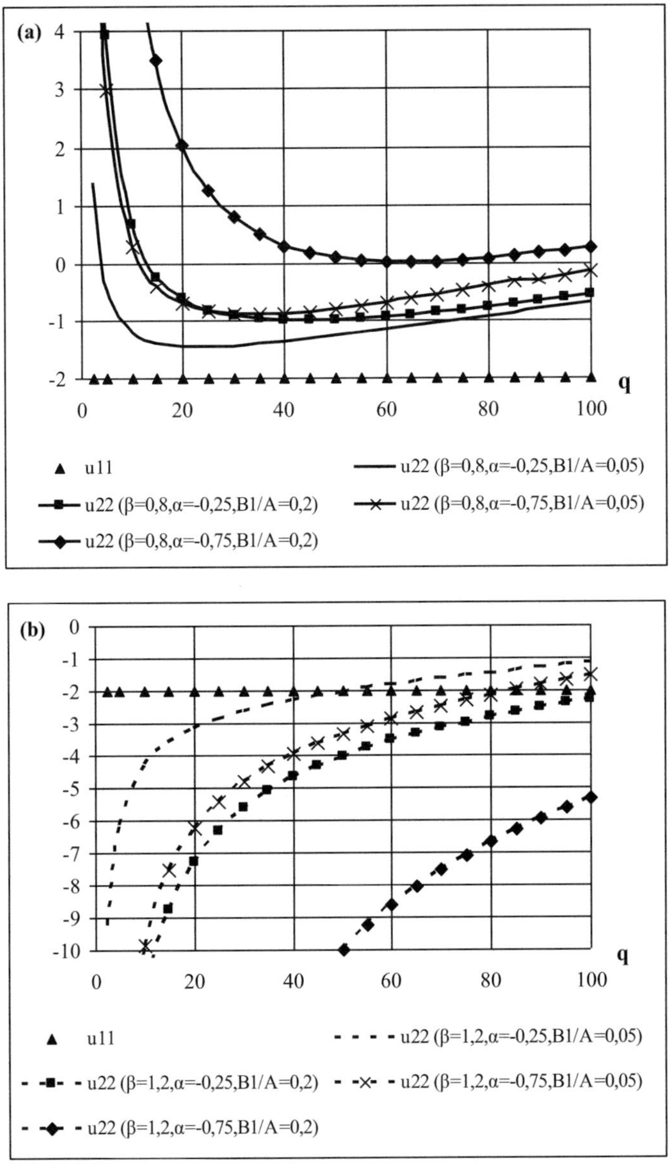


**Fig. 5** **a** Nonlinearity coefficient for power-law demand functions and costs, $\beta = 0.8$. **b** Nonlinearity coefficient for power-law demand functions and costs, $\beta = 1.2$

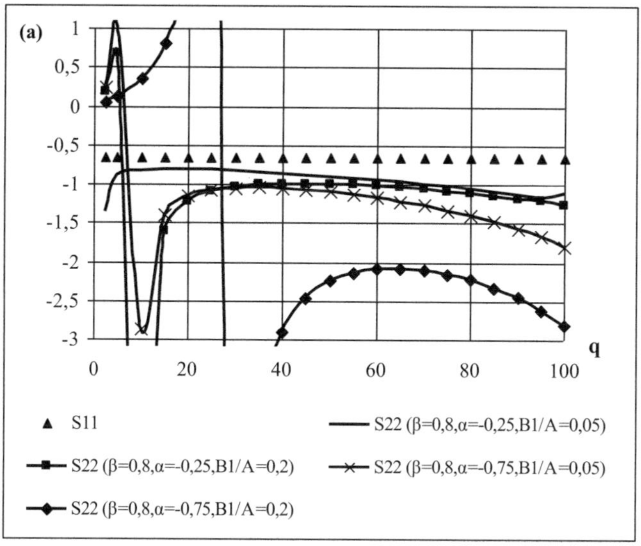


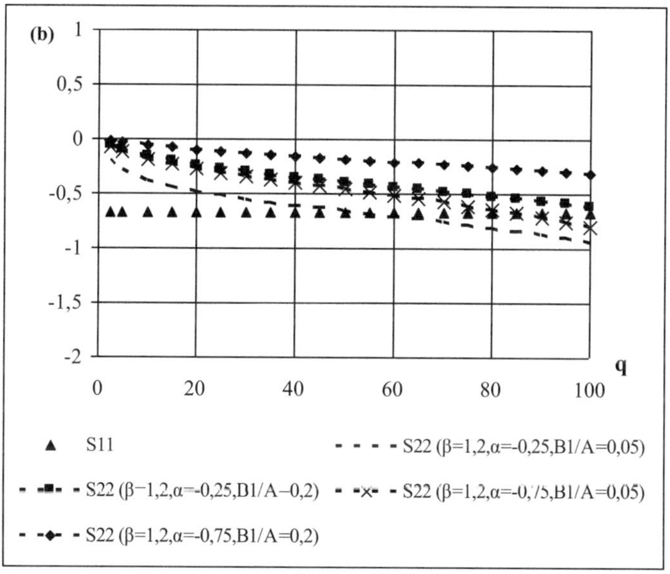


**Fig. 6** **a** Sum of variations for power-law demand functions and costs, $\beta = 0.8$. **b** Sum of variations for power-law demand functions and costs, $\beta = 1.2$

Fourthly, in the case of power demand function (9), we analyze the influence of quadratic cost function (12) on S, that is, we study the difference of $u_{23}$ from $-2$ and $S_{23}$ from $-0.67$ for different values of the demand elasticity $|\alpha| < 1$, the constant Q $= 100$, variable ratios $\frac{q}{Q} \in [0, 1]$ and $\frac{B_2}{A} \in [0; 0, 2]$ (Figs. 7 and 8). Conclusions:

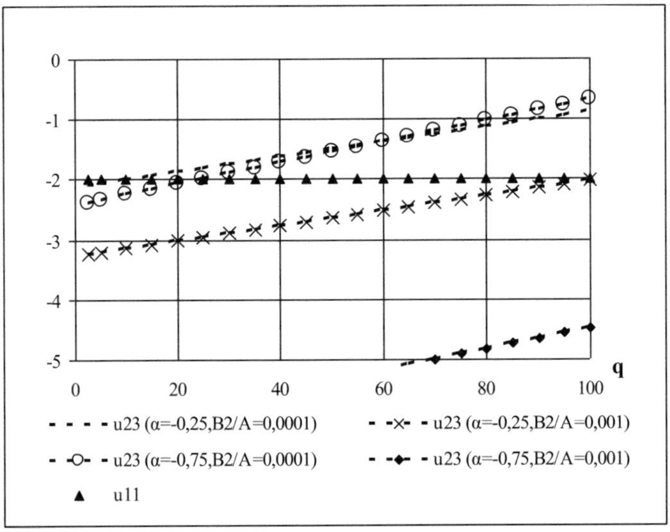


**Fig. 7** Nonlinearity coefficient for power demand function and quadratic costs

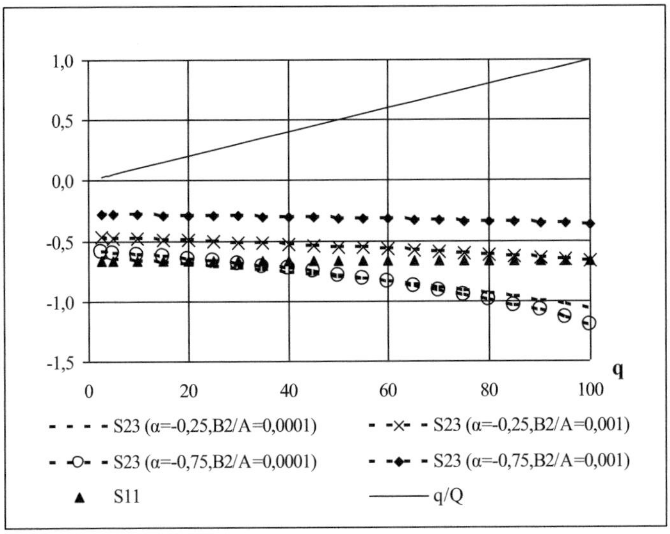


**Fig. 8** Sum of variations for power demand function and quadratic costs

(1) the nonlinearity coefficient $u_{23} < \vee > -2$, in addition, it increases with increasing in the parameter $\frac{q}{Q}$ and decreases with increasing in the coefficient $B_2$, and it can depend non-monotonically on the demand elasticity $|\alpha|$, in particular, for small values of $B_2$;
(2) the sum of variations $S_{23} < \vee > -0,67$, and it decreases with increasing in the parameter $\frac{q}{Q}$, increases with increasing in the coefficient $B_2$, and it can depend non-monotonically on the demand elasticity $|\alpha|$.

In general, these considerations mean that in the case under consideration (i.e., the power-law demand function and the quadratic cost function), the sum of the player's variations is negative. The modulus of this sum increases with the growth of the player's market share and decreases with the growth rate of costs. In addition, the modulus of this sum is smaller than for the linear case if the player's market share is small, and otherwise it is higher than in the linear case.

## 5 Conclusion

We summarize the results obtained on the basis of a comparison of the linear model (i.e., the linear demand function and the linear cost functions) and a variety of nonlinear models.

The linear-quadratic model (i.e., the linear demand function and the quadratic cost functions) shows a less radical player's assumption about a decrease in the environment's sales volume as a result of the player's increase in the volume than the linear model. The reason is that the cost level increases with the quadratic cost function compared to the linear case. Therefore, with a smaller decrease in sales, the environment achieves the same cost reduction that requires more effort in the linear case.

The model with the power-law demand function and the linear cost functions provides greater value for the modulus of reducing the environment's action relative to the linear model, therefore, the environment tends to reduce the aggregate price to greater extent. Probably, this is caused by sharper increase in the price due to a decrease in the action aggregate in the case of the power-law demand function compared to the linear one.

The linear-power model (i.e., the linear demand function and the power cost functions) exhibits different properties depending on the type of scale effect. With a positive scale effect (i.e., $\beta < 1$) the absolute value of the player's variations sum is larger, and with a negative scale effect (i.e., $\beta > 1$) it is smaller than in the linear case. The reason for this difference is similar to the aforementioned phenomenon: with a positive scale effect (i.e., the concave cost functions), the costs of the environment are smaller than in the linear case. Therefore, the environment can afford a larger reduction in sales. In the case of a negative scale effect, the situation is the opposite.

The power model (i.e., the case of the power functions of demand and costs) is the most multivariate among others, because the options of scale effects interfere with different rates of the demand function decline for small and large values of the player's market share. If this share is small, then with a positive scale effect (i.e., $\beta < 1$) the player can expect a positive sum of variations, that is, an increase in the volume of the environment sales due to the growth of the player's volume. If this share is large, then the player expects a decrease in the volume of the environment sales, but unlike the linear-power model, this decrease is greater than in the linear case. With a negative scale effect (i.e., $\beta > 1$) the player always expects a decrease in the volume of the environment sales, and this decrease is less than in the linear case if the player's market share is small, and vice versa otherwise.

The quadratic power-law model (i.e. the case of the power-law demand function and the quadratic cost functions) has the same characteristics as the power-law model with a negative scale effect (i.e., $\beta > 1$), because the quadratic cost function also has a negative scale effect. However, in this situation, the relationship between the cost growth rate and the elasticity of demand plays an important role. For highly elastic demand and low cost growth rates, the trend is similar to that in the power-law model above. However, for low-elastic demand and high cost growth rates, the player expects the environment to react as in the linear power-law model with a negative scale effect (i.e., $\beta > 1$): the reduction in the environment's sales volume is smaller compared to the linear model.

## References


1. Anderson, S.P., Erkal, N., Piccinin, D.: Aggregative games and oligopoly theory: short-run and long-run analysis. RAND J. Econ. **51**(2), 470–495 (2020)
2. Bowley, A.L.: The Mathematical Groundwork of Economics. Oxford Univers. Press, Oxford (1924)
3. Stackelberg, H.: Market Structure and Equilibrium, 1st edn. Translation into English, Bazin, Urch & Hill, Springer. (2011, Original 1934)
4. Novikov, D.A., Chkhartishvili, A.G.: Reflexion and Control: Mathematical Models. CRC Press, London (2014)
5. Julien, L.A.: On noncooperative oligopoly equilibrium in the multiple leader–follower game. Eur. J. Oper. Res. **256**(2), 650–662 (2017)
6. Geraskin, M.I.: The properties of conjectural variations in the nonlinear stackelberg oligopoly model. Autom. Remote. Control. **81**(6), 1051–1072 (2020)
7. Aizenberg, N.I., Zorkal'tsev, V.I., Mokryi, I.V.: Study of unsteady oligopoly markets. J. Appl. Industr. Math. **11**(1), 8–16 (2017)
8. Algazin, G.I., Algazina, Y.G.: To the analytical investigation of the convergence conditions of the processes of reflexive collective behavior in oligopoly models. Autom. Remote. Control. **83**(3), 367–388 (2022)
9. Fedyanin, D.N.: Monotonicity of equilibriums in Cournot competition with mixed interactions of agents and epistemic models of uncertain market. Procedia Comput. Sci. **186** (2021)
10. Lo, C.F., Yeung, C.F.: Quantum stackelberg oligopoly. Quantum Inf. Process. **21**(3), 85 (2022)
11. Ougolnitsky, G., Gorbaneva, O.: Sustainability of intertwined supply networks: a game-theoretic approach. Games **13**(3), 35 (2022)
12. Ugolnitsky, G.A., Usov, A.B.: Comparative analysis of the effectiveness of methods for organizing the interaction of economic agents in Cournot duopoly models, taking into account environmental conditions. Avtomatika i telemekhanika **2**, 150–168 (2023)

13. Filatov, A.Y.: Heterogeneity in the behavior of firms in an oligopolistic market: strategic firms and price takers. Izvestia Irkutsk State Univ. 13, 72–83 (2015)
14. Cornes, R., Fiorini, L.C., Maldonado, W.L.: Expectational stability in aggregative games. J. Evol. Econ. **31**(1), 235–249 (2021)
15. Geras'kin, M.I., Chkhartishvili, A.G.: Structural modeling of oligopoly market under the nonlinear functions of demand and agents' costs. Autom. Rem. Contr. **78**(2), 332–348 (2017)
16. Kanieski da Silva, B., Tanger, S., Marufuzzaman, M., Cubbage, F.: Perfect assumptions in an imperfect world: managing timberland in an oligopoly market. Forest Policy Econ. **137**, 102691 (2022)
17. Zhou, X., Pei, Z., Qin, B.: Assessing market competition in the Chinese banking industry based on a conjectural variation model. Chin. World. Econ. **29**(2), 73–98 (2021)
18. Mammadov, J.F., Abdullaev, K.S., Agaev, U.H., Aliev, I.R., Huseynova, G.G.: Searching and selection of a flexible manufacturing system by means of frame model. studies in systems, decision and control. Cyber-Physic. Syst. Model. Intel. Contr. 139–152 (2021)
19. Bolshakov, A., Veshneva, I., Lushin, D.: Mathematical model of integration of cyber-physical systems for solving problems of increasing the competitiveness of the regions of the Russian federation. In: Society 5.0: Cyberspace for Advanced Human-Centered Society. Studies in Systems, Decision and Control, pp. 129–139. Springer (2021)
20. Bolshakov, A.A., Kovalchuk, Y., Slobodyanyuk, L., Shashikhina, O.: A combined method for solving the problem of optimizing the production schedule of metal structure processing for use in a cyber-physical control system of a metallurgical enterprise. Cyber-Physic. Syst. Modell. Industr. Appl. 243–259 (2022)

# Airline Crew Scheduling System Based on Genetic Algorithm

**Oleg Protalinskiy, Vitaly Borodin, and Alexander Antyukhov**

**Abstract** The airline crew scheduling problem is considered. A multistage approach for constructing crew and aircraft schedules is described, as well as a two-stage approach for solving Airline Crew Pairing Problem. A mathematical model of the second sub-stage, the problem of efficient selection of pairings from the generated set, is constructed. The main constraints and the optimality criterion of the problem are formulated. The representation of the problem through the set covering problem is illustrated. The use of the genetic algorithm for constructing an efficient schedule is proposed. Also, a solution to the problem of schedule correctness using a minimal set of mappings is proposed. Different ways of using the minimum set were analyzed. An approach that adds correctness control at all stages of the genetic algorithm was chosen. The individual model, generation, and crossing algorithms considering correctness control are described. The results of the algorithm on data from five thousand flights are presented.

**Keywords** Airline crew scheduling · Pairing problem · Crew pairings · Genetic algorithm · Set covering problem

## 1 Introduction

Civil aviation is a very important transportation industry. According to the International Air Transport Association (IATA), total global air traffic in 2023 grew by 36.6% compared to 2022 and reached 94.1% of the pre-pandemic (2019) [1]. Airports Council International (ACI) estimates international passenger traffic at 8.7 billion in 2023 and forecasts 9.7 billion in 2024 [2].

O. Protalinskiy (✉)
National Research University "Moscow Power Engineering Institute", 14 Krasnokazarmennaya, Moscow 111250, Russia
e-mail: protalinskiy@gmail.com

V. Borodin · A. Antyukhov
Astrakhan State Technical University, 16 Tatishcheva, Astrakhan 414025, Russia



A. G. Kravets and A. A. Bolshakov (eds.), *Human-Centric Cyber-Society*, Studies in Systems, Decision and Control 639, https://doi.org/10.1007/978-3-032-06787-6_5

In the Russian Federation according to the Ministry of Transport in 2023 the volume of passenger transportation exceeded 105 million people [3]. Such volumes are provided by a large number of flights and flight hours of crews. Small and medium-sized airlines perform tens of thousands of flights per year, large airlines—hundreds of thousands and more. American Airlines Group performed about 2 million flights in 2022 [4], Turkish Airlines—more than 472 thousand flights in 2022 [5]. Nordwind performed 35 thousand flights in 2021 [6], Aeroflot Group—327 thousand flights [7]. To perform such a number of flights, a large staff is needed. According to the annual report of Aeroflot Group, the number of pilots in the airline amounted to more than 4 thousand people, the number of flight attendants—more than 11 thousand [8]. According to the report of UTair airline, the number of commanders in 2017 was 247, co-pilots 273, and flight attendants 1298 [9]. Crew schedules need to be planned in advance and also changed promptly.

Along with fuel costs, crew labor costs are the largest airline expense. In 1991, crew labor costs at American Airlines amounted to 1.3 billion dollars, at Norwest Airlines—1.05 billion dollars in 1989 [10]. Total labor costs for all employees amounted to 51,905 million rubles in Aeroflot Group in 2021 [8].

## 2 Problem Statement

Aircraft and crew scheduling is a very complex problem and usually consists of four steps [10] (Fig. 1):

1. Flight scheduling—at this stage, airline flights are planned, considering the requirements of the sales department and the capabilities of the company's fleet, as well as various constraints: for example, runway availability windows at airports.
2. Aircraft scheduling—at this stage, an aircraft from the company's fleet is assigned to each flight. Airlines try to reduce operating costs by taking into account various constraints—for example, the inability of certain airports to service certain aircraft models.
3. Anonymous crew scheduling (airline crew pairing problem)—at this stage flights are distributed among virtual crews in the form of pairings—cyclic routes starting and ending at the same base airport. The flights of the pairing are grouped into flight shifts separated by rest hours (Fig. 2). A correctly constructed pairing must satisfy a large number of legal restrictions [11, 12] and internal airline regulations.

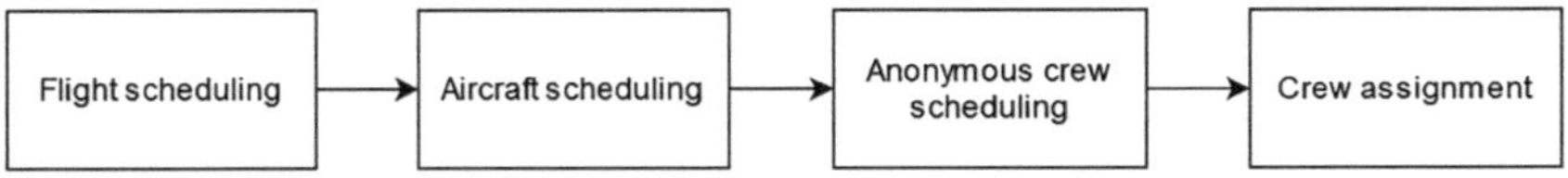


**Fig. 1** Aircraft and crew schedule planning

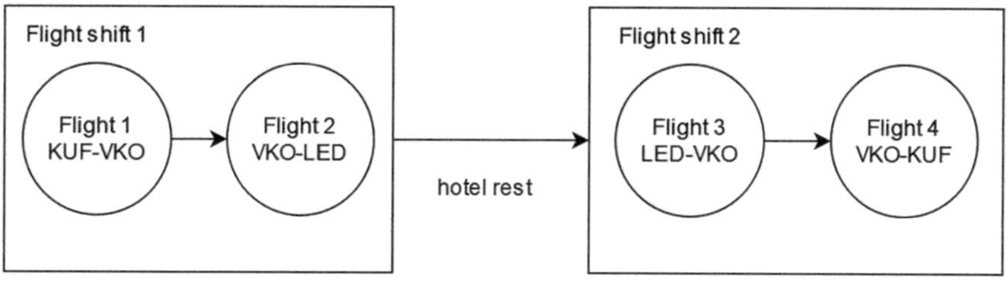


**Fig. 2** An example of pairing

4. Crew assignment (airline crew rostering problem)—at this stage, pairings are assigned to specific employees. it is necessary to consider the employee's position, schedule, personal vacation schedule, etc.

This chapter considers the third stage of the scheduling process, the airline crew pairing problem.

The aim of the work is to develop a system for scheduling routes of anonymous crews. Tasks of work are to develop a schedule model and scheduling algorithm. The input data of the system is the schedule of airline flights with assigned aircrafts. Output data is crew schedule. System constraints are the cyclic form of each crew route (a pairing should start and end at the same base airport) and schedule consistency—there should be no routes of different crews with the same flights in the schedule. Resource constraints—number of employees at the airport, prohibition of simultaneous departure of a certain number of employees from the airport—are not considered in this chapter.

A large number of papers have been devoted to ways of solving the airline crew pairing problem. Among the ways to solve the problem one can find the method of column generation [13, 14], branch and bound [10], heuristic methods such as ant algorithm [15, 16], particle swarm optimization [17], tabu search [18] and others. In some systems the problem is divided into 2 parts (Fig. 3):

1. Pairings generation—at this stage a large number of pairings are generated without taking into account the schedule consistency constraints and the number of employees at the base airport. In some cases, the stage is divided into two substages—the generation of shifts from flights and the generation of pairings from shifts [19]. In the generation stage, it's important to control satisfying legal and internal constraints of pairings. The number, quality and variety of pairings significantly influence the next stage.
2. Effective subset selection stage—at this stage the schedule is created by selecting pairings from the generated set.

In this chapter, the second stage is considered. As an input set could be used pairings generated by any route search algorithm, or with heuristic algorithms based on airline data, including previous versions of the calculation, updated for the required time range (in the case of a relatively unchanged aircraft schedule). Also manually calculated pairings could be used.

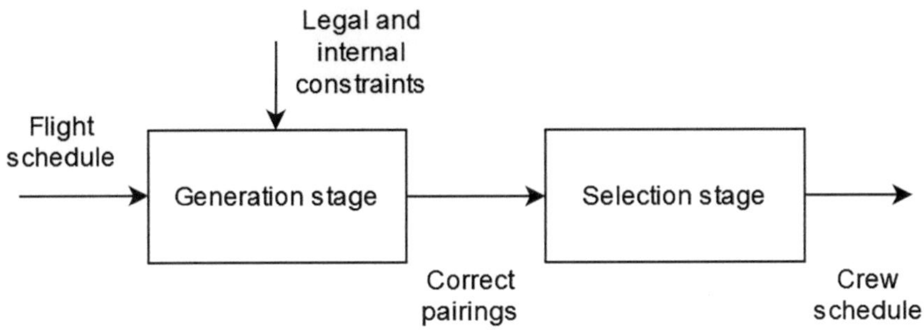


**Fig. 3** A two-stage algorithm for forming a crew schedule

The task of pairing selection could be formulated as follows. Let Flt = {$f_i$}, where *Flt* is set of airline flights *f,* for which the crew schedule should be built. Let Pr = {$p_j$}, where *Pr* is set of correct pairings *p* from generation stage. Every pairing consists of subset of flights: $p_j$ = {$f_k$...}, $f_k \in$ Flt. Every pairing could be assigned a *cost,* where cost ($p_i$ )> 0.

Let Sch = {p}, p ∈ Pr, where *Sch* is scheduling of pairings *p,* belonging to the total set of pairings *Pr*. To build a correct schedule, we can neither skip a flight nor assign more than one pairing to a flight (with some exceptions, two crews cannot do the same route). Let cpf(Sch, f) be a function, that returns a number of pairings of schedule *Sch,* containing a flight *f. T*he schedule is considered correct if ∀f ∈ Flt, cpf(Sch, f) = 1.

The optimality criterion is the total cost of the selected pairings. The scheduling algorithm should select the solution pairings to minimize it: $\sum(\text{cost } p_s) \rightarrow \min$, $p_s \in$ Sch. Usually, the cost of the pairing includes payment for the flight shift (daily or per flight), payment for hotel stays and payment for crew flights as passengers: $\text{cost}(p_s) = \text{cost}_{shift}(p_s) + \text{cost}_{hotel}(ps) + \text{cost}_{deadhead}(ps)$.

In this formulation, the selection problem could be represented as the set covering problem [10, 19, 20].

Figure 4 shows an example of the task. An airline needs to schedule six flights between Vnukovo (VKO), Pulkovo (LED) and Kurumoch (KUF) airports. At the first stage, pairings are generated, each pairing is a subset of a set of flights. The base airports are VKO and KUF, so pairings can only start with flights from these airports. The first departure airport of a pairing must be equal to the last arrival airport. The generation stage is followed by the selection stage, which can be represented through the set covering problem—it is necessary to select pairings so that the resulting schedule covers all elements without intersections. Different variants of schedules are possible, for example, the selection of three two-flight pairings, or one four-flight and one two-flight pairings.

The set covering problem is NP-hard, that's why it has no fast and accurate solution. The genetic algorithm [19–22] was chosen for the solution since it is quite flexible for further use of various heuristics specific to the subject area.

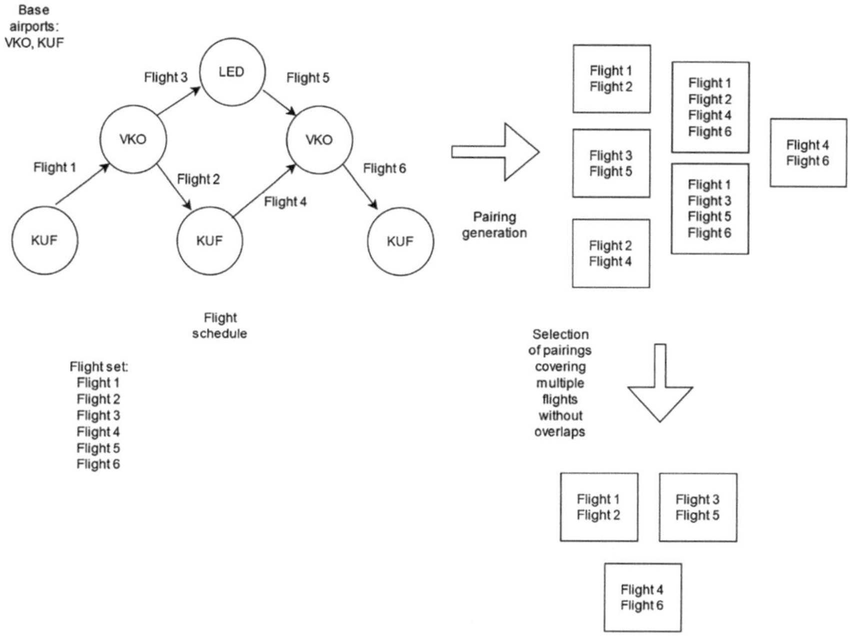


**Fig. 4** Representation of the problem through the set covering problem

## 3 Realization of the System of Pairing Selection Based on Genetic Algorithm

The schedule was chosen as the genetic algorithm's individual. The sum of the costs of the pairings was chosen as the value of adaptability. A binary genetic algorithm was chosen in terms of encoding the individual (Fig. 5). To encode an individual it is necessary to represent the set of all pairings *Pr* in the form of a bit array whose power is equal to the number of pairings in the total set. If the solution-individual contains a pairing *i*, then at position *i* in the bit array the value will be equal to one, if it does not, then zero. Mutations of an individual can be a change in the value of random bits to the reverse ones. Crossover can be performed by splitting the genome of parental individuals in half or into certain segments and joining the parts of two individuals into a single one. Knowing the adaptation values of individuals $\sum(\text{cost}(p_s))$, the genetic algorithm selects higher quality individuals-schedules generation by generation.

However, in such a representation the following difficulty arises: it is necessary to construct not only an efficient but also a correct schedule. Not every set of bit array values is a correct solution. A completely null bit array implies the absence of any pairings, an array of ones implies a schedule with a large number of repeated flights. Most randomly generated individuals will either lack flights or have an excessive number of flights.

| pairing 1 | pairing 2 | pairing 3 | pairing 4 | | pairing N-1 | pairing N |
|---|---|---|---|---|---|---|
| 1 | 0 | 1 | 0 | ... | 1 | 1 |

**Fig. 5** Binary representation of pairings in an individual

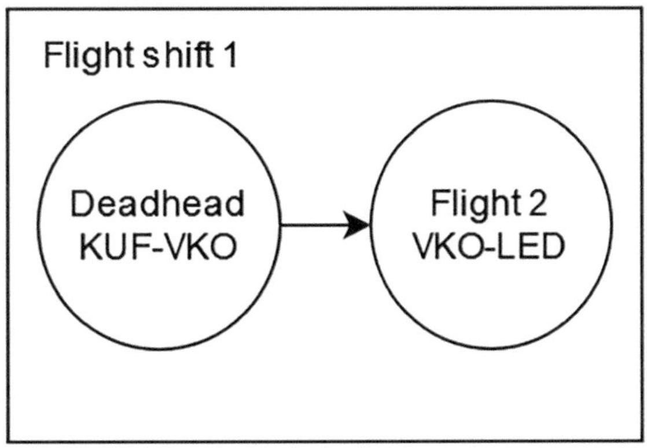


**Fig. 6** An example of pairing from a minimum set

First of all, to solve the problem, a *minimum set* was added to the set of pairings. A minimal set is a set of pairings, where each pairing contains only one flight to be performed by the crew. For the minimal set solution to make practical sense, the single-flight pairings are augmented with deadheads (Fig. 6). *Deadhead* is a flight where the crew flies as passengers.

For example, a Moscow-Samara flight may be supplemented by a Samara-Moscow flight in which pilots fly as passengers. This can be either a flight of the airline for which the schedule is built or a flight of another airline. The cost of tickets is included in the cost of such a pairing. Deadhead affects the cost of the pairing but is not taken into account when analyzing overlapping flights of schedule.

Thus, if for each flight from *Flt* a minimal pairing can be constructed, we avoid the situation where it is impossible to construct a schedule. In the worst and most expensive case, the schedule will consist only of minimum pairings.

Further, as part of the research process the following ways were tested to solve the problem of overlapping in individuals:

1. Penalties for intersecting or missing flights. This method showed rather low efficiency—the genetic algorithm did not converge to solutions without intersections.
2. Correcting intersections by creating new pairings—intersecting flights were replaced by deadheads. However, this significantly increased the cost of the individual and did not always lead to correct results—not every flight could be

replaced by a deadhead while keeping the other flights of the pairing (due to legal restrictions and internal rules of airlines, e.g. minimum time between flights).
3. Correction of individuals after generation and crossover. In this approach, a special algorithm tried to fix an already constructed individual. Missing values had to be supplemented with pairings that had no intersections with the current one or with pairings from the minimal set. This method allowed constructing correct schedules, however, they had a rather high cost due to a large number of expensive pairings from the minimal set. In addition, correcting all intersections of all individuals was quite time consuming.
4. Control of the correctness of individuals at all stages. In this method, the algorithm kept the correctness of individuals at the stage of their generation, crossing and mutations. This approach showed the best results.

To ensure the correctness, a boolean matrix *Imt* of pairing intersections by flight was created.

$$Imt = \begin{pmatrix} k_{0,0} \dots k_{0,m} \\ k_{1,0} \dots k_{1,m} \\ k_{m,0} \dots k_{m,m} \end{pmatrix}, k \in \{0, 1\}$$

If the pairing $p_i$ doesn't intersect with pairing $p_j$, then $Imt_{i,j} = 1$. If they intersect by any of the flights, then $Imt_{i,j} = 0$.

Let's consider an algorithm for creating a random individual (Fig. 7):

1. Create *Avb*, a bit array of ones. *Avb* is the set of allowable pairings.
2. Create *Slt*, a bit array of zeros. *Slt* is the current solution.
3. Choose random pairing index *ir*, from allowable pairings in *Avb*, whose value is equal to one.
4. Set the *ir* index bit in *Slt* to one.
5. Apply the bitwise AND operation to the bit array of allowable pairings *Avb* with the row *ir* of the matrix *Imt*.
6. Repeat operations 3–5 as long as there are ones in the *Avb* bit array.

Consider the random mutation algorithm:

1. Let *X* и *Y* are parent individuals.
2. Find *M* pairings that are not contained in *X* and *Y* individuals.
3. Create bit array *U*. For each pairing from *M* set the bit value to one.
4. Select random pairing from *U* by index *irm*.
5. Exclude intersecting pairings in *X, Y* and *U* using bitwise AND operation with row *irm* of matrix *Imt*.
6. Repeat the mutation algorithm for all *U* pairings.

Consider the crossover algorithm:

1. Select random pairing from *X* by index *irx*.
2. Remove the intersecting pairings in *Y* using a bitwise AND with *irx* row of the matrix *Imt*.

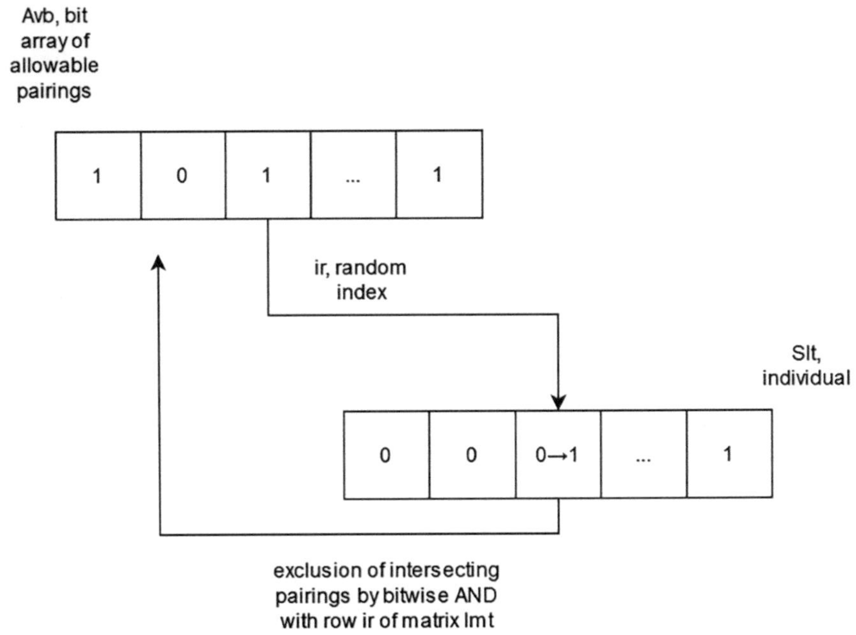


**Fig. 7** Algorithm of individual's generation

3. Similarly, select a random pairing from the remaining pairings in *Y* and remove the intersecting ones in *X*.
4. Repeat steps 1–3 until all pairings in *X* and *Y* will be non-overlapping. Create a descendant individual *Ch* and set the bits of the individual to match the remaining pairings in *X* and *Y*. At this stage, we have a correct but most likely incomplete pairing (Fig. 8).
5. Create *Avb,* a bit array from ones. *Avb* is the set of allowable pairings.
6. For the index *ich* of each pairing from *Ch*, apply a bitwise AND operation to *Avb* with a row of the matrix *Imt*.
1. 7. Select a random *ir* pairing from *Avb* similar to the individual generation algorithm and place it in the individual *Ch*.
7. Exclude intersecting pairings with *ir* in *Avb* using a bitwise AND with the row *ir* in the matrix *Imt*.
8. Repeat operations 6–8 until the *Avb* bit array becomes zero.

The initial results of the algorithm showed a variety of correct pairings with high cost. To reduce the cost, instead of creating completely random pairings, the generation algorithm was changed to a greedy (by pairing's cost) algorithm with elements of randomness.

Figure 9 shows the performance of the algorithm for input data of 5 thousand flights and more than 15 thousand pairings. The population size is 40 individuals.

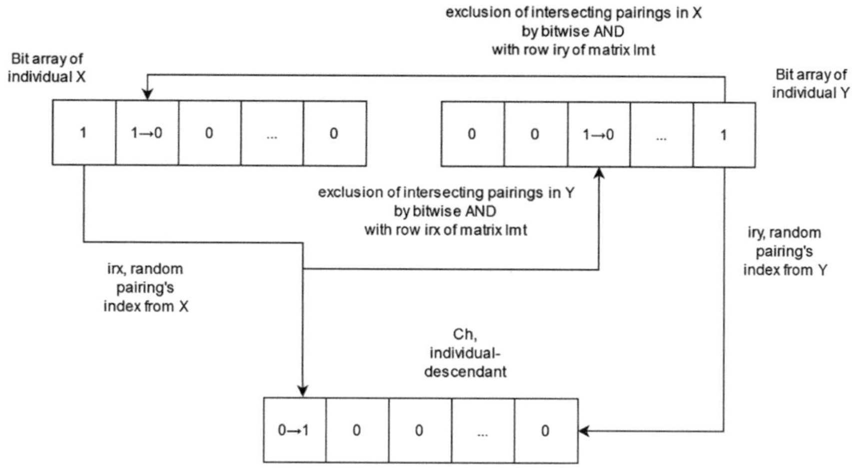


Fig. 8 Crossover of individuals taking into account intersections of pairings by flights

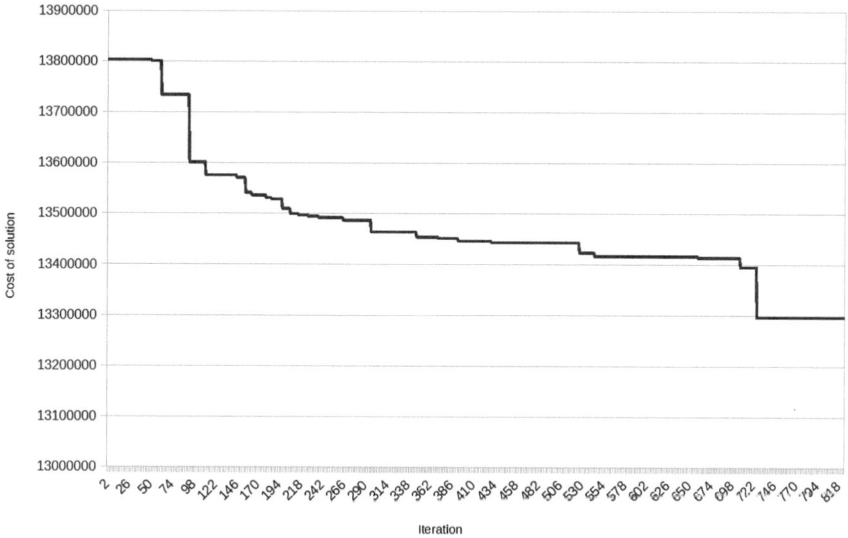


Fig. 9 Results of generic algorithm

# 4 Conclusion

As a result of this work, a system of schedule construction from the generated set of flight crew pairings was developed. The model of representation of pairings, flights and schedules in the system was developed, constraints and optimality criterion of the system were formulated. Different ways of ensuring the correctness of the schedule at the selection stage were investigated. A modification of the genetic algorithm was developed to ensure control of the correctness of an individual at each stage—using a minimal set of pairings. Runs of the system on test data have been carried out. The cost of schedules built by the system is 5–10% lower than manually built schedules for data sets of several thousand flights. In the future it is planned to extend the system with the help of different algorithms of schedule modification—for example, search or heuristic algorithms, as well as with genetic algorithm improvements.

## References

1. Global Air Travel Demand Continued Its Bounce Back in 2023. IATA. URL: https://www.iata.org/en/pressroom/2024-releases/2024-01-31-02/ (2023)
2. The trusted source for air travel demand updates. ACI World. URL: https://aci.aero/2024/02/13/the-trusted-source-for-air-travel-demand-updates
3. Vitaly, S.: told the senators of the Federation Council about the implementation of the Transport Strategy. Ministry of Transport of the Russian Federation. URL: https://mintrans.gov.ru/press-center/news/11097 (date of visit 03/20/2024)
4. American Airlines reports fourth-quarter and full-year 2023 financial results. American Airlines. URL: https://americanairlines.gcs-web.com/news-releases/news-release-details/american-airlines-reports-fourth-quarter-and-full-year-2023
5. Turkish Airlines. Annual Report 2022. URL: https://investor.turkishairlines.com/documents/yillik-raporlar/2022-yillik-faaliyet-raporu_en.pdf
6. Nordwind summed up the results of 2021 Official website of Nordwind Air-lines. URL: https://nordwindairlines.ru/ru/news/240 (in Russian)
7. Operational statistics Aeroflot. URL: https://ir.aeroflot.ru/ru/reporting/traffic-statistics/
8. Annual report 2021 Aeroflot. URL: https://ir.aeroflot.ru/fileadmin/user_upload/files/rus/common_info/gosa_doc_2022/Aeroflot_AR21_RUS_book_web.pdf
9. UTair Airlines—Personnel. URL: https://www.utair.ru/about/today/personal/
10. Andersson, E., Housos, E., Kohl, N., Wedelin, D.: Crew pairing optimization. In: Operations Research in the Airline Industry, pp. 228–258. Springer, US, Boston, MA (1998)
11. Order of the Ministry of Transport of the Russian Federation dated November 21, 2005 N 139 "On approval of the Regulations on the peculiarities of working hours and rest time for crew members of civil aviation aircraft of the Russian Federation" (with amendments and additions).System GARANT. URL: http://base.garant.ru/189086/#ixzz6vNh5yzp (2005)
12. CFR § 91.1059-Flight time limitations and rest requirements: One or two pilot crews. Cornell Law School. URL: https://www.law.cornell.edu/cfr/text/14/91.1059
13. Anbil, R., Forrest, J.J., Pulleyblank, W.R.: Column generation and the airline crew pairing problem. Doc. Math. **3**(1), 677 (1998)
14. Zeren, B., Özkol, I.: A novel column generation strategy for large scale airline crew pairing problems. Expert Syst. Appl. **55**, 133–144 (2016)
15. Crawford, B., Castro, C., Monfroy, E.: A hybrid ant algorithm for the airline crew pairing problem. In: MICAI 2006: Advances in Artificial Intelligence: 5th Mexican International Conference on Artificial Intelligence, Apizaco, Mexico, November 13–17, pp. 381–391, Proceedings 5. Springer Berlin Heidelberg (2006)
16. Saemi, S., Komijan, A.R., Tavakkoli-Moghaddam, R., Fallah, M.: Solving an integrated mathematical model for crew pairing and rostering problems by an ant colony optimisation algorithm. Euro. J. Industr. Eng. **16**(2), 215–240 (2022)
17. Azadeh, A., Asadipour, G., Eivazy, H., Nazari-Shirkouhi, S.: A unique hybrid particle swarm optimisation algorithm for simulation and improvement of crew scheduling problem. Int. J. Operat. Res. **13**(4), 406–422 (2012)

18. Caserta, M.: Tabu search-based algorithm for large scale crew scheduling problems. Contaduría y Administración **215**, 49–70 (2005)
19. Kornilakis, H., Stamatopoulos, P.: Crew pairing optimization with genetic algorithms. In: Hellenic Conference on Artificial Intelligence, pp. 109–120. Springer Berlin Heidelberg, Berlin, Heidelberg (2002)
20. Aggarwal, D., Saxena, D. K., Bäck, T., Emmerich, M.: Real-world airline crew pairing optimization: customized genetic algorithm versus column generation method. In: International Conference on Evolutionary Multi-Criterion Optimization, pp. 518–531. Springer Nature Switzerland, Cham (2023)
21. Simon, D.: Evolutionary optimization algorithms. John Wiley & Sons (2013)
22. Beasley, J.E., Chu, P.C.: A genetic algorithm for the set covering problem. Eur. J. Oper. Res. **94**(2), 392–404 (1996)

# Cargo Port Risk Management Ontological Model

**Irina Bondareva, Anna Khanova, and Mendes Alain August**

**Abstract** The possibility of using digital business models for managing the logistics processes of a cargo port to gain a competitive advantage through research and full development of the relationships between the risks of a transport logistics enterprise is considered. A method for cognitive modeling of a cargo port based on ontological modeling is proposed. The structure of the elements of the ontological model of cargo port risk management, based on the application of the concept of multi-level goal setting, is described and formalized. Risks are considered at three levels of management of the organizational system (cargo port): strategic, tactical and operational. Situations of failure to achieve goals, as well as failure of indicators to achieve various levels of standard values are considered as risks. The ontology allows you to trace all sorts of connections between risks, goals and indicators and, through queries, formulate management decisions to correct existing risk situations. A formalized description of the classes, object properties, and data properties of the ontological model of cargo port risk management is presented, and the main restrictions that define the connections between the classes of the ontological model and their instances are formulated and described. The possibility of using the Protégé toolkit is described, which allows, based on the constructed ontology structure, to identify implicit connections between its elements. The construction of this kind of ontology is possible for any organizational system, including cyber-physical.

**Keywords** Ontological model · Risk management · Balanced scorecard · Cargo port · Digital transformation

I. Bondareva (✉) · A. Khanova · M. A. August
Astrakhan State Technical University, Astrakhan, Russia
e-mail: i.o.bondareva@mail.ru

M. A. August
Université Catholique de L'Afrique de L'Ouest de Dakar, Dakar, Senegal



A. G. Kravets and A. A. Bolshakov (eds.), *Human-Centric Cyber-Society*, Studies in Systems, Decision and Control 639, https://doi.org/10.1007/978-3-032-06787-6_6

# 1 Introduction

In the Transport and Logistics industry, most companies have formed digital transformation strategies, and 55% of companies have begun to implement it [1]. The development and constant step-by-step use of digital business models for managing various business processes is the key to the stable development of companies in the market, as well as strengthening their competitive positions. At the same time, the Water Freight Transportation sector lags significantly behind other industry sectors in the application of digital business models (see Fig. 1).

A promising direction of development is the possibility of anticipating and anticipating undesirable events that are likely to occur, which can have a negative impact on the activities of the organization and lead to destructive consequences [2–7]. For most companies engaged in water cargo transportation, including cargo ports (CP), an urgent task is to develop an existing risk management mechanism.

Based on the Balanced Scorecard (BSS) in the form of a strategic map, the prospects and goals of the SOE were identified [8–15]. Failure to achieve a port's objective is the risk to the port associated with that specific objective. The structure of goals according to the BSC concept has a tree-like hierarchical structure; accordingly, failure to achieve any of the goals of a lower level necessarily has a negative impact on the general goal of the GP. The risk structure of a GP is based on the mutual influence of all structural elements.

Cause-and-effect relationships are supplemented with indicators for assessing the achievability of goals, the failure of which to achieve standard values is considered as risks at the tactical level of management. Operational level risks were created by detailing part of the assessment indicators according to a number of detail indicators: types of cargo, types of services provided and structural divisions.

The complex structure of the objectives of the GP BSC [16–18], the not always obvious mutual influence of many indicators, the different names of indicators within different divisions of one port and companies in the Water Freight Transportation

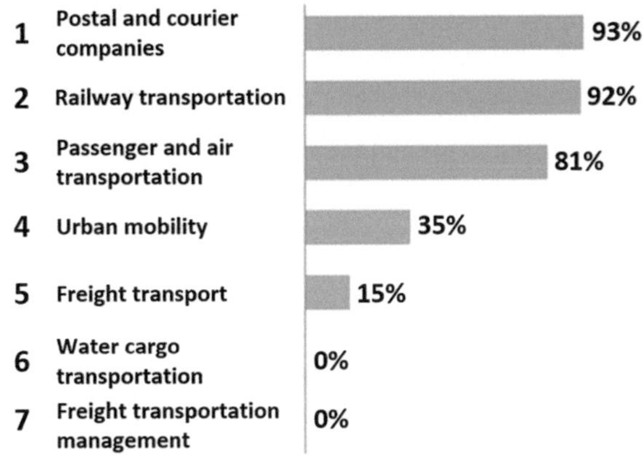


**Fig. 1** Application of digital models in the transport and logistics industry

sector as a whole, determined the semantic heterogeneity of the subject area and the choice of ontological modeling as a research tool [19–21].

The ontology model (OM) was implemented in the Protégé ontology editor using the ontology modeling language OWL (Web Ontology Language). The tools of this software package allow, based on the built connections between all the components of the ontology, to select various parameters according to certain conditions.

## 2 Ontology Model

The cargo port risk management OM is presented in the following form:

$$O_{CPRM} = \{OD^C, OD^O, OD^D, OD^A\},$$

where $OD^C$—are statements, corresponding to OM concepts (classes); $OD^O$—statements describing the object properties of OM concepts, used to indicate connections between concepts; $OD^D$—statements characterizing the properties of the data; $OD^A$—a set of axioms that define and formulate the main restrictions for all existing elements of the sets $OD^C$, $OD^O$, $OD^D$ [16]. The formalization of OM concepts $OD^C$ can be presented as:

$$OD^C = \{OD^C_G, OD^C_I, OD^C_R\},$$

where $OD^C_G$ represents the concept class "Goals" and contains information about the Goals formulated for the port, $OD^C_I$—the class "Indicators", contains information about the Indicators of the achievability of the designated goals; $OD^C_R$—class "Risks", contains information about Risks associated with goals and indicators.

Figure 2 shows the hierarchical structure of the ontological model classes as well as the designations of the remaining child concept classes.

The $OD^C_G$, representing the "Goals" concept class, is described by the following set of elements:

$$OD^C_G = \{GN, OD^C_{DG}, OD^C_{LGS}\},$$

includes an instance of $GN$ corresponding to the main goal of the cargo port (increasing competitiveness), and two child classes: $OD^C_{DG}$—"Subgoals", containing information about detailed (clarifying) Goals, as well as $OD^C_{LGS}$—"Goal Setting Level", containing information about the levels of goal setting, to which each of the detailed objectives relate to. According to the BSC and the LP model, four such levels were identified:

$$OD^C_{LGS} = \{OD^C_{1LGS}, OD^C_{2LGS}, OD^C_{3LGS}, OD^C_{4LGS}\},$$

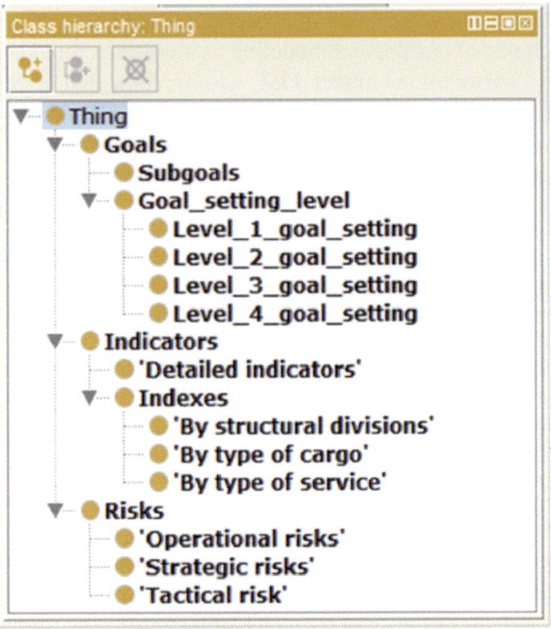


**Fig. 2** Ontology classes diagram

$OD_{1LGS}^{C}$, $OD_{2LGS}^{C}$, $OD_{3LGS}^{C}$, and $OD_{4LGS}^{C}$ include clarifying Goals of the appropriate level of goal setting (from 1 to 4).

All instances of the "Goals" class corresponding to the goals of the cargo port BSC are shown in Fig. 3.

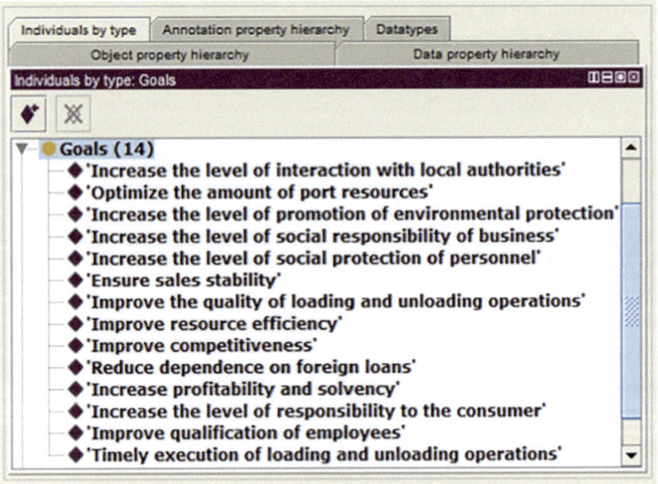


**Fig. 3** Instances of the "goals" class

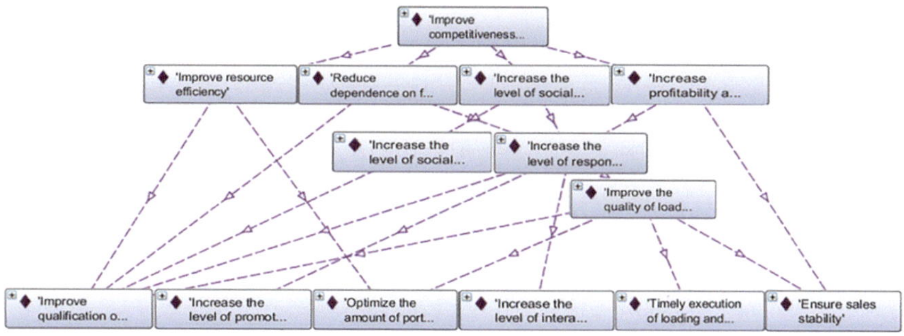


**Fig. 4** Cargo port goal tree

Let's consider the object properties features of the cargo port OM. To create a goal tree in the form of a multi-level structure of the OM, feedback and composite links were created. Reciprocal are the connections $OD^O_{DbG}$—"divided_by_goal", which allows you to subdivide the goal into subgoals that clarify it, and $OD^O_{CtG}$—"combined_into_goal", which allows you to combine subgoals into a large goal of a higher level. Based on the formulated goals, a tree of cargo port goals was built using the Protégé ontology editor (see Fig. 4).

Evaluation indicators are connected to indicators by the connection $DO^O_{DI}$—"details the indicator" and $DO^O_{DtI}$—"details on the indicator" (feedback to $DO^O_{DI}$) details all available indicators. The ontology contains three compound relations:

$$OD^O_{HI} = \langle OD^O_{SbI} \circ OD^O_{DtI} \rangle,$$
$$OD^O_{DoI} = \langle OD^O_{DbG} \circ OD^O_{SbI} \rangle,$$
$$OD^O_{DoDI} = \langle OD^O_{DbG} \circ OD^O_{SbI} \circ OD^O_{DtI} \rangle,$$

where the "o" operator defines property chains in owl terminology.

All of the above object properties of OM are described by the following set (see Fig. 5):

$$OD^O = \{OD^O_{DbG}, OD^O_{CtG}, OD^O_{EG}, OD^O_{SbI}, OD^O_{DI}, OD^O_{DtI}, OD^O_{F}, OD^O_{HI}, OD^O_{DoI}, OD^O_{DoDI}\}.$$

In the cargo port risk management OM, Data properties are specified to describe the properties of data instances of the "Indicators" class. The main components of this set are the following nested data properties:

$$OD^D = \{OD^D_{FI}\}, \quad OD^D_{FI} = \{OD^D_{I}\}, \quad OD^D_{I} = \{OD^D_{SI}\},$$

where $OD^D_{FI}$—represents the data property "Indicator Formula" and includes the composite auxiliary indicators $OD^D_{SI}$—included in the formula for calculating the indicator, by corresponding the data property "Indicator" $OD^D_{I}$ to the $OD^C_{I}$ class of the same name.

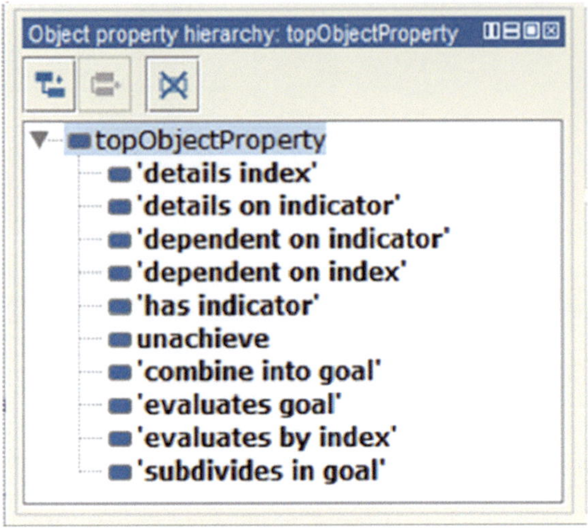


**Fig. 5** Object properties of the ontological model

The developed data property elements are presented in the Fig. 6.

The ontology allows you to work with numeric data types and perform calculations. Changes in the values of the components included in the calculation of a particular indicator are presented by selecting the values "increases", "decreases", or remains "without change". A description of all available properties of a specific instance of "Absolute liquidity ratio" of the "Indicators" class is given as an example in Fig. 7.

The $O_{CPRM}$ port risk management ontology model includes $OD^{A}$ constraints that define the relationships between classes and their instances. A total of 10 such restrictions were identified. Just a few of them are given as an example:

1. One goal (subgoal) can be assessed by several indicators (one-to-many relationship):

$$\mathrm{OD}_{G}^{C}\ \overrightarrow{1\mathrm{OD}_{\mathrm{SbI}}^{O}\infty}\ \mathrm{OD}_{I}^{C};$$

2. One indicator can evaluate several goals:

$$\mathrm{OD}_{I}^{C}\ \overrightarrow{1\mathrm{OD}_{\mathrm{EG}}^{O}\infty}\ \mathrm{OD}_{G}^{C};$$

3. One goal can be divided into several clarifying (detailed) goals. Or a goal can be related to another goal (including a subgoal) in a one-to-many relationship for the $OD_{DbG}^{O}$ object property:

$$\mathrm{OD}_{G}^{C}\ \overrightarrow{1\mathrm{OD}_{\mathrm{DbG}}^{O}\infty}\ \mathrm{OD}_{\mathrm{DG}}^{C};$$

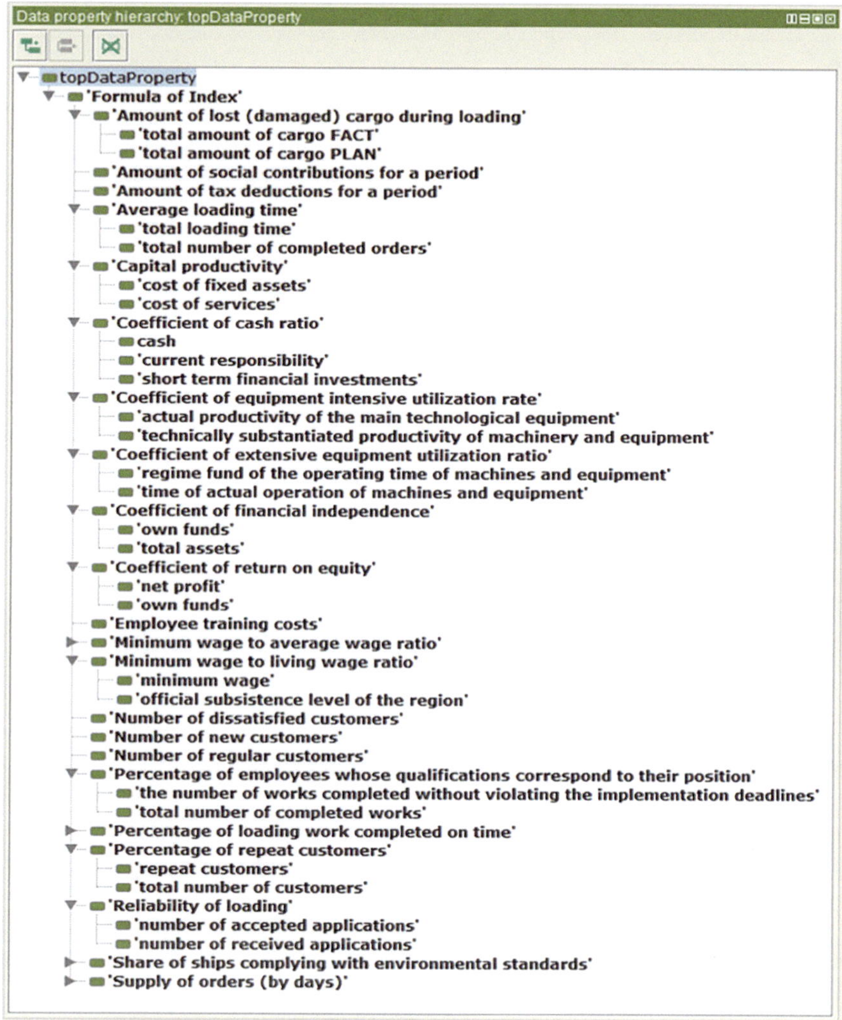


**Fig. 6** Data properties

4. One indicator can detail several indicators for assessing the achievement of port goals, i.e. any indicator has a one-to-many relationship with an indicator for an object property $OD_{DI}^{O}$:

$$\mathrm{OD}_{\mathrm{DI}}^{C}\ \overrightarrow{1\mathrm{OD}_{\mathrm{DI}}^{O}\infty}\ \mathrm{OD}_{I}^{C};$$

5. Any indicator can be detailed only by one of the three presented indicators of detail, i.e. any indicator has a one-to-one relationship with the detail indicator for an object property $OD_{DtI}^{O}$:

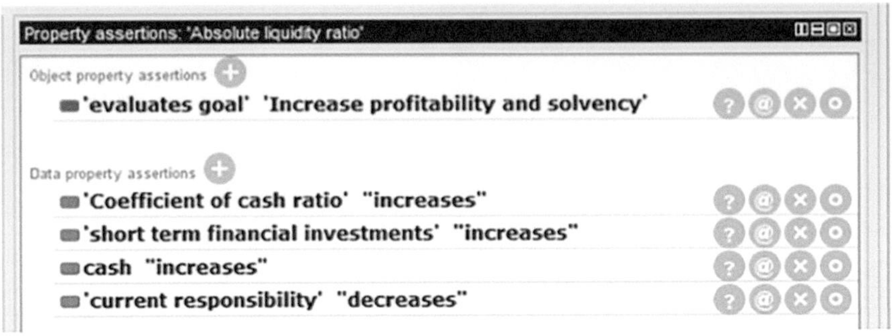


**Fig. 7** Description of the properties of the "absolute liquidity ratio" instance of the "indicators" class

$$\mathrm{OD}_{\mathrm{DI}}^{C}\ 1\overrightarrow{\mathrm{OD}_{\mathrm{DtI}}^{O}}\infty\ \mathrm{OD}_{I}^{C};$$

the same statement in a logical expression for a specific indicator will contain a strict disjunction:

$$\mathrm{OD}_{I}^{C} = \left(\mathrm{OD}_{\mathrm{DtI}}^{O}\mathrm{OD}_{\mathrm{ICT}}^{C}\right) \otimes \left(\mathrm{OD}_{\mathrm{DtI}}^{O}\mathrm{OD}_{\mathrm{ITS}}^{C}\right) \otimes \left(\mathrm{OD}_{\mathrm{DtI}}^{O}\mathrm{OD}_{\mathrm{ISS}}^{C}\right).$$

The OM of cargo port risk management includes all cause-and-effect relationships between elements defined by the concept of multi-level goal setting based on the BSC. Reasoner produces results based on logical transformations; it also gives an explanation (or several, depending on the degree of elaboration of the logical statements) to each of them, for example, a logical conclusion that the goal "Increase competitiveness" is dependent on the indicator "Tally department" explained in 36 different ways.

The Reasoner tool [22] allows you to trace all kinds of connections formed on the basis of constructing an ontology and setting up its elements and their properties. The Fig. 8 shows the results of the Reasoner tool for the "Improve resource efficiency" goal (implicit connections are highlighted in yellow).

Let it be necessary to find out the goals that depend on the indicator "Absolute liquidity ratio" and "Coefficient of extensive equipment utilization ratio". Let's imagine a situation in which it is necessary to find out which goals will be affected by these indicators when they change together. To do this, DL Query requires sampling based on several indicators. The result of the query shows that the joint change of these two coefficients affects the goals "Increase profitability and solvency" and "Increase competitiveness" (Fig. 9).

Next, using the SPARQL Query tool [23, 24], a request was made for the values of the indicators. As a result of the request, it was found that in order to increase the "Equipment Intensive Use Coefficient" indicator, it is necessary to reduce the "technically justified productivity of machinery and equipment" parameter and increase the "actual productivity of the main technological equipment".

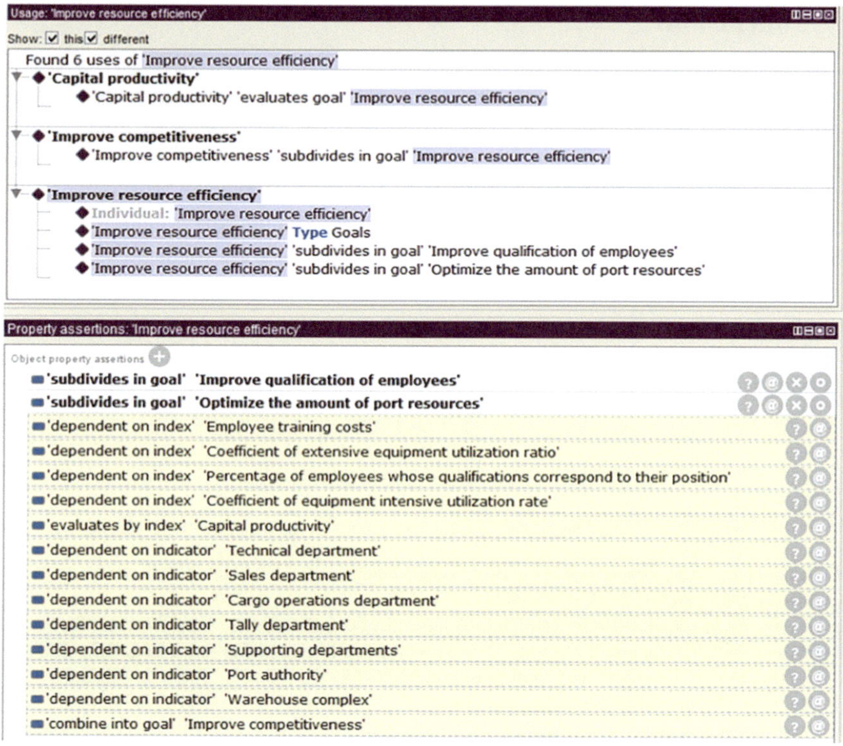


**Fig. 8** Reasoner tool inferences (for the "improve resource efficiency" goal)

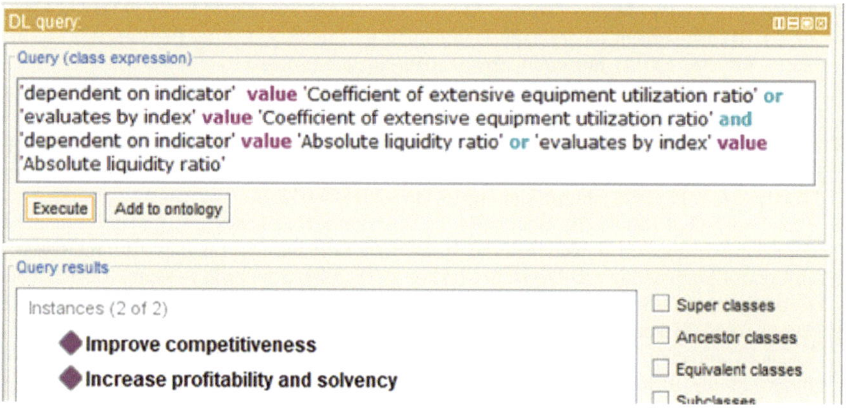


**Fig. 9** DL query and the result of the selection

## 3 Conclusion

The most significant advantage of using ontology in managing organizational systems is the integrity of the approach to managed business processes. By presenting a holistic view of the subject area under consideration, the ontological model allows us to comply with the principle of consistency. By unifying the material, i.e. By describing it in a single specified format, ontology implements the principle of uniformity, making the perception of ontology material more convenient for perception. Compliance by the ontology with the principle of complexity is manifested in the possibility of restoring and supplementing the missing logical connections between the elements of the ontology. The results of requests to a risk-based OM are essentially options for management decisions aimed at reducing risks in a cargo port.

## References

1. Gasparian, M.S., Kiseleva, I.A., Titov, V.A., Olenev, L.A.: Emergency risk assessment and management. Nexo Rev. Científ. **35**(1), 165–173 (2022)
2. Ho, M.W., Ho, K.H.: Risk management in large physical infrastructure investments: the context of seaport infrastructure development and investment. Maritime Econom. Log. **8**(2), 140–168 (2006). https://doi.org/10.1057/palgrave.mel.9100153
3. Ramírez-Nafarrate, A., González-Ramírez, R.G., Smith, N.R., et al.: Impact on yard efficiency of a truck appointment system for a port terminal. Ann. Oper. Res. **258**, 195–216 (2017). https://doi.org/10.1007/s10479-016-2384-0
4. Wee, H., Blos, M., Yang, W.-H.: Risk management in logistics. Intell. Syst. Ref. Libr. **33**, 285–305 (2012). https://doi.org/10.1007/978-3-642-25755-1_15
5. Khalil, M., Abdou, M.A., Mansour, M.S., Farag, H.A., Ossman, M.E.: A cascaded fuzzy-LOPA risk assessment model applied in natural gas industry. J. Loss Prevent. Process Ind. **25**(6), 877–882 (2012). https://doi.org/10.1016/j.jlp.2012.04.010
6. Cullinane, K., Bergqvist, R., Wilmsmeier, G.: The dry port concept: theory and practice. Maritime Econ. Log. **14**(1), 1–13 (2012). https://doi.org/10.1057/mel.2011.14
7. Peng, Y., Wang, W., Xu, X., Chen, M., Song, X., Li, X.: A simulation-based dynamic programming method for interchange scheduling of port collecting and distributing network. J. Adv. Transp. **54**, 1–17 (2018). https://doi.org/10.1155/2018/4805250
8. Protalinsky, O., Khanova, A., Bondareva, I., Averianova, K., Khanova, Y.: Cognitive model of the balanced scorecard of manufacturing systems. In: Dolinina, O., et al. (eds.) Recent Research in Control Engineering and Decision Making. ICIT 2020. Studies in Systems, Decision and Control, vol. 337. Springer, Cham (2021). https://doi.org/10.1007/978-3-030-65283-8_47
9. Strohhecker, J.: Factors influencing strategy implementation decisions: an evaluation of a balanced scorecard cockpit, intelligence, and knowledge. J. Manag. Control. **27**, 89–119 (2016). https://doi.org/10.1007/s00187-015-0225-y
10. Sandkuhl, K., Seigerroth, U.: Method engineering in information systems analysis and design: a balanced scorecard approach for method improvement. Softw. Syst. Model. **18**, 1833–1857 (2019). https://doi.org/10.1007/s10270-018-0692-3
11. Bondareva, I., Khanova, A.A.: Multi-level management of organizational systems on the basis of risk cascading, logical-probabilistic modeling and simulation. In: Kravets, A.G., Bolshakov, A.A., Shcherbakov, M. (eds.) Society 5.0: Human-Centered Society Challenges and Solutions. Studies in Systems, Decision and Control, vol. 416. Springer, Cham (2022). https://doi.org/10.1007/978-3-030-95112-2_13

12. Solozhentsev, E.: Logic and probabilistic risk models for management of innovations system of country. Int. J. Risk Assess. Manag. **18**(3–4), 237–255 (2015). https://doi.org/10.1504/IJRAM.2015.071211
13. Solozhentsev, E., Mityagin, S.: Logical and probabilistic risk models for assessment and analysis of the drug addiction problem in a region. Int. J. Risk Assess. Manag. **18**, 1–20 (2015). https://doi.org/10.1504/IJRAM.2015.068153
14. Protalinskiy, O., Andryushin, A., Shcherbatov, I., Khanova, A., Urazaliev, N.: Strategic decision support in the process of manufacturing systems management. In: Proceedings of the 2018 Eleventh International Conference on Management of Large-Scale System Development, pp. 1–4. MLSD, Moscow (2018). https://doi.org/10.1109/MLSD.2018.8551760
15. Solozhentsev, E., Karasev, V.: Hybrid logical and probabilistic models for management of socioeconomic safety. Int. J. Risk Assess. Manag. **21**(1–2), 89–110 (2018). https://doi.org/10.1504/IJRAM.2018.090258
16. Solozhentsev, E., Karaseva, E.: Data structures, logical-probabilistic models and digital management of the safety and quality of systems in the economics. Int. J. Risk Assess. Manag. **23**(1), 27–53 (2020). https://doi.org/10.1504/IJRAM.2020.106162
17. Bondareva, I.O., Shendo, M.V., Luneva, T.V., Khanova, A.A.: Logical-probabilistic and simulation modeling as a toolkit for complex analysis and risk management of a cargo port. E3S Web Confer. **224**, 02027 (2020). https://doi.org/10.1051/e3sconf/202022402027
18. Bondareva, I., Khanova, A., Khanova, Y.: Configuring systems based on petri nets, logic-probabilistic, and simulation models. In: Kravets, A.G., Bolshakov, A.A., Shcherbakov, M. (eds.) Cyber-Physical Systems: Modelling and Intelligent Control. Studies in Systems, Decision and Control, vol. 338. Springer, Cham (2021). https://doi.org/10.1007/978-3-030-66077-2_21
19. Sarkar, D.: Simulation application in project risk management for infrastructure transportation project. Int. J. Project Organ. Manag. **3**(3/4), 374–392 (2010). https://doi.org/10.1504/IJPOM.2011.042039
20. Gómez-Fuster, J.M., Jimenez, P.R.: Probabilistic risk modelling for port investments: a practical approach. Case Stud. Transp. Policy **8**, 822–831 (2020). https://doi.org/10.1016/j.cstp.2020.06.001
21. Dong, S., Yu, T., Farahmand, H., Mostafavi, A.: Probabilistic modeling of cascading failure risk in interdependent channel and road networks in urban flooding. Sustain. Cit. Soc. **62**, 102398 (2020). https://doi.org/10.1016/j.scs.2020.102398
22. Uschold, M., Gruninger, M.: Ontologies: principles, methods and applications. Knowl. Eng. Rev. **11**(2), 93–136 (1996)
23. Noy, N.F., McGuinness, D.L.: Ontology development 101: a guide to creating your first ontology. In: Technical Report. Stanford Knowledge Systems Laboratory, p. 25 (2001)
24. Gómez-Pérez, A., Fernández-López, M., Corcho, Ó.: Ontological engineering: with examples from the areas of knowledge management, e-commerce and the semantic web. Adv. Inform. Knowl. Process. **54**, 404 (2004)

# Environment and Ecology

# Multidimensional Statistical Analysis as a Tool for Analyzing the Yield of Grain Crops

**A. N. Tyrsin and Hasan Ahmed Mohammed Hasan**

**Abstract** The yield of agricultural crops is an integral indicator of the biological productivity of agro landscapes. Therefore, it is considered as a measure of the efficiency of the agricultural sector. The yield of the most important agricultural crops in Kirkuk, wheat, and barley, has been analysed. The influence of climatic and socio-economic factors on the yield of winter crops in Kirkuk for the period from 2000 to 2022 has been studied. Multidimensional methods of correlation and regression analysis and data classification were used for data analysis. Significant indicators included population engaged in crop production, gross domestic product per capita, normalized vegetation index, soil moisture profile, surface pressure, and wind speed. The study revealed the influence of socio-economic and climatic factors on the yield of winter grain crops.

**Keywords** Model · Yield · Grain crop · Climate · Correlation · Regression

## 1 Introduction

Agriculture is one of the most important components of the economic prosperity of the Republic of Iraq. Therefore, studying the impact of various factors on the efficiency of the agricultural sector is undoubtedly a relevant task [1, 2]. Many authors [3–6] note a decrease in soil fertility due to climate change and military actions,

A. N. Tyrsin (✉) · H. A. M. Hasan
Ural Federal University, Yekaterinburg, Russia
e-mail: at2001@yandex.ru

H. A. M. Hasan
e-mail: ahha.susu@gmail.com

A. N. Tyrsin
Engineering Center "Reliability and Resource of Large Systems and Machines", Ural Branch, Russian Academy of Sciences, Yekaterinburg, Russia

H. A. M. Hasan
Tikrit University, Tikrit, Saladin Province, Iraq

 
A. G. Kravets and A. A. Bolshakov (eds.), *Human-Centric Cyber-Society*, Studies in Systems, Decision and Control 639, https://doi.org/10.1007/978-3-032-06787-6_7

leading to reduced crop yields in crop production. However, crop production is a complex activity that depends not only on climatic conditions but also on several other components—economic level, social sphere, technological development [7, 8]. The yield of agricultural crops is an integral indicator of the biological productivity of ago-landscapes [9–11]. Therefore, it is precisely this indicator that is used as a measure of the efficiency of the agricultural sector. Establishing the most important factors influencing the yield of agricultural crops will allow for the adoption of well-founded programs and plans for implementing investment agricultural projects.

The purpose of the chapter is to use the methods of correlation-regression analysis and data classification to study, using the example of Kirkuk province, the influence of climatic and socio-economic factors on the agricultural sector of Iraq.

## 2 Data and Region of Research

The analyzed period is from 2000 to 2022. As factors influencing yield, we will consider both climatic and socio-economic factors. Climatic factors will include vegetation indices, soil moisture, air humidity and precipitation, surface pressure and wind characteristics, solar flux behavior, temperature indicators, with values generated monthly. Let's specify climatic factors by groups:

1. Vegetation indices: normalized difference vegetation index (NDVI), vegetation health index (VHI), agricultural stress index (ASI);
2. Soil wetness: surface soil wetness, profile soil moisture, root zone soil wetness;
3. Humidity and precipitation: specific humidity at 2 m, relative humidity at 2 m, adjusted precipitation;
4. Surface pressure and wind characteristics: surface pressure, wind direction at 2 m, wind speed at 2 m, wind direction at 10 m, wind speed at 10 m, wind speed at 2 m maximum, wind speed at 2 m minimum, wind speed at 10 m maximum, wind speed at 10 m minimum, wind speed at 2 m range, wind speed at 10 m range;
5. Solar fluxes and related: cloud amount, top-of-atmosphere shortwave downward irradiance, ultraviolet radiation of the entire sky surface, intensity of ultraviolet radiation over the entire sky surface, albedo of the entire sky surface, normal illumination of the entire sky surface with short-wave radiation downward, short-wave downward radiation of the entire sky surface, shortwave downward clear sky radiation, total area of the entire sky surface, diffuse shortwave radiation of the entire sky surface, total clear sky surface area;
6. Temperature indicators: earth skin temperature, temperature at 2 m, dew/frost point at 2 m, wet bulb temperature at 2 m, earth skin temperature maximum, earth skin temperature minimum, temperature at 2 m maximum, temperature at 2 m minimum, temperature at 2 m range, frost days, all sky surface long-wave downward irradiance. The main socio-economic indicators are the number

of people employed in crop production and gross domestic product (GDP) per capita.

Kirkuk is a city in Iraq and the capital of the Kirkuk Governorate (formerly known as Tameem), located 236 km north of the capital Baghdad (Fig. 1). It is built on the Khasa River and lies on the geographical coordinates of Latitude 35° 28′ 5″ Northing, Longitude 44° 23′ 31″ Easting, approximately 350 m above sea level, Kirkuk has a climate with extremely hot and dry summers and cool, rainy winters. The average annual temperature is daytime +28.1 °C, nighttime +19.6 °C. In the hottest month of August, the average temperature is 43.2 °C during the day and +32.0 °C at night. In the coldest month of January, the average temperature is 15.0 °C during the day and 7.8 °C at night. Observational data (climatic data) for the Kirkuk stations, utilized in the development of models' multivariate correlation and regression analysis techniques, were acquired from NASA's POWER project at the Langley Research Center, funded by the NASA Earth Science/Applied Science Program. GDP and per capita income data were obtained from the World Bank national accounts data, and OECD National Accounts data files. Data for winter agricultural crops (wheat and barley) were obtained from the Planning and Follow-up Department in Ministry of Agriculture (Iraq).

Population proportions data were obtained from the annual reports of the Iraqi Ministry of Planning. NDVI, VHI, and ASI data were procured from the United Nations Food and Agriculture Organization (FAO). Annual socio-economic data and crop yields were obtained from the Ministry of Agriculture of Iraq.

In Kirkuk, the main agricultural crops are wheat and barley. Therefore, in this study, we focus on these winter crops. Wheat and barley are winter crops, and their life cycle occurs from November to April. We will use the average values of climatic indicators over their life cycle as annual values.

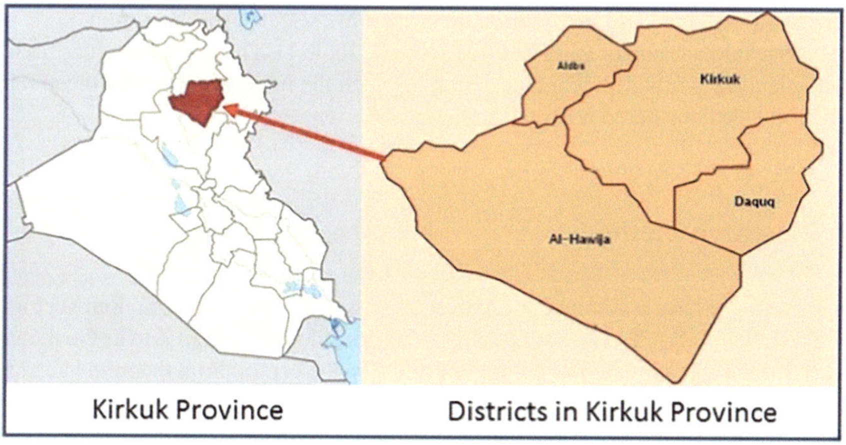


**Fig. 1** The location of Kirkuk

**Table 1** Analysed indicators

| Groups of indicators | Indicators |
|---|---|
| $\mathbf{X}_1$—population and GDP | $X_{11}$—population of Kirkuk, million people |
| | $X_{12}$—gross domestic product (GDP) per capita in Iraq in 2023 prices (to account for inflation), thousand US dollars |
| $\mathbf{X}_2$—vegetation indices | $X_{21}$—NDVI, unit |
| | $X_{22}$—VHI, unit |
| $\mathbf{X}_3$—soil and air humidity | $X_{31}$—soil moisture profile, unit |
| | $X_{32}$—specific humidity at 2 m, g/kg |
| $\mathbf{X}_4$—pressure and wind | $X_{41}$—surface pressure, kPa/10 |
| | $X_{42}$—wind speed at a distance of at least 10 m, m/s |
| $\mathbf{X}_5$—solar fluxes and temperature | $X_{51}$—downward shortwave radiation of clear sky, MJ/m$^2$/day |
| | $X_{52}$—temperature at 2 m, degrees celsius |
| $\mathbf{Y}$—grain crops | $Y_1$—wheat yield, tons/ha |
| | $Y_2$—barley yield, tons/ha |

When compiling the data, clarifications were made. Firstly, due to the prolonged armed conflict resulting from the US invasion of Iraq in 2003 and military clashes with ISIS in 2014–2015, statistical data for several years of certain socio-economic indicators are missing for individual provinces. Therefore: (1) instead of the population engaged in crop production, we use the population of Kirkuk; (2) instead of Kirkuk, we use the gross domestic product per capita correlation in Iraq. The proposed replacement indicators are closely correlated with the original ones, making this substitution appropriate. Secondly, a pronounced multicollinearity effect is observed in all groups of climatic indicators. To eliminate multicollinearity, the indicators most correlated with the yield of winter grain crops were selected in each group.

As a result, a group of variables was formed (Table 1). The absolute values of the coefficients of pairwise linear correlation between the indicators do not exceed 0.8, which can be considered acceptable.

## 3 Research Methods

In Table 1, we have a system $\mathbf{Z}$ of indicators from six subsystems (random vectors) $= \mathbf{Y} \cup \mathbf{X}, \mathbf{X} = \bigcup_{k=1}^{5} \mathbf{X}_k, \mathbf{X}_k = \left(X_{k,1}, X_{k,2}\right)$. We consider the system $\mathbf{Z}$ to be Gaussian: (1) the use of the normal distribution law relies on the central limit theorem [12]; (2) the analyzed sample is small, which does not allow us to estimate the distribution law of the random vector $\mathbf{Z}$; (3) testing the significance of linear correlation between all pairs of components based on the Fisher $F$-test [13] did not reject the null hypothesis.

Let's consider known indicators of the tightness of the correlation connection, which we will use for such a system. For Gaussian random vectors **Z**, a coefficient of tightness of the joint linear correlation connection, is proposed in [14], equal to

$$d_m(\mathbf{Z}) = 1 - |\mathbf{R_Z}|^{1/m}, \tag{1}$$

where $\mathbf{Z} = (Z_1, ..., Z_m)$ is a multidimensional random variable with a joint normal distribution and correlation matrix $\mathbf{R_Z}$.

Let's define $n$ continuous random vectors of arbitrary dimensions with no coincident components. Let's denote them as $\mathbf{Z}^{(1)} = \left(Z_1^{(1)}, ..., Z_{m_1}^{(1)}\right), \ldots, \left(Z_1^{(n)}, ..., Z_{m_n}^{(n)}\right)$. We assume that the components of these random vectors have variances and can be correlated. For this set of random vectors, let's introduce their union

$$\mathbf{Z} = \bigcup_{j=1}^{n} \mathbf{Z}^{(j)} = \left( \underbrace{Z_1, ..., Z_{m_1}}_{\mathbf{Z}^{(1)}}, \ldots, \underbrace{Z_{m_1+m_{n-1}+1}, ..., Z_{m_1+\cdots+m_n}}_{\mathbf{Z}^{(n)}} \right).$$

In [15], a scalar measure of the tightness of the relationship between random vectors $\mathbf{Z}^{(1)}, \ldots, \mathbf{Z}^{(n)}$ is introduced

$$d\left(\bigcap_{j=1}^{n} \mathbf{Z}^{(j)}\right) = 1 - \frac{\left|\mathbf{R}_{\bigcup_{j=1}^{n}\mathbf{Z}^{(j)}}\right|}{\prod_{j=1}^{n} |\mathbf{R}_{\mathbf{Z}^{(j)}}|}. \tag{2}$$

To consider the dimensionality of vector **Z**, similar to (1), we express (2) as

$$d_m\left(\bigcap_{j=1}^{n} \mathbf{Z}^{(j)}\right) = 1 - \left(\frac{\left|\mathbf{R}_{\bigcup_{j=1}^{n}\mathbf{Z}^{(j)}}\right|}{\prod_{j=1}^{n} |\mathbf{R}_{\mathbf{Z}^{(j)}}|}\right)^{1/m}. \tag{3}$$

The coefficient $d_m\left(\bigcap_{j=1}^{n} \mathbf{Z}^{(j)}\right)$ takes values from 0 to 1. If each of the elements of any subsystem functions independently of all elements of the other subsystems, then $d_m\left(\bigcap_{j=1}^{n} \mathbf{Z}^{(j)}\right) = 0$. If at least one element of at least one subsystem is linearly dependent on at least one element of at least one other subsystem, then $d_m\left(\bigcap_{j=1}^{n} \mathbf{Z}^{(j)}\right) = 1$. In the presence of correlation dependencies between elements of different subsystems $0 < d_m\left(\bigcap_{j=1}^{n} \mathbf{Z}^{(j)}\right) < 1$. Obviously, $d_m\left(\bigcap_{j=1}^{n} \mathbf{Z}^{(j)}\right) \le d_m(\mathbf{Z})$.

Formula (3) allows for a standardized quantitative measure of the tightness of correlation relationships. Let's consider its special cases:

1. If all $m_j = 1$, so $\mathbf{Z}^{(j)} = Z_j, j = 1, \ldots, m$, then formula (3) takes the form (1).
2. Let $n = 2$: $\mathbf{Z} = \mathbf{Z}^{(1)} \cup \mathbf{Z}^{(2)}$. Then the tightness of the correlation relationship between random vectors $\mathbf{Z}^{(1)}$ and $\mathbf{Z}^{(2)}$ is equal to

$$d_m\left(\mathbf{Z}^{(1)} \cap \mathbf{Z}^{(2)}\right) = 1 - \left(\frac{|\mathbf{R}_{\mathbf{Z}^{(1)} \cup \mathbf{Z}^{(2)}}|}{|\mathbf{R}_{\mathbf{Z}^{(1)}}| \bullet |\mathbf{R}_{\mathbf{Z}^{(2)}}|}\right)^{1/m}. \tag{4}$$

3. The tightness of the correlation relationship between the random vector $\mathbf{Z} = (Z_1, \ldots, Z_m)$ and the Gaussian random variable $U$, considering (4), is equal to

$$d_{m+1}(\mathbf{Z} \cap U) = 1 - \left(\frac{|\mathbf{R}_{\mathbf{Z} \cup U}|}{|\mathbf{R}_{\mathbf{Z}}|}\right)^{1/m+1}. \tag{5}$$

Let's define the marginal contribution of any component $X_j$ to the increase in the tightness of the correlation relationship between the random vectors $\mathbf{X}$ and $\mathbf{Y}$ as

$$\Delta d_m\left(\mathbf{X} \cap \mathbf{Y} | X_j\right) = \left(\frac{\left|\mathbf{R}_{(\mathbf{X} \backslash X_j) \cup \mathbf{Y}}\right|}{\left|\mathbf{R}_{\mathbf{X} \backslash X_j}\right| \bullet |\mathbf{R}_{\mathbf{Y}}|}\right)^{1/m} - \left(\frac{|\mathbf{R}_{\mathbf{X} \cup \mathbf{Y}}|}{|\mathbf{R}_{\mathbf{X}}| \bullet |\mathbf{R}_{\mathbf{Y}}|}\right)^{1/m}. \tag{6}$$

Also, due to the large number of variables, it is useful to employ the partial coefficient of pairwise linear correlation. For instance, for a Gaussian random vector $\mathbf{Z}$, the partial coefficient of pairwise linear correlation between variables $Z_j$ and $Z_l$ is [16]

$$R_{jl}(\mathbf{Z}) = -A_{jl}/\sqrt{A_{jj} \bullet A_{ll}}, \tag{7}$$

where $A_{jl}$, $A_{jj}$, $A_{ll}$ are algebraic complements of the elements $r_{jl}$, $r_{jj}$, $r_{ll}$ of the correlation matrix $\mathbf{R}_{\mathbf{Z}}$.

For the data from Table 1, there is no strong multicollinearity between the input indicators and there are no dominant variables strongly correlated with the $Y_j$ variables. There are also no significant errors in the measurement of input variables [13, 17]. Therefore, standard linear multivariate regression analysis can be used [18].

Along with multivariate correlation-regression modeling, statistical methods of data classification with training are of interest, for example, a combination of discriminant analysis [19] with binary logistic regression [20]. It makes it possible to reduce the computational errors inherent in directly finding logistic regression coefficients using the maximum likelihood criterion [21, 22]. First, a separating hyperplane equation is constructed from the data sample, from which a binary logistic regression model is then obtained. Based on this model, it can be used for monitoring tasks and developing management decisions to increase the yield of grain crops [23].

The equation for the separating hyperplane for two clusters is given by [19]

$$W(\mathbf{x}) = C - f(\mathbf{x}) = 0, \tag{8}$$

where $f(\mathbf{X}) = \sum_{j=1}^{M} a_j X_j$ is the discriminant function, $M$ is the number of variables (analysed climatic and socio-economic indicators), $C$ is the discrimination constant (if $W(\mathbf{x}) > 0$, then the observation under consideration $\mathbf{x}_i = (x_{i,1}, \ldots, x_{i,M})^{\mathrm{T}}$ is assigned to the first group, if $W(\mathbf{x}) < 0$ to the second group).

Vector **a** of the coefficients of the discriminant function and the discrimination constant $C$ in (8) are calculated using formulas $\mathbf{a} = \mathbf{S}_*^{-1}(\overline{\mathbf{X}}_1 - \overline{\mathbf{X}}_2)$, $C = (\overline{f}_1 + \overline{f}_1)/2$, where $\mathbf{X}_k$, $\overline{\mathbf{X}}_k = (\overline{\mathbf{X}}_{k,1}, ..., \overline{\mathbf{X}}_{k,M})^{\mathrm{T}}$ are the matrix of values and the vector of average values of indicators of the $k$-th group, $\overline{f}_k = \frac{1}{n_k}\sum_{i=1}^{n_k} f(\mathbf{x}_{ki})$, $\mathbf{S}_* = \frac{\mathbf{X}_1^T\mathbf{X}_1 + \mathbf{X}_2^T X_2}{n_1+n_2-2}$ is the combined covariance matrix, $n_k$ is the number of observations in the $k$-th group ($k = 1, 2$).

## 4 Results

Consideration of the annual yield indicators for wheat and barley showed their steady growth with a reliability of more than 99%. Figure 2 shows the dynamics of wheat and barley yields for 2000–2022, with regression equations for crop yields relative to time and coefficients of determination $R^2$ of the constructed dependencies. Table 2 presents the partial correlation coefficients between yields $Y_k$ and variables $X_j$ from 2000 to 2022, calculated according to (7).

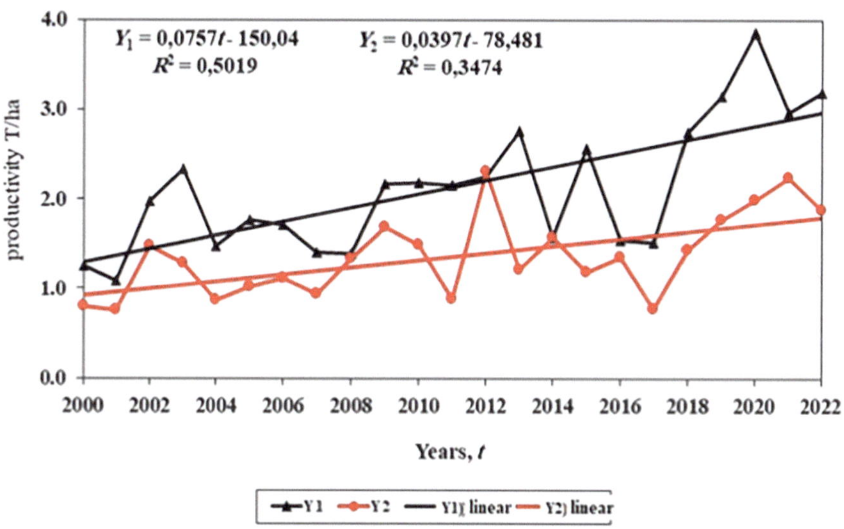


**Fig. 2** Illustrates the dynamics of wheat yield ($Y_1$) and barley yield ($Y_2$)

**Table 2** Matrix of partial correlation coefficients for the years 2000–2022

| | $X_{11}$ | $X_{12}$ | $X_{21}$ | $X_{22}$ | $X_{31}$ | $X_{32}$ | $X_{41}$ | $X_{42}$ | $X_{51}$ | $X_{52}$ |
|---|---|---|---|---|---|---|---|---|---|---|
| $Y_1$ | 0.71 | −0.64 | 0.40 | −0.27 | 0.15 | −0.20 | −0.68 | 0.40 | 0.29 | −0.11 |
| $Y_2$ | 0.36 | −0.10 | 0.00 | 0.08 | −0.21 | −0.10 | −0.12 | −0.03 | −0.11 | −0.12 |

Regression modeling of wheat and barley yields did not yield dependencies that satisfied all quality indicators. Therefore, due to the small sample size, variables with a $p$-level of <0.2 were included in the model. The best regression dependencies of wheat and barley yields are presented in Table 3 and 4, respectively. The low value of the coefficient of determination $R^2$ for Table 4 is explained by the fact that barley yield values for several years were based on incomplete statistical data. The results in Tables 3 and 4 correspond to the partial correlation coefficients from Table 2.

Let us now proceed to the multivariate correlation analysis of the data. The input and output variables are represented by vectors $\mathbf{X}$ and $\mathbf{Y}$. Therefore, in conducting the study, we will use the formulas (1)–(7) of multivariate correlation analysis as provided above. To examine the dynamics of the indicators, we will use subsets of 14 years: from 2000 to 2013, …, from 2009 to 2022. Reducing the size of the subsets leads to instability of the results. A comparative analysis of the correlation matrices showed that increasing the size of these subsets has little effect on the results. Therefore, to maintain the possibility of studying dynamics, we will focus on 14-year subsets. In the graphs below, the date implies that the result is calculated on a subset of 14 years, including the current year. For example, in 2013, the period from 2000 to 2013 is considered, and so on.

In Fig. 3 shows the dynamics of the tightness coefficients of the joint linear correlation relationship $d_m(\mathbf{Z})$, $d_m(\mathbf{X} \cap \mathbf{Y})$, $d_m\left(\bigcap_{j=1}^{6}\mathbf{Z}^{(j)}\right)$ from 2000 to 2022 are presented. Here $\mathbf{X}_1 = \mathbf{Z}^{(1)}$, …, $\mathbf{X}_5 = \mathbf{Z}^{(5)}$, $\mathbf{Y} = \mathbf{Z}^{(6)}$. Table 5 presents the values of coefficients $d_m(\mathbf{Z})$, $d_m(\mathbf{X} \cap \mathbf{Y})$, $d_m\left(\bigcap_{j=1}^{6}\mathbf{Z}^{(j)}\right)$ for the entire sample.

**Table 3** Regression model for wheat yield

| Variables | Coefficients | Standard errors | $t$ | $p$-value |
|---|---|---|---|---|
| Intercept | 706.8863 | 146.2285 | 4.8341 | 0.0002 |
| $X_{11}$ | 2.3741 | 0.4065 | 5.8400 | 0.0000 |
| $X_{12}$ | −0.2204 | 0.0598 | −3.6847 | 0.0018 |
| $X_{21}$ | 1.6164 | 1.1003 | 1.4691 | 0.1601 |
| $X_{41}$ | −72.8578 | 15.0553 | −4.8393 | 0.0002 |
| $X_{42}$ | 3.3434 | 2.1005 | 1.5917 | 0.1299 |

*Note* $R^2 = 0.8421$, $F(5,17) = 18.131$, $p < 0.0000$, standard error $s_e = 0.3275$

**Table 4** Regression model for barley yield

| Variables | Coefficients | Standard errors | $t$ | $p$-value |
|---|---|---|---|---|
| Intercept | 4.4661 | 2.3456 | 1.9040 | 0.0722 |
| $X_{11}$ | 0.8669 | 0.2336 | 3.7112 | 0.0015 |
| $X_{31}$ | −3.8344 | 2.1276 | −1.8022 | 0.0874 |
| $X_{52}$ | −0.0921 | 0.0638 | −1.4436 | 0.1651 |

*Note* $R = 0.4746$, $F(3,19) = 5.721$, $p < 0.0058$, standard error $s_e = 0.3563$

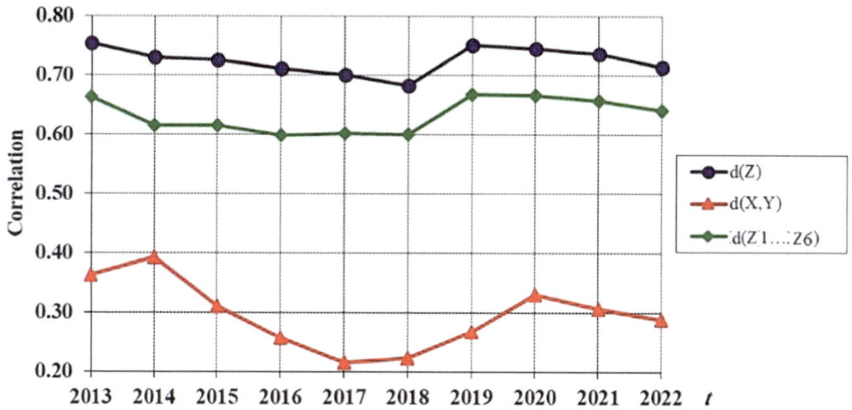


**Fig. 3** Coefficients $d_m(\mathbf{Z})$, $d_m(\mathbf{X} \cap \mathbf{Y})$, $d_m\left(\bigcap_{j=1}^{6} \mathbf{Z}^{(j)}\right)$

**Table 5** Coefficients $d_m(Z)$, $d_m(X \cap Y)$, $\boldsymbol{d_m}\left(\bigcap_{j=1}^{6} \mathbf{Z}^{(j)}\right)$

| $d_m(\mathbf{Z})$ | $d_m(\mathbf{X} \cap \mathbf{Y})$ | $d_m(\cap_{j=1}^{6} \mathbf{Z}^{(j)})$ |
|---|---|---|
| 0.609 | 0.180 | 0.468 |

We observe a strong joint correlation $d_m(\mathbf{Z})$ among all components of the $\mathbf{Z}$ set, and a correlation $d_m\left(\bigcap_{j=1}^{6} \mathbf{Z}^{(j)}\right)$ between sets $\mathbf{X}_1, \ldots, \mathbf{X}_5, \mathbf{Y}$. The significantly lower closeness of correlation $d_m(\mathbf{X} \cap \mathbf{Y})$) between sets $\mathbf{X}$ and $\mathbf{Y}$ is explained, according to (3), by the too high correlation between input variables from different groups.

Next, we will evaluate, using formula (6), the marginal contributions of input variables to the increase in the closeness of the correlation relationship between random vectors $\mathbf{X}$ and $\mathbf{Y}$ over the period from 2000 to 2022 (see Table 6). The greatest marginal contribution was found among the variables statistically significant in Tables 3 and 4.

Let's determine, using formulas (1) and (4), the coefficients $d_m(\mathbf{X}_k \cup \mathbf{Y})$, $d_m(\mathbf{X}_k \cap \mathbf{Y})$ over the period from 2000 to 2022 (Table 7). The large difference between them is due to a closer correlation within groups compared to intergroup correlations of variables. This was particularly evident for sets $\mathbf{X}_3$ and $\mathbf{X}_4$.

The results of the calculation according to formula (5) of the dynamics of coefficients $d_m(\mathbf{X} \cap \mathbf{Y})$ representing the tightness of the correlation relationship between yields $Y_k$ and the set of input variables $\mathbf{X}$ are presented in Fig. 4. The values of the

**Table 6** Limit contributions $\Delta d_m(X \cap Y|X_j)$ of input variables to the increase in the tightness of the correlation between random vectors X and Y

| $X_j$ | $X_{11}$ | $X_{12}$ | $X_{21}$ | $X_{22}$ | $X_{31}$ | $X_{32}$ | $X_{41}$ | $X_{42}$ | $X_{51}$ | $X_{52}$ |
|---|---|---|---|---|---|---|---|---|---|---|
| $\Delta d_m$ | 0.049 | 0.042 | 0.015 | 0.008 | 0.009 | 0.003 | 0.049 | 0.016 | 0.011 | 0.001 |

**Table 7** Coefficients of tightness of the correlation relationship between sets of input variables $X$ and yields $Y$

| $\mathbf{X}_k$ | $\mathbf{X}_1$ | $\mathbf{X}_2$ | $\mathbf{X}_3$ | $\mathbf{X}_4$ | $\mathbf{X}_5$ |
|---|---|---|---|---|---|
| $d_m(\mathbf{X}_k \cup \mathbf{Y})$ | 0.449 | 0.473 | 0.335 | 0.191 | 0.254 |
| $d_m(\mathbf{X}_k \cap \mathbf{Y})$ | 0.195 | 0.164 | 0.068 | 0.040 | 0.083 |

coefficients remain approximately at the same level. The low values compared to the results of regression analysis are explained by the small subsets.

The results of linear discriminant analysis for wheat yield are presented in Tables 8 and 9. Due to the small size of the data sample, all indicators with a $p$-level $<0.2$ were included in Table 8. Excluding any of the indicators from Table 8 leads to a significant change in the variance ratio, indicating that all indicators make an important contribution to group discrimination.

According to the obtained results, informative diagnostic indicators include population size ($X_{11}$), gross domestic product per capita ($X_{12}$), relative humidity at 2 m ($X_{32}$), surface pressure ($X_{41}$), and temperature at 2 m ($X_{52}$). It is worth noting that the

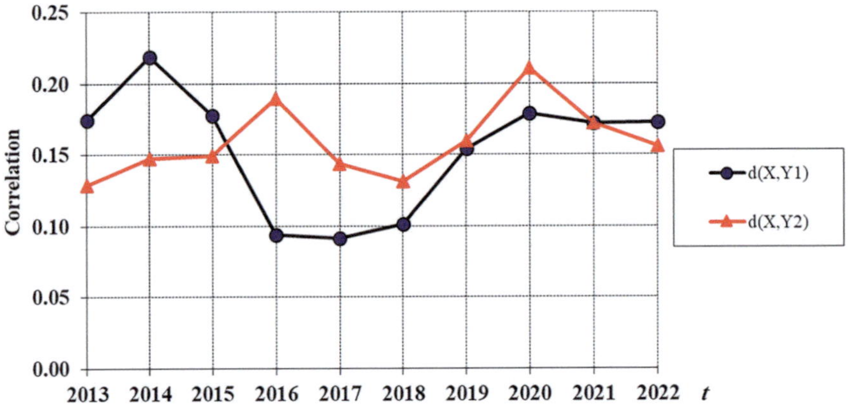


**Fig. 4** Coefficients $d_m(\mathbf{X} \cap Y_k)$ of tightness of the correlation relationship between yields $Y_k$ and the set of input variables $\mathbf{X}$

**Table 8** Results of discriminant analysis for wheat yield

| $N = 23$ | Wilks' lambda | Partial lambda | $F$-remove (1, 17) | $p$-level | Toler |
|---|---|---|---|---|---|
| $X_{11}$ | 0.521787 | 0.508585 | 16.42610 | 0.000827 | 0.216964 |
| $X_{12}$ | 0.295648 | 0.897596 | 1.93948 | 0.181674 | 0.373917 |
| $X_{32}$ | 0.299139 | 0.887121 | 2.16312 | 0.159625 | 0.590850 |
| $X_{41}$ | 0.582060 | 0.455920 | 20.28728 | 0.000313 | 0.469035 |
| $X_{52}$ | 0.298303 | 0.889609 | 2.10952 | 0.164593 | 0.722470 |

*Note* $Z = -1$ и $Z = 1$; $\lambda = 0.26537$; $F(5,17) = 9.4122$ $p < 0.0002$

**Table 9** Classification results of two clusters $Z = -1$ и $Z = 1$

| Group | Percent correct | $G_1: Z = -1@@p = 0.47826$ | $G_2: Z = 1@@p = 0.52174$ |
|---|---|---|---|
| $G_1: Z = -1$ | 100,0000 | 11 | 0 |
| $G_2: Z = 1$ | 91,6667 | 1 | 11 |
| Total | 95,6522 | 12 | 11 |

formal division into two groups based on the training sample was statistically significant—the calculated value of the $F$-statistic is 9.412, and the $p$-level is <0.0002. The equation of the separating hyperplane (8) will take the form:

$$W(\mathbf{X}) = a_0 + a_1X_{11} + a_2X_{12} + a_3X_{32} + a_4X_{41} + a_5X_{52} = 0, \quad (9)$$

where $a_0 = 8324.31$, $a_1 = 19.695$, $a_2 = -1.003$, $a_3 = -3.0197$, $a_4 = -855.005$, $a_5 = -1.0228$.

Let's analyse the Table 8. Parameter $\lambda$ (Wilks statistics, Wilks lambda) takes values from 0 to 1—the higher $\lambda$, the better the data is divided into clusters. Here $\lambda = 0.265$.

Wilks' Lambda parameter is the value of Wilks' statistic with the corresponding variable excluded. The greater it differs from $\lambda$, the more desirable the presence of the variable in the discrimination procedure. It should be noted that excluding surface pressure ($X_{41}$) most significantly deteriorates the quality of cluster recognition. This result is explained by the reaction of yield to changes in surface pressure.

Partial Lambda statistic is the value of Wilks' statistic for one variable (the individual contribution of the corresponding variable to discrimination between clusters). The smaller it is, the greater the contribution of the variable to overall discrimination. As evident from Table 8, none of the variables allow for the division of the dataset into two groups. The next two columns show the F-statistics ($F$-remove) of the analysis of variance and $p$-levels, characterizing the contribution of each variable to the difference between the two groups. According to the results, $X_{41}$ and $X_{11}$ were also the most significant.

The tolerance value (Toler.) of the variable is calculated as $1 - R^2$, where $R^2$ is the coefficient of determination of the linear regression of the $X_i$ feature on the remaining set of statistically significant features used in the analysis. The tolerance of the variable is a measure of its redundancy in the model; the smaller it is, the more redundant the variable is, i.e., the least correlated variable, and consequently the least redundant, is the temperature at 2 m ($X_{52}$). If the statistic is not greater than the lower tolerance threshold (0.05), then the corresponding variable carries little additional information and should not be included in the model.

According to Table 9, the average probability of correct recognition was 0.957, which exceeds the required reliability level of 0.95. This means that there is a statistically significant difference between the values of socio-economic and climatic indicators in years with low ($Z = -1$) and high ($Z = 1$) wheat yield. The only misclassified data was the yield in the year 2015.

Thus, it can be asserted that socio-economic and climatic conditions have a statistically significant impact on wheat yield. The classification will be performed using the logistic function [23]

$$h(\mathbf{x}) = \frac{1}{1 + \exp\{-\mathbf{a}^T\mathbf{x}\}} \in (0;\, 1), \tag{10}$$

This function can be interpreted as the probability of belonging to clusters $G_1$ or $G_2$. In this case, the vector $\mathbf{a}^T = (a_0, a_1, \ldots, a_5)$ defines the separating boundary in the form of a hyperplane $\Omega$: $\mathbf{a}^T\mathbf{x} = 0$. Then $\forall\mathbf{x} \in G_1\ h(\mathbf{x}) < 0{,}5$, $\forall\mathbf{x} \in G_2\ h(\mathbf{x}) > 0{,}5$, $\forall\mathbf{x} \in \Omega\ h(\mathbf{x}) = 0{,}5$ and $\forall\mathbf{x}\ P(\mathbf{x} \in G_1) = 1 - h(\mathbf{x})$ and $P(\mathbf{x} \in G_2) = h(\mathbf{x})$.

It is known [23], that $\mathrm{grad}h(\mathbf{x}) = \left(\frac{\partial h(\mathbf{x})}{\partial x_1}, \ldots, \frac{\partial h(\mathbf{x})}{\partial x_m}\right) = \frac{1+\exp\{-\mathbf{a}^T\mathbf{x}\}}{(1+\exp\{-\mathbf{a}^T\mathbf{x}\})^2} \bullet \mathbf{a}^T$, i.e. $\mathrm{grad}h(\mathbf{x})$ always orthogonal to the separating hyperplane $\Omega$.

The increase in yield according to (10) is equivalent to the increase in the probability $h(\mathbf{x})$ or the change in the indicator vector x in the direction of $\mathrm{grad}h(\mathbf{x})$. This means that to increase the value of $h(\mathbf{x})$, the increments $a_k$ must be positive when the corresponding $\Delta x_k$ coefficients are positive, and negative when the coefficients are negative. That is, to increase wheat yield, it is necessary to decrease the indicators $X_{12}, X_{32}, X_{41}, X_{52}$ and increase $X_{11}$. There is no contradiction here—the signs of the coefficients in (9) coincide with the regression coefficients of the corresponding variables in Table 3. It is worth noting that the increase in the population of Kirkuk is accompanied by a simultaneous increase in the rural population engaged in agriculture (rural population was not included as an indicator due to the lack of reliable statistical data for several years). The negative coefficient for gross domestic product can be explained by the negative correlation between it and real investments in agriculture.

In the future, for practical use of the model (9), (10) in the form of forming management recommendations based on optimization tasks aimed at increasing wheat yield, it is necessary to refine the socio-economic indicators.

## 5 Conclusion

Using multivariate statistical analysis methods, a study was conducted on the influence of several climatic and socio-economic factors on the agricultural sector based on statistical data from the past 23 years. The study showed the influence of socio-economic and climatic factors on the yield of winter grain crops. It demonstrated the possibility of forming management recommendations aimed at increasing the yield of grain crops. The influence of climatic conditions, as the research results showed, turned out to be less significant than a priori assumed.

## References


1. Hussein, A.J.A., Abd, M.J., Shuaib, R.Z.: Estimation and analysis of the impact of the contribution ratio of the oil and agricultural sectors to the gross domestic product in Iraq for the period 1990–2016. J. Kufa Stud. Center **1**, 55 (2019). https://doi.org/10.36322/jksc.v1i55.4994.
2. Saleh Joaani, S.M., Al-Douri, B.F.: An economic and measurement study of the impact of government support and some other variables on wheat crop production in Iraq for the period (1992–2019). Tikrit J. Agricult. Sci. **21**(4), 79–90 (2021). https://doi.org/10.25130/tjas.21.4.9
3. Bai, Z.G., Dent, D.L., Olsson, L., Schaepman, M.E.: Proxy global assessment of land degradation. Soil Use Manag. **24**(3), 223–234 (2008). https://doi.org/10.1111/j.1475-2743.2008.00169.x
4. Fadhil, A.M.: Land degradation detection using geo-information technology for some sites in Iraq. J. Al Nahrain Univ. Sci. **12**(3), 94–108 (2009). https://doi.org/10.22401/JNUS.12.3.13
5. Awadh, W.R.: The problem of utilization of the water resources of the Republic of Iraq under progressive desertification conditions. J. Geogr. Nat. Resour. **35**(4), 373–379 (2014). https://doi.org/10.1134/S1875372814040106
6. Hamad, K.O., Surucu, A.: Land degradation sensitivity and desertification risk in Harrir region, northern Iraq. Heiyion **10**(5), 123 (2024). https://doi.org/10.1016/j.heliyon.2024.e27123
7. Imad, A.R., El-Din, Q.S.: Policies to support the prices of wheat crop and its impact on the planted area in Iraq for the period (1970–2002). Anbar Univ. J. Econ. Admin. Sci. **5**(10), 118–135 (2013)
8. Oborin, M.S., Chernikova, S.A.: The influence of agriculture on the socio-economic development of the region. Vestnik Zabaykal'skogo Gosudarstvennogo Universiteta **8**, 137–146 (2018)
9. Park, S.E., Benjamin, L.R., Watkinson, A.R.: Comparing biological productivity in cropping systems: a competition approach. J. Appl. Ecol. **39**(3), 416–426 (2002). https://doi.org/10.1046/j.1365-2664.2002.00732.x
10. Gollin, D.: Agricultural productivity and economic growth. In: Evenson, R., Pingali, P. (eds.) Handbook of Agricultural Economics, pp. 3825–3866. Elsevier (2010). https://doi.org/10.1016/S1574-0072(09)04073-0
11. Voronina, V.P., Ruleva, O.V.: Phytocenotic structure and productivity of cereal associations in agricultural landscapes. Izvestiya Nizhnevolzhskogo agrouniversitetskogo kompleksa: nauka i vyssheye professional'noye obrazovaniye **4**(52), 57–64 (2018). https://doi.org/10.32786/2071-9485-2018-04-7
12. Gnedenko, B.V.: Theory of Probability, 6th edn., p. 520. Gordon and Breach Science Publishers, London (1998)
13. Kendall, M.G., Stuart, A.: The Advanced Theory of Statistics: Inference and Relationship, vol. 2, p. 684. Hafner Publishing Company, New York (1961)
14. Pena, D., Rodriguez, J.: Descriptive measures of multivariate scatter and linear dependence. J. Multivar. Anal. **85**(2), 361–374 (2003). https://doi.org/10.1016/S0047-259X(02)00061-1
15. Tyrsin, A.N.: Scalar measure of the relationship between several random vectors. Fact. Lab. Diagn. Mater. **88**(3), 73–80 (2022). https://doi.org/10.26896/1028-6861-2022-88-3-73-80
16. Baba, K., Shibata, R., Sibuya, M.: Partial correlation and conditional correlation as measures of conditional independence. Austr. N. Zeal. J. Stat. **46**(4), 657–664 (2004). https://doi.org/10.1111/j.1467-842X.2004.00360.x
17. Tyrsin, A.N., Makarova, T.A.: Estimation of linear structural relationships between random variables in small samples. Fact. Lab. Diagn. Mater. **79**(5), 73–76 (2013)
18. Seber, G.A.F., Lee, A.J.: Linear Regression Analysis, p. 572. Wiley, New York (2003)
19. Hardle, W.K., Simar, L.: Applied Multivariate Statistical Analysis, 5th edn. Springer, New York (2019). https://doi.org/10.1007/978-3-030-26006-4
20. Hilbe, J.M.: Logistic Regression Models, p. 637. CRC Press, New York (2009)
21. Febrianti, R., Widyaningsih, Y., Soemartojo, S.: The parameter estimation of logistic regression with maximum likelihood method and score function modification. J. Phys. Confer. Ser. **1725**, 012014 (2021). https://doi.org/10.1088/1742-6596/1725/1/012014

22. Tyrsin, A.N., Kostin, K.K.: Consideration of estimation of logistic regression as an optimization problem. Bull. Tomsk State Univ. Manag. Comput. Inform. Sci. **3**(40), 52–60 (2017). https://doi.org/10.17223/19988605/40/6
23. Tyrsin, A., Chistova, E., Kostin, K.: Logistic regression as a diagnostic model for stochastic systems: applied methods of statistical analysis. In: Nonparametric Methods in Cybernetics and System Analysis: AMSA'2017, Krasnoyarsk, Russia, 18–22 September 2017: Proceedings of the International Workshop. Novosibirsk: NSTU, 2017, pp. 355–362. http://www.amsa.conf.nstu.ru/amsa2017/proceedings/AMSA2017-proceedings.pdf

# Small Data Samples Modeling of Well Interaction for Analyzing the Effectiveness of Flooding Systems

**A. N. Tyrsin and S. E. Kashcheev**

**Abstract** At the final stage of oil field development, the problem of maintaining acceptable levels of oil production through operational management of flooding is an urgent one. The complexity increases due to the growing number of wells in the basin and the variability of their interaction. This requires new approaches that take into account these trends in oil production. Two algorithms focused on large dimensions and small data samples are proposed. They were tested on model data, in which 60 injection wells, 160 production wells and 17 observations, random errors were present. These algorithms have demonstrated acceptable performance both in terms of accuracy and speed, and in terms of the possibility of using them for predicting oil production.

**Keywords** Material balance model · Flooding system · Well interaction mask · Entropy · Forecast · Capacitance-resistive model

## 1 Introduction

In recent years, there has been a trend for most oil fields under development to reach the final stage. This stage is characterized by an increase in the amount of water in extracted products up to 60–70% and above [1]. This poses a significant challenge in maintaining acceptable levels of oil production through the management of flooding. Operational optimization of flooding systems has become a priority in mature, often low-efficiency fields. However, there are several problems here. Earlier

A. N. Tyrsin (✉)
Ural Federal University, Yekaterinburg, Russia
e-mail: at2001@yandex.ru

Engineering Center "Reliability and Resource of Large Systems and Machines", Ural Branch, Russian Academy of Sciences, Yekaterinburg, Russia

S. E. Kashcheev
South-Ural State University, Chelyabinsk, Russia
e-mail: kascheevstas@yandex.ru

 
A. G. Kravets and A. A. Bolshakov (eds.), *Human-Centric Cyber-Society*, Studies in Systems, Decision and Control 639, https://doi.org/10.1007/978-3-032-06787-6_8

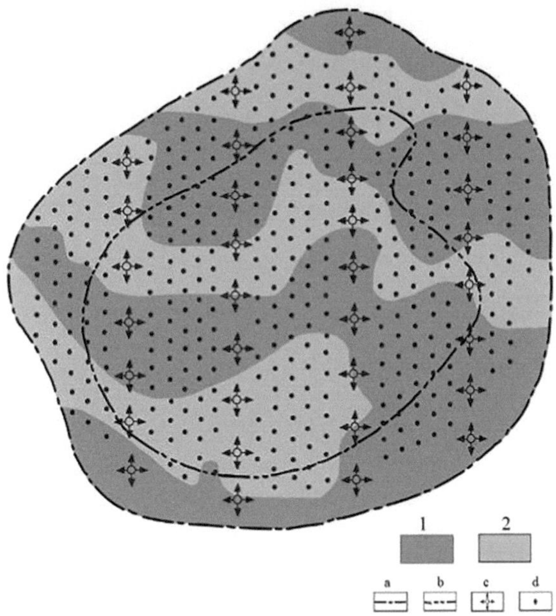


**Fig. 1** Various types of flooding systems. Area with reservoir thickness and reservoir properties: (1) high, (2) low. Oil bearing contours: (a) external, (b) internal, (c) injection, (d) extractive

studies have examined the problems associated with the lack of reservoir data. In [2, 3], a model is being developed to detect patterns in the parameters of the flooding system. In a study by Moreno and Lake [4], the general characteristics of an optimal injection scheme are described. However, there is another challenge. This is due to the particular nature of the numerical solutions of differential equations that describe reservoir processes. It is necessary to use simpler mathematical models rather than complex hydrodynamic models [5, 6]. Figure 1 shows a diagram of a variety of flooding systems.

One of the promising areas of support for the development of oil fields is the use of capacity-resistive analytical models (CRM). These models represent an analytical solution to the equation of dynamic material balance in the form of a time series [7–12]. The CRM was first described in 2005 as a differential equation describing the dynamics of oil flow rate [7].

The general type of the CRM model in the presence of $L > 1$ producing wells and $M > 1$ injection well is investigated. The oil flow rate for each producing well at $k$th time point is described as [13]

$$q_l(k) = q_l(k-1)e^{-1/\tau_l} + \left(1 - e^{-1/\tau_l}\right)\sum_{j=1}^{M} f_{jl} I_j(k) + \varepsilon(k) \tag{1}$$

where $q_l = q_l(k)$, $I_j(k)$—flow rate of the $l$th production well and injectivity of the $j$th injection well in the $k$th time period, respectively, $\tau_l > 0$—time (time associated with the elastic energy reserve in the reservoir and determines the rate of propagation of disturbances in the reservoir); $f_{jl} \geq 0$ is the coefficient of mutual influence of the injection well $I_j$ and the production well $P_l$. $k$ is the number of the period (for example, a month), $l = 1, ..., L$, $j = 1, ..., M$, $k = 1, ..., N$. The term $\varepsilon(k)$ in (1) reflects random errors that are always present in real data. In (1), it is assumed that the bottom-hole pressure in injection wells remains constant. In real cases, this parameter is usually ignored or considered constant, since measuring bottom-hole pressure is a very time-consuming process.

The popularity of CRM models is mainly due to the fact that they do not require knowledge of the properties of the inter-well space. These properties are difficult to determine with certainty. As noted in [12, p. 14], the parameters $\tau_l$ and $f_{jl}$ of model (1) allow to take into account the characteristics of the reservoir-well system. In practice, these parameters are quite sufficient to obtain a high-quality forecast of the dynamics of technological indicators based on the CRM model.

Support for the development of oil wells includes monitoring the flooding system and forecasting oil production. To do this, it is necessary to determine the interference coefficients and time constants based on the available data on the injectivity of injection wells and the oil flow rate of production wells. Then, using the found parameters and planned injectivity of injection wells, the oil flow rate in production wells is predicted. The presence of a large number of wells, the nonconvex nature of the functional dependence, and the presence of restrictions on the domain of values lead to a not unique solution and difficult to find [12].

In [13] a new approach for CRM modeling is proposed. It based on determining the mask of interference. This significantly reduces the number of determined interference coefficients. The mask of interference is understood as a set of injection wells that affect the formation of the oil flow rate and the corresponding values of the interference coefficient $f_{jl}$. However, when testing this approach, shortcomings were revealed. Firstly, it requires relatively large amounts of training data, about 30 observations, for each well. This number of observations is not always feasible, because the model parameters may change during this time period. Secondly, the issue of the scale of actual flooding systems, where the number of wells can reach several hundred, remains unresolved.

The aim of this work is to increase the effectiveness of monitoring of flooding systems by reducing the size of the training data sample and expanding the scale of the analyzed systems from several tens to hundreds of wells.

## 2 Research Methods

One of the problems of evaluating the parameters of the CRM model is the presence of a large number of production and injection wells. The number of such wells can approach hundreds and thousands, respectively. A large number of wells leads to an increase in the number of possible interaction pairs. This leads to computational instability in determining the interaction coefficients. In fact, only part of the injection wells influences to the oil flow rate of the producing wells $P_l$, $l = 1, ..., L$. The definition and use of such a set of variables avoid reducing the accuracy of estimates of interference coefficients [12]. The set of such variables that affect the flow rate formation will be called the mask of interference.

In the CRM model (1), we consider the flooding system as a network structure. In this network structure, injection and production wells are stochastically connected to each other through interference coefficients. The network of structures [14] is a set of stable connections between the elements of the system that ensure reproducibility under changing conditions. In addition, the network structure may contain subsystems (subnets), which should also be reflected as relationships at the subsystem level [15, 16].

The task of evaluating the CRM model into 3 stages: (1) forming a preliminary mask of interference; (2) refining the mask of interference and forming an initial approximation for the CRM model; (3) determining the parameters of the CRM model for the entire flooding system.

## 2.1 Formation of a Preliminary Mask of Interference

The lack of information about the reservoir properties and the presence of a large number of production and injection wells leads to the need to use statistical methods for preliminary determination of the mask of interference.

The mask is a matrix of significant interference coefficient of producing and injection wells. The minimum set of input data for forming the mask is the coordinates of wells, values of injectivity and oil flow rates. Additional information for forming the matrix can be information about the structure of reservoirs, the development system.

Dimensionality reduction is achieved in two ways. First way—zeroing out the interference coefficients $f_{jl}$ for injection wells located sufficiently far from the production well and for which there is no information about the effect of the well injectivity $I_j$ on the oil flow rate well $P_l$. Second way—zeroing out the interference coefficients $f_{jl}$ for injection wells, whose values below the specified threshold level formed by the production well. It based on a priori information about the well system [13]. As a result, there is a preliminary mask of interference $G^* = \{(j(l), l)\}$, $l = 1, ..., L$, $j(l) \in \{1, ..., M\}$—a set of pairs of numbers corresponding to two injection wells for each of the producing wells $P_l$ with non-zero interference coefficients, that is $\forall (j, l) \notin G^* f_{jl} = 0$.

It is worth noting the number of significant injection wells for each production well, which form the interference coefficients, is limited by the number of observations of the training data sample. In particular, for the following example of 17 observations, for each production well, the set of injection wells is limited to 15 nearby wells.

## 2.2 Refining the Mask of Interference and Forming an Initial Approximation for the CRM Model

For each production well, the parameters of the CRM-model is estimated by minimizing the sum of squares of discrepancies between the actual and model values with physical restrictions on the model parameters [6]

$$\sum_{k=2}^{N}\left(q_l^*(k) - q_l(k)\right)^2 \to, \quad l = 1, \ldots, L, j(l) \in \{1, \ldots, M\} \tag{2}$$

$$q_l^*(k) = a_l q_l(k-1) + \sum_{j(l)} b_{j(l),l} I_{j(l)}(k), q_l^*(1) = q_l(1) \tag{3}$$

$$b_{jl} \geq 0, 0 < a_l < 1 \tag{4}$$

where $a_l = e^{-1/\tau_l}$, $b_{jl} = \left(1 - e^{-1/\tau_l}\right) f_{jl}$, $\tau_l = -1/\ln \ln a_l, f_{jl} = b_{jl}/(1 - a_l)$.

As a result of solving $L$ minimization problems (2)–(4), the vector of parameters $a_1^*, \ldots, a_L^*$ and the set of interaction coefficients $f_{j(l),l}^*$, $j(l) \in \{1, \ldots, M\}$, $l = 1, \ldots, L$ is obtained. The values of the interference coefficient are too small values that are close to zero (for example, if $f_{jl} < 10^{-4}$) are removed from consideration.

Next, for each producing well, three sub-systems are considered:

1. a production well $P_l$, the oil flow rates of which are formed as $P_l(k) = q_l(k) - a_l^* q_l(k-1)$, where $a_l^*$ is the estimate of the parameter $a_l$ as a result of solving problem (2)–(4),
2. the set of all injection wells $I_l = \left(I_{1,l}, \ldots, I_{M,l}\right)$,
3. a set of injection wells, from which an arbitrary $j$th well is removed $I_l^j = I_l / I_{j,l} = \left(I_{1,l}, \ldots, I_{j-1,l}, I_{j+1,l}, I_{M,l}\right)$.

As an indicator, the entropy indicator of the relationship [13] is used. It is expressed in terms of the determinants of correlation matrices as

$$\delta_{jl} = \frac{G(P_l \cap I_l) - G\left(P_l \cap I_l^j\right)}{G\left(P_l \cap I_l^j\right)} \cdot 100\% \tag{5}$$

where $G(P_l \cap I_l) = -\frac{1}{2} \ln \ln\left(\left|R_{P_l \cup I_l}\right| / \left|R_{I_l}\right|\right)$, $G\left(P_l \cap I_l^j\right) = -\frac{1}{2} \ln \ln\left(\left|R_{P_l \cup I_l^j}\right| / \left|R_{I_l^j}\right|\right)$.

The coefficient $_{jl}$ characterizes the percentage decrease in the entropy index of the relationship between the system $I_l$ and the producing well $P_l$ when an arbitrary well $I_l$ is removed from the system $I_j$.

From the mask of interference, we exclude the mutual interference coefficient $f_{jl}$ of injection wells for which $_{jl}$ is less than the threshold level. This threshold level is set based on a priori information about the well system and the required accuracy of estimating the parameters of the CRM model. Then, once again, for each producing well, we solve problems (2)–(4) with a refined set of injection wells that are significant for it, and determine estimates of the interaction coefficients $f_{jl}^0$ and $a_l^0$.

Solving $L$ problems (2)–(4) avoids the procedure for calculating the reduced injectivity of injection wells. This significantly reduces the required sample size for determining the parameters of the flooding system, which in [13] depends on the maximum value of the time constant $\tau_l$ of producing wells. For example, at $\tau_l = 6$, the calculation of the reduced injectivity of injection wells will require at least 15 additional observations. As a result, the minimum allowable data sample size is about 30 observations, which is not always possible in practice.

## 2.3 *Defining Parameters of the CRM Model for the Entire Flooding System*

Next, the parameters of the CRM-model for the entire flooding system are defined. Consider two algorithms: (1) direct solution of the optimization problem (algorithm A); (2) iterative algorithm (algorithm B).

Algorithm A solves the problem

$$\sum_{k=2}^{N}\sum_{l=1}^{L}\left(q_l^*(k) - q_l(k)\right)^2 \to \min_{a_l, f_{jl}}, \tag{6}$$

$$q_l^*(k) = a_l q_l(k-1) + (1-a_l)\sum_{j(l)} f_{j(l),l} I_{j(l)}(k),\ q_l^*(1) = q_l(1), \tag{7}$$

$$f_{jl} \geq 0,\ 0 < a_l < 1,\ l = \underline{1, L},\ j = \underline{1, M} \tag{8}$$

$$\sum_{l=1}^{L} f_{jl} \leq 1,\ j = \underline{1, M} \tag{9}$$

Algorithm B includes step 0 and cyclically repeats the steps 1 and 2 ($k = 1, 2, \ldots$):

*Step 0.* Calculation of the interference coefficients $f_{j(l),l}^{(0)}$ for the entire CRM model according to the parameters $a_1^{(0)}, \ldots, a_L^{(0)}$ found when solving $L$ problems (2)–(4):

$$\sum_{k=2}^{N}\sum_{l=1}^{L}\left(q_l^*(k) - q_l(k)\right)^2 \to \min_{f_{j(l),l}}, \tag{10}$$

$$q_l^*(k) = a_l^{(0)} q_l(k-1) + \left(1 - a_l^{(1)}\right) \sum_{j(l)} f_{j(l),l} I_{j(l)}(k),\ q_l^*(1) = q_l(1), \tag{11}$$

$$f_{j(l)l} \geq 0, \sum_{l=1}^{L} f_{jl} \leq 1, j = \underline{1, M}. \tag{12}$$

*Step 1*($k$). Calculation of the delay coefficients $a_1^{(k)}, \ldots, a_L^{(k)}$ for the entire CRM model based on the coefficients of mutual influence found in the previous step $f_{j(l),l}^{(k-1)}$:

$$\sum_{k=2}^{N} \sum_{l=1}^{L} \left(q_l^*(k) - q_l(k)\right)^2 \rightarrow \min_{a_l}, \tag{13}$$

$$q_l^*(k) = a_l \left( q_l(k-1) - \sum_{j(l)} f_{j(l),l}^{(k-1)} I_{j(l)}(k) \right) + \sum_{j(l)} f_{j(l),l}^{(k-1)} I_{j(l)}(k), \tag{14}$$

$$q_l^*(1) = q_l(1), 0 < a_l < 1, l = \underline{1, L}. \tag{15}$$

*Step 2*($k$). Calculation of the interference coefficients $f_{j(l),l}^{(k)}$ for the entire CRM model based on the parameters $a_1^{(k)}, \ldots, a_L^{(k)}$:

$$\sum_{k=2}^{N} \sum_{l=1}^{L} \left(q_l^*(k) - q_l(k)\right)^2 \rightarrow \min_{f_{j(l),l}}, \tag{16}$$

$$q_l^*(k) = a_l^{(k)} q_l(k-1) + \left(1 - a_l^{(k)}\right) \sum_{j(l)} f_{j(l),l} I_{j(l)}(k),\ q_l^*(1) = q_l(1) \tag{17}$$

$$f_{j(l)l} \geq 0, \sum_{l=1}^{L} f_{jl} \leq 1, j = \underline{1, M}. \tag{18}$$

The loop stops if the decrease in the objective function becomes insignificant.

Algorithms for solving optimization problems (2)–(4), (6)–(18) realized in Python. The globally convergent method of moving the asymptotes (MMA) for local gradient-based optimization from the NLOPT nonlinear optimization library [17] is used. MMA was first described in [18]. MMA is a nonlinear programming algorithm from the family of conservative convex separable approximations (CCSA). This method is used in the field of structural optimization, where calculations of functions and gradients take a lot of time. The method also has global convergence and high-speed [19]. The essence of the method is that at each current point, a local approximation is generated using the gradient of the function, constraint conditions, and a quadratic penalty term to make the approximation conservative.

Algorithm the Improved Stochastic Ranking Evolution Strategy (ISRES) algorithm from the NLOPT library [20, 21] does not show equally stable performance for small data samples with a large problem dimension [13] and was replaced by the MMA method. The ISRES method is an evolutionary algorithm based on a combination of the mutation rule (with log-normal step size update and exponential smoothing) and differential variation (update rule similar to the Nelder-Mead rule). In optimization problems (2)–(4), (6)–(18) there are no non-linear constraints, so the fitness rating is performed only using the objective function. The ISRES method uses a large number of randomly generated parameter sets, which reduces the convergence rate of the algorithm.

The estimation of parameters the CRM-model was carried out on the basis of the indicators root-mean-square deviations (RMS) of errors in the interference coefficients, delay errors, and average absolute errors and forecast.

## 3 Numerical Experiments

In this section, an example of a model of a real oil field flooding system is considered. The system consists of 60 injection wells and 160 production wells. Injectivity of injection wells are real data from the flooding system. For each well $I_j, j = 1, \ldots, M$, there are $N = 17$ observations recorded at equal time intervals. The matrix of interference coefficients $f_{jl}$ is formed randomly, taking into account the normalization $(f_{j(l),l} \geq 0, \sum_{l=1}^{L} f_{jl} \leq 1, j = 1, ..., M)$. The number of non-zero interference coefficients for each producing well varied from 1 to 7, and there were 554 of them in total. The time constants $_l$ is formed randomly in the range from 1 to 6. The oil flow rate of production wells $P_l$ is formed according to the formula (1).

Algorithms A and B is tested for resistance to random errors. Several data sets of flow rates with different and RMS random error $\varepsilon(k)$ are formed: $(\sigma_\varepsilon = 0; 0, 1; 0, 2; 0, 3)$. Algorithms is estimated by such parameters as the rate of convergence of the algorithm (calculation time), the value of the objective function, and the RMS errors of estimates of interference coefficient and delay parameters. The RMS errors rate of interference coefficient is calculated as

$$RMSE_f = \sqrt{\frac{\sum_{j=1}^{M} \sum_{l=1}^{L} (\hat{f}_{jl} - f_{jl})^2}{LM}}, l = 1, \ldots, L, j = 1, \ldots, M \tag{19}$$

where $\hat{f}_{jl}(k)$ is the predicted value of the oil flow rate for the well $P_l$ in the $k$th time period, $f_{jl}$ is the actual oil flow rate for the well $P_l$ in the $k$th time period.

At the first step, the preliminary mask is formed according to formulas (2)–(4). Then, for each producing well, a set of injection wells is specified using the entropy indicator (5). The threshold level for $\delta_{jl}$ is set to 2.5%. The indicator $\delta_{jl}$ allowed about $30 \div 35$ errors of the 1st kind. There are no errors of the 2nd kind.

Additionally, algorithm A is implemented without the use of mask of interference, i.e., minimization is performed over the entire matrix of interference coefficients. Accordingly, 9760 parameters are used in the minimization problem: 160 delay coefficients and 9600 interference coefficients. Table 1 shows a comparison of implemented algorithms.

Algorithm A without using a mutual influence mask shows a significant deterioration in quality and is inferior to the other two in all indicators with an increase RMS in the random error rate.

Algorithm A without using a mutual influence mask shows a significant deterioration in quality and is inferior to the other two in all indicators with an increase RMS in the random error rate.

Algorithm A with the ISRES method performs worse than the same algorithm with the MMA method. So, ISRES runs 10 ÷ 16 times longer, and is inferior to MMA in estimation accuracy (the minimum value of the objective function) by 6 ÷ 11 times. Such results may be caused by the algorithm's specific features—the

**Table 1** Algorithm results

| Performance indicators | RMS of random error $\int \Sigma$ | Algorithm A, MMA | Algorithm A without cross-influence mask, MMA | Algorithm B, MMA | Algorithm A, ISRES |
|---|---|---|---|---|---|
| Convergence rate, sec | 0 | 33.4 | 97.9 | 56.5 | 369.3 |
| | 0.1 | 48.1 | 140.7 | 82.6 | 877.0 |
| | 0.2 | 57.4 | 281.0 | 76.1 | 875.3 |
| | 0.3 | 59.5 | 198.9 | 86.8 | 844.9 |
| Objective function value | 0 | 0.0 | 12,462.5 | 0.0 | 0.0 |
| | 0.1 | 48.0 | 11,674.0 | 31.4 | 241.1 |
| | 0.2 | 116.6 | 10,835.9 | 109.3 | 1476.55 |
| | 0.3 | 293.5 | 13,361.7 | 257.3 | 3296.6 |
| RMS $f_{jl}$ | 0 | 0.000 | 0.025 | 0.000 | 0.000 |
| | 0.1 | 0.002 | 0.025 | 0.002 | 0.002 |
| | 0.2 | 0.005 | 0.025 | 0.005 | 0.004 |
| | 0.3 | 0.006 | 0.025 | 0.006 | 0.005 |
| Average absolute error of the forecast, $m^3$ | 0 | 1.34 | 6.12 | 1.34 | 1.34 |
| | 0.1 | 1.36 | 6.17 | 1.34 | 1.51 |
| | 0.2 | 1.47 | 6.25 | 1.46 | 2.04 |
| | 0.3 | 1.65 | 5.99 | 1.63 | 2.59 |
| Average absolute error of the forecast, % | 0 | 3.45 | 15.76 | 3.45 | 3.45 |
| | 0.1 | 3.50 | 15.89 | 3.45 | 3.89 |
| | 0.2 | 3.78 | 16.09 | 3.76 | 5.25 |
| | 0.3 | 4.25 | 15.42 | 4.20 | 6.67 |

population size in problems is $20 \times (n + 1)$, where $n$ is the number of optimization parameters.

Algorithms A and B show similar results and are resistant to random errors. Algorithm A is $1.3 \div 1.7$ times faster, but loses to algorithm B in estimation accuracy (the minimum value of the objective function) by $1.1 \div 1.5$ times. However, discrepancies can be caused by a small sample—a total of three implementations of random errors were considered.

The quality of predicting is evaluated, so additional $NP = 17$ observations are formed. The average absolute predict error is equal to

$$D = \frac{1}{L(NP - N)} \sum_{k=N+1}^{N+NP} \sum_{l=1}^{L} |\hat{q}_l(k) - q_l(k)|, l = 1, \ldots, L, j = 1, \ldots, M \quad (20)$$

where $\hat{q}_l(k)$ is the predicted oil flow rate for the well $P_l$ in the $k$th time period $q_l(k)$ is the actual oil flow rate for the well $P_l$ in the $k$th time period. For the implemented algorithms, $D$ is 1.34 $m^3$ or 3.45% of the average actual oil flow rate for all wells in the forecast period. The prediction error without using a mask is 6.61 $m^3$, or 17.02% of the average actual forecasted rate for all wells in the forecast period. Thus, algorithms using a mask of interference are preferable for monitoring and predicting the level of oil production. Figure 2 shows an example of actual and prediction oil flow rates for wells $P_5$ and $P_{133}$.

The actual and prediction values differ slightly in the figure. However, for solving problems with real data, this difference is insignificant.

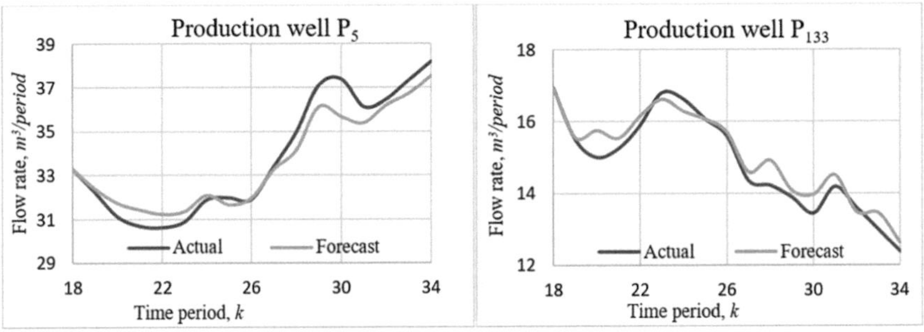


**Fig. 2** Actual and forecast oil flow rates for wells $P_5$, $P_{133}$

## 4 Conclusion

In this research, the advantage of using a mask of interference for evaluating the CRM model is shown. A method for preliminary estimation of the mask using entropy indicators of the relationship is described. This method allows to specify for each producing set of influencing injection wells.

Defining the mask of interference allows to significantly reduce the size of optimization problems and improve the quality of CRM modeling. So, in the considered example, this made it possible to reduce the number of minimization parameters from 9760 to 712, reduce the convergence time of the algorithm, and increase the accuracy of estimating the parameters of the CRM model.

The solution allowed a practical two-fold reduction of the training sample compared to the known solution. This is achieved by forming a preliminary mask of interference and an initial approximation for the CRM model by considering all production wells separately. The algorithms have shown resistance to random errors and fairly accurate forecast results. The dimensions considered correspond to real flood systems. The obtained algorithms, after successful testing on real data, can be used for monitoring the flooding system and predicting oil production.

## References

1. Yushkov, I.R., Khizhnyak, G.P., Ilyushin, P.Y.: Development and Operation of Oil and Gas Fields, p. 177. PNRPU Publ, Perm (2013)
2. Hernandez-Mejia, J.L., Pisel, J., Jo, H., et al.: Dynamic time warping for well injection and production history connectivity characterization. Comput. Geosci. **27**, 159–178 (2023). https://doi.org/10.1007/s10596-022-10188-w
3. Salehian, M., Çınar, M.: Reservoir characterization using dynamic capacitance–resistance model with application to shut-in and horizontal wells. J. Petrol Explor. Prod. Technol. **9**, 2811–2830 (2019). https://doi.org/10.1007/s13202-019-0655-4
4. Moreno, G.A., Lake, L.W.: Input signal design to estimate interwell connectivities in mature fields from the capacitance-resistance model. Pet. Sci. **11**, 563–568 (2014). https://doi.org/10.1007/s12182-014-0372-z
5. Aziz, K., Settari, E.: Mathematical Modeling of Reservoir Systems: Translation from English. M.: Nedra, p. 408 (1982)
6. Pospelova, T.A., Stepanov, S.V., Strekalov, A.V., Sokolov, S.V.: Mathematical Modeling for Decision-Making on the Development of Deposits, p. 427. Nedra Publishing House, Moscow (2021)
7. Yousef, A.A., Gentil, P.H., Jensen, J.L., Lake, L.W.: A capacitance model to infer interwell connectivity from production and injection rate fluctuations. In: SPE Annual Technical Conference and Exhibition, 9–12 October 2005, Dallas (2005)
8. Kim, J.S., Lake, L., Edgar, T.F.: Integrated capacitance-resistance model for characterizing waterflooded reservoirs. In: Proceedings of the 2012 IFAC Workshop on Automatic Control in Offshore Oil and Gas Production, Norwegian University of Science and Technology, Trondheim, Norway, May 31–June 1, pp. 19–24 (2012)
9. Khatmullin, I.F., Tsanda, A.P., Andrianova, A.M., Budyonny, S.A., Margarit, A.S., Lushpeev, V.A., Simonov, M.V., Peretz, D.S.: Semi-Analytical Models for Calculating Well Interference Based on the CRM Model Class, pp. 38–41 (2018)

10. Danko, M.Y., Brilliant, L.S., Zavyalov, A.S.: Application of the dynamic material balance method and CRM-method to calculating reserves. Achimov Bazhenov Reserv **4**(80), 76–85 (2019)
11. Afanaskin, I.V., Kryganov, P.V., Glushakov, A.A., Yalov, P.V.: Using CRM models of well interference to evaluate reservoir filtration and capacity properties based on development data **1**(1), 17–27 (2020)
12. Stepanov, S.V., Bekman, A.D., Ruchkin, A.A., Pospelova, T.A.: Support of Oil Field Development Using CRM Models, p. 300. CPI "Express", Tyumen (2021)
13. Tyrsin, A.N., Stepanov, S.V., Ruchkin, A.A., Beckman, A.D.: Increasing the reliability of well interaction modeling for the analysis of the efficiency of the flooding system. Math. Models Comput. Simul. **15**(6), 1092–1103 (2023). https://doi.org/10.1134/S2070048223060170
14. Pullan, W.: Structure. Cambridge University Press, Cambridge (2000)
15. Novikov, D.A.: Network Structures and Organizational Systems. IPU RAS, Moscow (2003)
16. Lindenlaub, I., Prummer, A.: Network structure and performance get access arrow. Econ. J. **131**(634), 851–898 (2021)
17. Johnson, S.G.: The NLopt Nonlinear-Optimization Package. http://github.com/stevengj/nlopt://github.com/stevengj/nlopt. Accessed 13 March 2025
18. Svanberg, K.: The method of moving asymptotes: a new method for structural optimization. Int. J. Numer. Methods Eng. **24**(2), 359–373 (1987)
19. Svanberg, K.: A class of globally convergent optimization methods based on conservative convex separable approximations. SIAM J. Optim. **12**(2), 555–573 (2002)
20. Runarsson, T.P., Yao, X.: Search biases in constrained evolutionary optimization. IEEE Trans. Syst. Man. Cybern. C Appl. Rev. **35**(2), 233–243 (2005)
21. NLopt algorithms. https://nlopt.readthedocs.io/en/latest/NLopt_Algorithms/. Accessed on 22 March 2025

# Method for Determining River Flow Characteristics Based on Gauging Stations Data and Machine Learning

**Alexander Khoperskov, Anna Klikunova, Sergey Khrapov, and Illarion Popov**

**Abstract** A method is being developed for determining the parameters of hydraulic resistance to water flow in a riverbed by matching the results of direct hydrodynamic modeling with measurement data at gauging stations using machine learning methods. The numerical hydrodynamic model is based on shallow water equations using Combined Smoothed Particle Hydrodynamics—Total Variation Diminishing on the GPU. The hydraulic resistance model contains three free parameters, calculated using a neural network with a Long Short-Term Memory architecture. An additional free parameter determines the boundary conditions at the exit of the river flow from the computational domain. Determination of these four characteristics provides the best match between time series of water levels measured at several gauging stations and numerical modeling data using the example of the Lower Volga River. Two methods for determining characteristics were investigated. In the first case, the parameters maintained their constant values throughout the year. The second approach is based on calculating the non-stationary characteristics of hydraulic resistance for each day.

**Keywords** Computational fluid dynamics · Parallel computing · Hydrological regime · Shallow water model · Neural network

A. Khoperskov (✉) · A. Klikunova · S. Khrapov · I. Popov
Volgograd State University, Volgograd, Russia
e-mail: khoperskov@volsu.ru

A. Klikunova
e-mail: klikunova@volsu.ru

S. Khrapov
e-mail: khrapov@volsu.ru

I. Popov
e-mail: popov.larion@volsu.ru



A. G. Kravets and A. A. Bolshakov (eds.), *Human-Centric Cyber-Society*, Studies in Systems, Decision and Control 639, https://doi.org/10.1007/978-3-032-06787-6_9

# 1 Introduction

Hydrological research is of enormous practical importance in various fields of engineering, ecology, economics, agriculture, and for improving meteorological and climate forecasts [1]. Qualitative modeling of the hydrological regime of large river systems underlies the design and development of engineering projects and water resource planning [2–5].

The problem of verifying numerical models to describe the hydrological regime of water bodies and adjacent territories plays a key role in making certain decisions related to flash flooding [5–8], creating evacuation plans [9, 10], designing various kinds of hydraulic structures [11], forecasting the state of water bodies and nearby territories [12, 13], etc.

Traditionally used shallow water models are based on vertically integrated hydrodynamic equations, which leads to the appearance of a number of free parameters, in particular, characterizing the hydraulic resistance to fluid flow. Usually the so-called roughness coefficient or Manning parameter $n_M$ is used, which is one of the most important characteristics in hydrology [14–16]. This parameter is phenomenological, since it includes the total effect of a variety of physical factors that inhibit the water flow in the channel. Therefore, the quality of hydrodynamic models depends on the choice of $n_M$ and requires the use of observational data [17–20]. The active use of statistical and intelligent methods for processing hydrological data contributes to obtaining new important results [15, 21–23], especially in conditions of poor knowledge of bathymetry [24].

The hydraulic resistance in our study is divided into contributions due to the classical roughness coefficient $n_M$ and turbulent viscosity, which dominates as the river flow depth increases during the spring flooding period. These two factors have different dependences on the depth and velocity of the flow and they dominate in different parts of the river system at different time intervals. Separate inclusion of turbulent viscosity, as part of hydraulic resistance, provides better agreement with measurement data at gauging stations for the lower reaches of the Volga River from the Volzhskaya hydroelectric power station (HPP) dam. This model of hydraulic resistance makes it possible to qualitatively simulate river flow for a highly variable water charge, both for low-water intervals and at the stage of spring flooding, when the water charge $Q$ increases several times.

# 2 Results of Hydrodynamic and Neural Network Modeling

## *2.1 Model of Hydraulic Resistance in the Shallow Water Approximation*

We use software EcoGIS-Simulation [25, 26] to simulate the dynamics of surface water on a given topography $b(x, y)$ in the shallow water model:

$$\frac{\partial H}{\partial t} + \frac{\partial (Hu_x)}{\partial x} + \frac{\partial (Hu_y)}{\partial y} = q(\vec{\mathbf{r}}, t), \tag{1}$$

$$\frac{\partial (Hu_x)}{\partial t} + \frac{\partial (Hu_x^2)}{\partial x} + \frac{\partial (Hu_x u_y)}{\partial y} = -gH\frac{\partial \eta}{\partial x} + Hf_x^{(q)} + Hf_x^{(HR)}, \tag{2}$$

$$\frac{\partial (Hu_y)}{\partial t} + \frac{\partial (Hu_x u_y)}{\partial x} + \frac{\partial (Hu_y^2)}{\partial y} = -gH\frac{\partial \eta}{\partial y} + Hf_y^{(q)} + Hf_y^{(HR)}, \tag{3}$$

where the function $q(x, y, t)$ specifies the surface density of water sources and drains, $\eta = b + H$, $\eta(x, y, t)$ is the water level, $b(x, y)$ is the topography of the riverbed, $H(x, y)$ is the depth of the flow, $\mathbf{f}^{(q)}$ is the specific force due to the sources. The charge of water $Q(t)$ is defined as $Q(t) = \int q(x, y, t)\, \mathrm{d}x\mathrm{d}y$. The specific hydraulic resistance force $\mathbf{f}^{(HR)}$ is determined by the action of two factors. The first is the term $\mathbf{f}^{(M)}$ with the roughness parameter $n_M$. The second factor is the internal friction force $\mathbf{f}^{(turb)}$ due to turbulent viscosity $\nu^{(turb)}$[19], which gives the following general expression

$$\mathbf{f}^{(HR)} = \mathbf{f}^{(M)} + \mathbf{f}^{(turb)} = g\frac{n_M^2|\mathbf{u}|}{H^{4/3}}\mathbf{u} + \hat{\alpha}H^{\gamma_H}|\mathbf{u}|^{1-\gamma_H}\mathbf{u}, \tag{4}$$

where the free parameters $\hat{\alpha}$ and $\gamma_H$ characterize viscous friction. The boundary conditions at the flow exit from the computational domain are determined by an additional free parameter $\Theta^{(b)}$, which specifies the slope of the water level at the boundary of the river channel. Stream flow near the outlet boundary depends on the topography outside the computational domain, and this effect on water dynamics is determined by the quantity $\Theta^{(b)}$. Numerical integration of the system of Eqs. (1)–(3) is based on the Combined Smoothed Particle Hydrodynamics–Total Variation Diminishing (CSPH–TVD) algorithm [25, 26]. Parallel calculations were performed on the NVIDIA Volta V100 GPU.

We analyze the method of reconstructing hydraulic resistance in form (4) for the section of the Volga River from the dam of the Volga hydroelectric station to Raigorod village (Volgograd Region). For each specific annual cycle, there is a time series of water charge $Q(t)$ (or the so-called hydrograph) through the Volzhskaya HPP dam and the results of measurements of the dynamics of water levels at three gauging stations: $\eta_{obs}^{(k)}(t)$, $k = 1$ is for the gauging station "Dam Downstream of the Volga HPS", $k = 2$—the gauging station "Volgograd" (River port), $k = 3$—the gauging station "Svetly Yar". The simulation results give the corresponding time series $\eta_{mod}^{(k)}(t)$ for each fixed set of parameters $n_M$, $\hat{\alpha}$, $\gamma_H$, $\Theta^{(b)}$. We use the following measurement units in this chapter: $[n_M] = \mathrm{s}\cdot\mathrm{m}^{-1/3}$, $[\Theta^{(b)}] = \mathrm{m/m}$, which we omit in this chapter.

Let's define the following functions

$$\sigma_\eta^{(k)} = \sqrt{\frac{1}{N_d}\sum_{n=1}^{N_d}\left(\eta_{obs}^{(k)}(t_n) - \eta_{mod}^{(k)}(t_n)\right)^2}, \quad \sigma_\eta = \frac{1}{3}\left(\sigma_\eta^{(1)} + \sigma_\eta^{(2)} + \sigma_\eta^{(3)}\right), \tag{5}$$

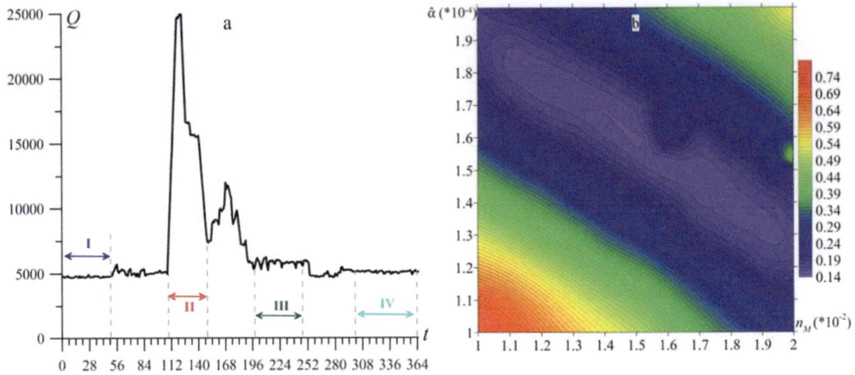


**Fig. 1** **a** Hydrograph of the Volzhskaya HPP dam in 2022, $[Q] = \mathrm{m}^3\mathrm{s}^{-1}$. **b** Function $\sigma_\eta(n_M, \hat{\alpha})$ for fixed parameter values $\gamma_H = 0.5245$, $\Theta_b = 6.23 \cdot 10^{-6}$ for the entire year 2022

where $N_d$ is the length of the time series for the $k$-th gauging station, the interval $N_d = 365$ defines the entire annual series, or we consider shorter time intervals, for example, the spring stage of flooding with a high water charge level $Q$. Figure 1a shows an example of the relationship c in 2022. Minimizing the function $\sigma_\eta(n_M, \hat{\alpha}, \gamma_H, \Theta^{(b)})$ makes it possible to achieve agreement between model and measured data

$$\sigma_\eta(n_M, \hat{\alpha}, \gamma_H, \Theta^{(b)}, ...) \to \min, \tag{6}$$

## 2.2 *Calculation of Stationary Parameters of Hydraulic Resistance*

The solution to the problem (6) is based on the results of training the Long Short-Term Memory (LSTM) neural network. The LSTM architecture allows you to include the influence of previous states of the hydrological system at a certain interval of duration $\tau_m$ $(t_{n-1}, t_{n-2}, ..., t_{n-m})$ on the state at time $t_n$. This is necessary, since in a large river system there are hysteresis phenomena [19, 27], when the properties of the flow at a given time depend on the prehistory over the entire interval $[t_{n-m}; t_n]$. The interval $\tau_m = m$ days characterizes the length of memory at the learning stage. Two classes of hydraulic resistance models are considered. The first is based on the assumption of constancy of the parameters $n_M, \hat{\alpha}, \gamma, \Theta_b$. The second class is based on the nonstationary behavior of all four quantities $n_M(t), \hat{\alpha}(t), \gamma(t), \Theta_b(t)$. Thus, there is the tuple of free parameters $\mathbf{H} = \{n_M, \hat{\alpha}, \gamma_H, \Theta_b\}$, calculated from the condition (6).

We built dataset based on the results of a large number of computational fluid dynamics experiments for training the neural network. Each numerical experiment for an annual interval is carried out for a fixed set of $\mathbf{H}$ with hydrograph $Q(t)$. The total number of such experiments is 32, and they differ in the values of $n_M$, $\hat{\alpha}$, $\gamma_H$, $\Theta_b$. We divided the model annual series $\eta_{mod}^{(k)}$ into subsets of various durations from 3 to 25 days, which limits the memory length to $\tau_m = 25$ days. As a result, the sample size for training model with memory length $\tau_m$ is $32 \times (365 - \tau_m)$ elements. The resulting dataset was randomly divided into training and validation subsets in a ratio of 7:3.

Minimizing the function (6) with constant parameters gives the solution $\mathbf{H}_c = \{n_M = 0.01531, \hat{\alpha} = 1.535 \cdot 10^{-4}, \gamma_H = 0.5245, \Theta_b = 6.23 \cdot 10^{-6}\}$ in the case of memory length $\tau_m = 5$ days [19]. Let us highlight some features of the solutions in the case $\mathbf{H} = \text{const}$. Figure 1b shows the distributions of $\sigma_\eta$ on the parameter plane $(n_M, \hat{\alpha})$ calculated for the entire year 2022. Functions $\sigma_\eta$ are also built for four time intervals: I ($t = 1 - 50$ days), II ($t = 110 - 150$ days), III.

($t = 200 - 245$ days), IV ($t = 300 - 365$ days) according to formula (5) (Fig. 2). Each specified period of time is characterized by a physically special hydrological flow regime, which is dictated by the variability of the hydrograph (See Fig. 1a).

The form of functions $\sigma_\eta^{(k)}(n_M, \hat{\alpha})$ have a qualitatively similar form for all time intervals (See Figs. 1b, 2). A special feature is the presence of a characteristic canyon profile in $\sigma_\eta(n_M, \hat{\alpha})$ (See color purple). Moreover, several local minima lying at the

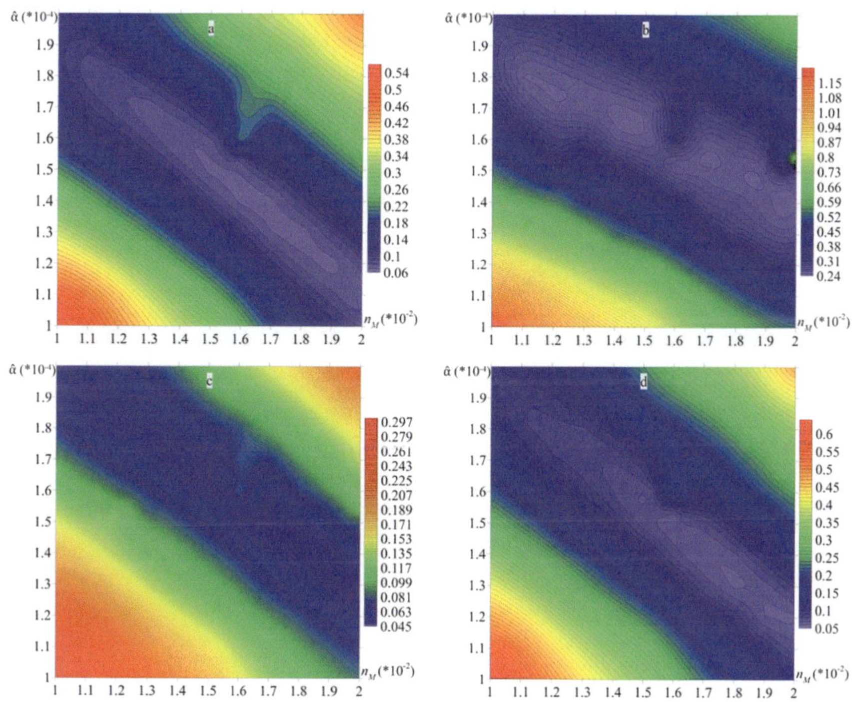


**Fig. 2** Functions $\sigma_\eta(n_M, \hat{\alpha})$ at $\gamma_H = 0.5245$, $\Theta_b = 6.23 \cdot 10^{-6}$ for various time intervals: **a** I in Fig. 1a, b II in Fig. 1a, c III, **d** IV

bottom are observed. The presence of several almost optimal solutions is associated with the structure of the formula (4), when close values of the resistance force can be achieved by different combinations of contributions from the first and second terms in (4).

Intervals I–IV are selected in accordance with the physically distinguished behavior of the water charge. The first low-water interval lasts 50 days ($1 \leq t \leq 50$) with an almost constant value of $Q \simeq 5000$ m/s. The second interval covers the spring flood ($110 \leq t \leq 150$), when the hydraulic resistance changes due to a fivefold increase in the water charge and the behavior of $Q(t)$ is significantly non-monotonic. A comparison of two low-water intervals I and IV indicates certain differences in the functions $\sigma_\eta(n_M, \hat{\alpha})$ (See Fig. 2a and d).

Apparently, this is due to certain changes in the structure of the riverbed topography due to the spring flood, when about 40% of the annual water flow passes along the channel very quickly.

The bottom position of such a canyon can be approximated by a linear function of the form

$$\hat{\alpha}^{(can)} = B^{(can)} - A^{(can)} n_M^{(can)}, \tag{7}$$

where $A^{(can)}$ and $B^{(can)}$ depend on the choice of time intervals (Table 1, See Fig. 2). The decreasing character (7) is due to the opposite influence of the two considered factors (roughness and turbulence) in the hydraulic resistance model (4). Approximately the same level of hydraulic resistance can be achieved either by greater roughness and less turbulence, or vice versa.

The total hydraulic resistance, taking into account roughness and turbulence, can be characterized by the effective Manning coefficient

$$n_M^{(eff)} = \sqrt{n_M^2 + \frac{\hat{\alpha}}{g} \frac{H^{4/3+\gamma_H}}{u^{\gamma_H}}} \tag{8}$$

which in the limit of very weak turbulence ($\hat{\alpha} = 0$) gives $n_M^{(eff)} = n_M$. Figure 3 shows the results of calculations of the effective Manning coefficient for ~100 km of the Volga River for low-water conditions (a) and at the peak of the flood (b). The value of $n_M^{(eff)}$ increases ~1.5 times during the spring flood period compared to the low water period due to an increase in depth, which is not compensated by an increase in flow velocity. The contribution of turbulence to hydraulic resistance in the formula (8) varies widely (20–300%) depending on the date and channel section. The blue

**Table 1** Coefficients in the formula (7) for four intervals and the whole year (See Fig. 1)

| | I | II | III | IV | Whole year |
|---|---|---|---|---|---|
| $A^{(\mathrm{can})}$ | 0.81 | 0.47 | 0.71 | 0.8 | 0.67 |
| $B^{(\mathrm{can})}$ | 2.73 | 2.33 | 2.7 | 2.77 | 2.61 |

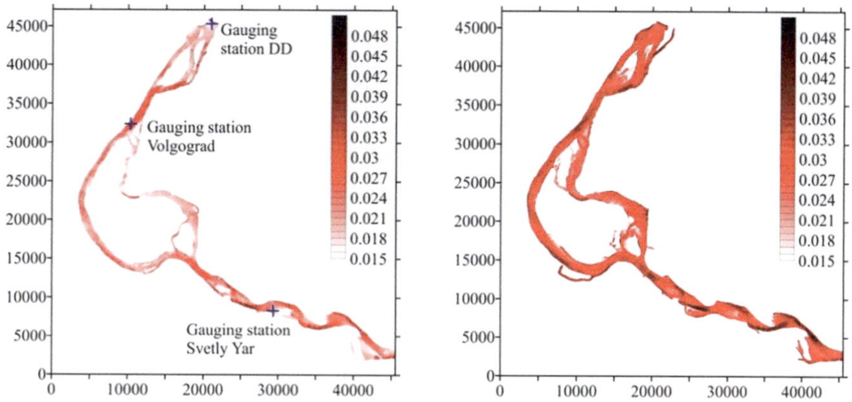


**Fig. 3** Distributions of the coefficient $n_M^{(eff)}$ in the Volga channel during low water (**a**) and at the peak of flood (**b**). The computational domain covers the Northern part of the Volga-Akhtuba floodplain with a size of 45 km × 45 km

crosses in this Figure show the positions of the three gauging stations at which the time series $\eta_{obs}^{(k)}(t)$ are measured.

## 2.3 *Calculation of Non-Stationary Parameters of Hydraulic Resistance*

The physical meaning of the Manning coefficient dictates the understanding that $n_M$ can be a dynamic quantity [14, 20]. For example, a study of the hydrological regime of the Yellow River basin shows improved forecasts when using spatiotemporal distributions of $n_M$ [14].

Figure 4 shows the results of training a neural network using different memory lengths $\tau_m$. The training of the neural network at each interval of $\tau_m = 15$ days is based on a time series of hydrodynamic simulations with a constant set of four parameters **H**. The stage of recovering parameters **H** using a neural network assumes that each of the four parameters is a time function

$$\mathbf{H}(t) = N^{(n)}(Q(t, \tau_m), \eta_{obs}^{(1)}(t, \tau_m), \eta_{obs}^{(2)}(t, \tau_m), \eta_{obs}^{(3)}(t, \tau_m)), \tag{9}$$

where $N^{(n)}$ is the neural network's objective function, which assigns time series a set of parameters **H** that changes every day. All time series are considered on the interval $\tau_m$.

Color saturation denotes the frequency of occurrence of parameter values $n_M, \hat{\alpha}, \gamma_H, \Theta_b$. The distribution width for all four parameters for low-water intervals is noticeably smaller than during floods. Moreover, all three parameters of hydraulic resistance $n_M, \hat{\alpha}, \gamma_H$ deviate in both directions from the low-water values.

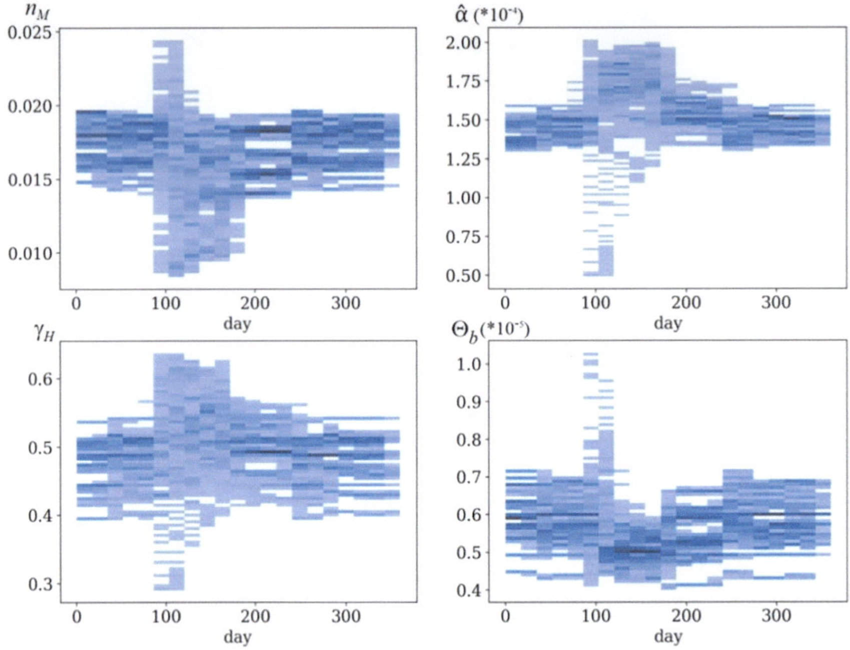


**Fig. 4** Distribution density of four parameters on an annual interval for different values of memory length $\tau_m$ based on the results of training a neural network

Figure 5a shows the dependence of mean squared error (MSE) on the epoch:

$$\text{MSE} = \frac{1}{M_{ss}} \sum_{i=1}^{M_{ss}} (\mathbf{H}_i - \widetilde{\mathbf{H}}_i)^2, \tag{10}$$

where $M_{ss}$ is the number of elements in the subset, $\mathbf{H}_i$ is the actual value obtained using numerical simulation, $\widetilde{\mathbf{H}}_i$ is the predicted parameters of $i$-th element. There is a rapid decrease in MSE over the first 150 epochs across all three subsets. Then the error on training and validation subsets decreases very slowly until the end of training (epoch = 8000). The function MSE(epoch) for model subset has a minimum in the vicinity of epoch = 1000, after which an overfitting effect is observed.

Note that the error on the validation set does not begin to grow during the same period. This is explained by the fact that the time series of elements of the same models from the training and validation subsets are highly similar. Therefore, it is necessary to compile sets from elements of different models.

The effect of overfitting is expressed rather weakly and the value of MSE remains almost unchanged in the interval epoch = 3000–8000. This is ensured by the choice of the simple neural network architecture with a small number of hidden layers.

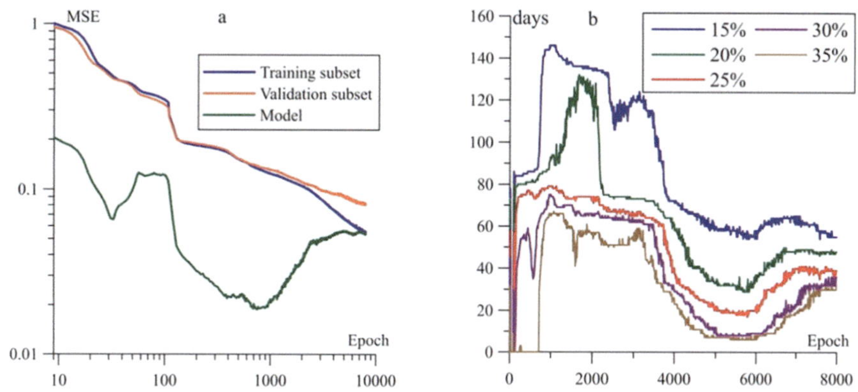


**Fig. 5** **a** Dependencies of the MSE on the epoch for the training, validation and model subsets with memory length $\tau_m = 15$ days. **b** Number of days during which the function $\hat{\alpha}(t)$ had high values versus epoch of learning

Figure 5b shows the length of time (LOT) with a higher level of turbulence, which is determined by the parameter $\hat{\alpha}$ at different threshold coefficients ($th$):

$$\mathrm{LOT} = \sum_{i=1}^{350} \begin{cases} 1, \hat{\alpha}_i > \hat{\alpha}_{LW} * th \\ 0, otherwise \end{cases}, \tag{11}$$

where $\hat{\alpha}_i$ is the value of parameter $\hat{\alpha}$ at $i$-th day, $\hat{\alpha}_{LW}$ is the mean value of $\hat{\alpha}$ in low water period (See, for example, intervals I and IV in Fig. 1a).

The flood begins on $t = 110$ day and ends approximately on the 185th day. This interval of ~70–80 days is characterized by high water with non-monotonic behavior of the hydrograph $Q(t)$(Fig. 1a).

The neural network responds weakly to changes in water charge at the initial stage of learning (epoch < 800). The NN-model becomes capable of detecting the presence of a flood only after approximately the 800th training epoch. However, the model begins to respond less strongly to floods after about epoch 3000. Analysis of these data indicates the advantage of a neural network model built at the 1000th epoch.

Figure 6 shows the time dependences of the parameters $n_M$, $\hat{\alpha}$, $\gamma_H$, which determine the hydraulic resistance in the model (4) for the studied section of the Volga River in 2022. The hydrograph of the dam $Q(t)$ is also shown in Figure. This approach allows us to study the specific effect of two factors in (4). The time dependences of $Q(t)$ clearly highlight significantly different intervals at which changing physical conditions of the river flow are realized.

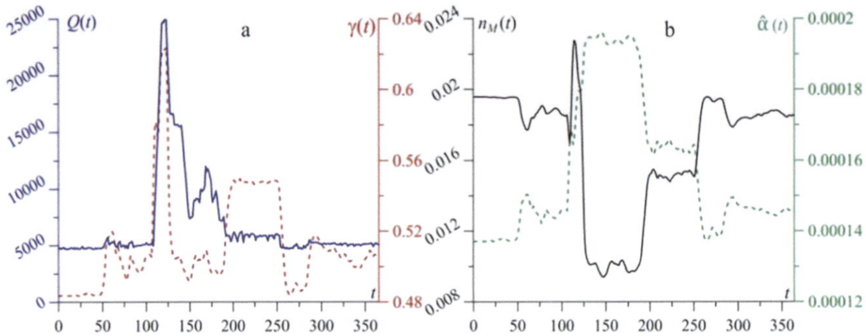


**Fig. 6** Results of reconstruction of hydraulic resistance parameters in the model (4). **a** The solid line is the hydrograph of the Volzhskaya HPP dam $Q$, the dashed line is the parameter $\gamma_H$. **b** The solid line is the roughness coefficient $n_M$, the dashed line is the parameter $\hat{\alpha}$

## 3 Conclusion

New model of hydraulic resistance for the Lower Volga channel has been studied, providing good agreement with the data of measurements of water levels $\eta_{obs}^{(k)}(t_n)$ at three gauging stations below the Volzhskaya HPP. The hydrodynamic model of the river system is based on numerical simulations in the shallow water approximation. The hydraulic resistance model contains three free parameters in addition to other free parameters of the numerical model, which are calculated by minimizing the differences between the time series of water levels $\eta_{mod}^{(k)}(t_n)$ and $\eta_{obs}^{(k)}(t_n)$. The presence of memory in a non-stationary river flow is modeled by an LSTM neural network.

The effective Manning coefficient $n_M^{(eff)}$ is proposed, which includes the contribution of internal friction due to turbulent viscosity in the flow in addition to the standard roughness coefficient $n_M$. Turbulent viscosity depends both on the distribution of water depths and on its velocity. For real conditions in 2022, it is shown that the contribution of turbulent viscosity must be significant to reproduce experimental data at gauging stations. Data from water level measurements at gauging stations for an extended section of the Volga River bed can be reproduced using a hydraulic resistance model with constant parameters $(n_M, \hat{\alpha}, \gamma_H)$. There is an extended region of “good” solutions that provide a standard deviation of the numerical simulation results from the experimental data within 20 cm for water levels.

The model of hydraulic resistance in the lower riverbed of the Volga River was also built for non-stationary parameters $(n_M, \hat{\alpha}, \gamma_H)$. An analysis of the influence of the memory of the neural network model and its training time on the quality of reproduction of natural data was carried out. The limits of variability of hydraulic resistance parameters at various stages of flooding of the northern part of the Volga-Akhtuba Floodplain have been determined.

**Acknowledgements** This work supported by the Russian Science Foundation (grant no. 23-71-00016, https://rscf.ru/project/23-71-00016/). The research is carried out using the equipment of the shared research facilities of HPC computing resources at Lomonosov Moscow State University.

## References

1. Platov, G.A., Golubeva, E.N., Krupchatnikov, V.N., Kraineva, M.V.: Numerical study of interaction between the climate system components and their role in arctic amplification of climate change. Water Resour. **50**, 664–674 (2023)
2. Voronin, A.A., Vasilchenko, A.A., Klikunova, A.Y., Vatyukova, O.Y., Khoperskov, A.V.: The problem of safe evacuation of large floodplains population during flooding. Adv. Syst. Sci. Appl. **22**(4), 65–78 (2022)
3. Isaeva I.I., Voronin A.A. (2024) Models for managing hydraulic projects in floodplain areas, considering the activity of economic entities. Math. Phys. Comput. Simul. 2024. Vol. 27, No. 1, 45–61. https://doi.org/10.15688/mpcm.jvolsu.2024.1.4.
4. Sukhinov, A.I., Chistyakov, A.E., Sidoryakina, V.V., Kuznetsova, I.Y.: Numerical simulation of dumping consequences during dredging in coastal marine systems. Math. Phys. Comput. Simul. **26**(3), 37–52 (2023)
5. Isaeva, I.I., Voronin, A.A., Khoperskov, A.V., Kharitonov, M.A.: Modeling the territorial structure dynamics of the northern part of the Volga-Akhtuba floodplain. Computation. **10**(4), 62 (2022)
6. Hu, H., Yu, M., Zhang, X., Wang, Y.: Urban hydrological model (UHM) developed for an urban flash flood simulation and analysis of the flood intensity sensitivity to urbanization. Geomat. Nat. Hazards Risk. **15**(1), 2302561 (2024)
7. Mikhailova, M.V.: Floods in the Vistula mouth area and their effect on changes in the delta hydrographic network. Water Resour. **50**, 816–827 (2023). https://doi.org/10.1134/S00978078 23700148
8. Kowalczuk, Z., Świergal, M., Wróblewski, M.: River flow simulation based on the HEC-RAS system. In: Advanced Solutions in Diagnostics and Fault Tolerant Control, pp. 253–266. Springer, Berlin (2018)
9. Vatyukova, O.Y., Klikunova, A.Y., Vasilchenko, A.A., Voronin, A.V., Khoperskov, A.V., Kharitonov, M.A.: The problem of effective evacuation of the population from floodplains under threat of flooding: algorithmic and software support with shortage of resources. Computation. **11**(8), 1–26 (2023)
10. Yin, J., Yang, Y., Yu, D., Lin, N., Wilby, R., Lane, S., Sun, B., Bricker, J., Wright, N., Yang, L., Guan, M.: Strategic storm flood evacuation planning for large coastal cities enables more effective transfer of elderly populations. Nat. Water. **2**(3), 274–284 (2024)
11. Girgidov, A.A.: Hybrid modeling in the design of hydraulic structures and FLOW-3D as a means of its implementation. Mag. Civil Eng. **3**(21), 21–27 (2011)
12. Kaur, H., Alam, M.A., Mariyam, S., Alankar, B., Chauhan, R., Adnan, R.M., Kisi, O.: Predicting water availability in water bodies under the influence of precipitation and water management actions using VAR/VECM/LSTM. Climate. **9**(9), 144 (2021). https://doi.org/10.3390/cli909 0144
13. Simonov, Y.A., Borsch, S.V., Semenova, N.K., Khristoforov, A.V.: Short-range streamflow forecasting for russian rivers using the HBV-96 model and the COSMO-Ru system. Russian Meteorol. Hydrol. **48**(12), 1011–1018 (2023)
14. Ye, A., Zhou, Z., You, J., Ma, F., Duan, Q.: Dynamic Manning's roughness coefficients for hydrological modelling in basins. Hydrol. Res. **49**(5), 1379–1395 (2018)
15. Mohanta, A., Patra, K.C., Sahoo, B.B.: Anticipate Manning's coefficient in meandering compound channels. Hydrology. **5**(3), 47 (2018)

16. Demir, V., Keskin, A.Ü.: Obtaining the Manning roughness with terrestrial-remote sensing technique and flood modeling using FLO-2D: a case study Samsun from Turkey. Geofizika **37**(2), 131–156 (2020)
17. Bandini, F., Lüthi, B., Peña-Haro, S., Borst, C., Liu, J., Karagkiolidou, S., et al.: A drone-borne method to jointly estimate discharge and Manning's roughness of natural streams. Water Resour. Res. **57**(2), e2020WR028266 (2021)
18. Tuozzolo, S., Langhorst, T., de Moraes Frasson, R.P., Pavelsky, T., Durand, M., Schobelock, J.J.: The impact of reach averaging Manning's equation for an in-situ dataset of water surface elevation, width, and slope. J. Hydrol. **578**, 123866 (2019)
19. Klikunova, A.Y., Polyakov, M.V., Khrapov, S.S., Khoperskov, A.V.: Problem of building high-quality predictive model of river hydrology: the combined use of hydrodynamic simulations and intelligent computing. Commun. Comput. Inform. Sci. **1909**, 191–205 (2023)
20. Khrapov, S., Pisarev, A., Kobelev, I., Zhumaliev, A., Agafonnikova, E., Losev, A., Khoperskov, A.: The numerical simulation of shallow water: estimation of the roughness coefficient on the flood stage. Adv. Mech. Eng. **5**, 787016 (2013)
21. Gaitan, C.F., Balaji, V., Moore, B.: Can we obtain viable alternatives to Manning's equation using genetic programming. Artif. Intell. Res. **5**(2), 92–101 (2016)
22. Azamathulla, H.M., Ahmad, Z., Ghani, A.: An expert system for predicting Manning's roughness coefficient in open channels by using gene expression programming. Neural Comput. Appl. **23**, 1343–1349 (2013)
23. Palchevsky, E.V., Antonov, V.V.: Decision support system based on application of the second generation neural network. Programmnaya Ingeneria. **13**(6), 301–308 (2022)
24. Garrote, J., González-Jiménez, M., Guardiola-Albert, C., Díez-Herrero, A.: The manning's roughness coefficient calibration method to improve flood hazard analysis in the absence of river bathymetric data: application to the urban historical Zamora city centre in Spain. Appl. Sci. **11**(19), 9267 (2021)
25. Khrapov, S.S., Khoperskov, A.V.: Application of graphics processing units for self-consistent modelling of shallow water dynamics and sediment transport. Lobachevskii J. Math. **41**(8), 1475–1484 (2020)
26. Khrapov, S.S.: Numerical modeling of hydrodynamic accidents: erosion of dams and flooding of territories. Vestnik St. Petersburg Univ. Math. **56**(2), 261–272 (2023)
27. Muste, M., Lee, K., Kim, D., Bacotiu, C., Oliveros, M.R., Cheng, Z., Quintero, F.: Revisiting hysteresis of flow variables in monitoring unsteady streamflows. J. Hydr. Res. **58**(6), 867–887 (2020)

# Method for Determining Turbulent Diffusion in the Atmosphere Boundary Layer Based on Simulation of the Dynamics of Fine Dust Particles

**Egor Savin and Alexander Khoperskov**

**Abstract** A method is proposed for determining the parameters of the vertically inhomogeneous diffusion coefficient in the atmospheric boundary layer based on matching data from numerical modeling of the dynamics of fine dust fractions and measurement results from quasi-periodic dust sources. The fitting of modeled and measured dust concentration time series is carried out using neural network modeling. An analysis of the influence of the neural network architecture choice and training procedure on the efficiency of atmospheric state reconstruction is presented. Data fitting is performed for several time series of dust concentrations of different particle diameters. We include the dispersion composition of dust into consideration when modeling its dynamics, which allows us to improve the reconstruction of the vertical profile of turbulent diffusion.

**Keywords** Computational experiments · Turbulent diffusion · Aerosol transport · Disperse composition · Machine learning

## 1 Introduction

Modeling the dynamics of pollutants and aerosols in the surface layer of the atmosphere is relevant for various applications in the field of ecology, technospheric safety, and atmospheric physics [1–3]. Numerous studies are aimed at analyzing the dispersion of fine aerosols in the air from various sources, primarily $PM_{10}$ and $PM_{2.5}$ [4, 5]. Features of the dynamics of pollutants depend on a large number of environmental parameters within the framework of certain atmospheric models used. Let us highlight the difficulties of measuring convection rate, boundary layer height,

---

E. Savin · A. Khoperskov (✉)
Volgograd State University, Volgograd, Russia
e-mail: khoperskov@volsu.ru

E. Savin
e-mail: e.s.savin@volsu.ru



A. G. Kravets and A. A. Bolshakov (eds.), *Human-Centric Cyber-Society*, Studies in Systems, Decision and Control 639, https://doi.org/10.1007/978-3-032-06787-6_10

Monin–Obukhov scale, heat flux density, vertical profile of temperature and wind and other characteristics in real atmospheric conditions on small time scales [4, 6, 7].

Determining the characteristics of polluting gas and aerosol components in the atmosphere and forecasting their dynamics significantly depend on uncertainties in the parameters that determine meteorological conditions [8, 9]. The properties of the air environment under specific conditions for a given territory are characterized by great uncertainty. The most significant factors influencing the dynamics of gaseous pollutants or aerosols are noted below [10–12].

1. Wind transfer is determined by the average wind speed vector $\mathbf{W}_0$, which is the result of averaging over an interval of 1–10 min [13] (the specific time varies in different countries) with a typical error within $\pm 0.5\ \text{m}\cdot\text{s}^{-1}$. The vertical dependence $W_0(z)$ is always important for describing the pollutants transport. Dependences on horizontal coordinates $x, y$ and time $t$ are determined by the size and topography of the area under study, as well as the characteristic time of the processes under consideration. Vertical air jets $W_0(z)$ can be caused by gas-dynamic instability and stratification of the atmosphere.
2. The second factor is diffusive transfer, determined by the flow turbulence [14, 15], which can be characterized by the pulsation velocity $\tilde{\mathbf{W}}(x, y, x, t)$ against the background of wind (averaged) motion $\mathbf{W}_0$:

$$\mathbf{W} = \mathbf{W}_0(\mathbf{r}) + \tilde{\mathbf{W}}(\mathbf{r}, t) \tag{1}$$

Pulsations of gas-dynamic characteristics are of a random nature and are characterized by a frequency and spatial spectrum. The coefficients of turbulent viscosity $\nu^{(turb)}$ and turbulent diffusion $\hat{D}^{(turb)}$ are expressed through the parameters of turbulent flow in a complex way, see, for example, [15, 16]. The characteristics of turbulence depend both on the height above the surface and the type of atmospheric stratification [16], which in the general case requires consideration of the tensor $\hat{D}^{(turb)}$.

There are various approaches for estimating the turbulent diffusion coefficient, including those based on fitting observational data on a chemical contaminant and the results of modeling its dynamic [17].

The process of dust formation and further transfer of particles has a significantly non-stationary character. This is due, firstly, to the variable regime of sources generating dust, secondly, to the spatial heterogeneity of environmental characteristics, and thirdly, to differences in the behavior of dust particles depending on their size [18]. If the source is under controlled conditions with known characteristics and we can measure the dynamics of the concentration of different diameters aerosols $d^{(a)}$, then this allows us to calculate restrictions on turbulent diffusion. The agreement between measurement data on the concentration of dust particles and the results of modeling the process of aerosol transport can provide information about the state of the atmosphere.

## 2 Numerical Model of Aerosol Dynamics

We consider a periodic source of dust particles $q(t)$ with a given dispersion composition $\phi(d^{(a)})$ at point $\mathbf{r}_s$. Particle concentration $n_\ell^{(aer)}(t) = n^{(aer)}(t; \mathbf{r}_\ell)$ is measured at certain points $\mathbf{r}_\ell$, where measuring complexes are located. Concentration time series are determined by the state of the atmosphere and, first of all, by the values of $\mathbf{W}_0$ and $D^{(turb)}$.

Our model is based on simple vertical profiles of wind and diffusion coefficient [17, 19, 20]:

$$U_0^{(w)}(z) = \mathbf{W}_0\mathbf{e}_x = U_{01}^{(w)} \frac{\ln\left(z/z_0^{(w)}\right)}{\ln\left(z_1^{(w)}/z_0^{(w)}\right)} \tag{2}$$

$$D^{(turb)}(z) = \frac{z\sqrt{e}D_{max}^{(turb)}}{z_{max}^{(turb)}} \cdot \exp\left[-\frac{1}{2}\left(\frac{z}{z_{max}^{(turb)}}\right)^2\right] \tag{3}$$

where $z_0^{(w)}$ is the roughness length of the underlying surface, free parameters $U_{01}^{(w)}$, $z_1^{(w)}$, $z_{max}^{(turb)}$, $D_{max}^{(turb)}$ determine the state of the atmosphere under specific conditions. The used vertical profile $D^{(turb)}$ ensures that the maximum is achieved at the height $z_{\max}^{(turb)}$.

The pollutants transport in the diffusion approximation is described by the equation

$$\frac{\partial\rho}{\partial t} + \nabla\left(\rho\mathbf{U}^{(w)}\right) = \nabla\left(\hat{D}^{(turb)}(\nabla\rho)\right) + q(x, y, z, t) + w_g\frac{\partial\rho}{\partial z} \tag{4}$$

$\rho$ is the mass density of dust, $q$ is the non-stationary source that determines the distribution of the dust entry rate into the atmosphere, $w_g$ is the effective rate of dust falling under the influence of gravity and the Stokes force. Figure 1 shows typical vertical profiles of $U_0^{(w)}(z)$ and $\hat{D}^{(turb)}(z)$ for some values of the free parameters in (2), (3). The turbulent diffusion coefficient $\hat{D}^{(turb)}$ is a tensor, which for the first term on the right in (4) gives

$$\nabla\left(\hat{D}^{(turb)}(\nabla\rho)\right) = \frac{\partial}{\partial x}D_x^{(turb)}\frac{\partial\rho}{\partial x} + \frac{\partial}{\partial y}D_y^{(turb)}\frac{\partial\rho}{\partial y} + \frac{\partial}{\partial z}D_z^{(turb)}\frac{\partial\rho}{\partial z} \tag{5}$$

where $D_{x,y,z}^{(turb)}$ are the diagonal elements of the tensor $\hat{D}^{(turb)}$.

There is a periodic source of dust particles with a known dispersion spectrum along the diameters $d^{(a)}$(Fig. 2). Such an aerosol is transported in the atmosphere with the characteristics $D_{max,obs}^{(turb)}$, $z_{max,obs}^{(turb)}$, $U_{01}^{(w)}$, $z_0^{(w)}$, which are unknown. Measurements at distance $L_{obs}$ yield time series $\rho_{obs}^{(d)}(t)$ for each diameter $d^{(a)}$. The most important

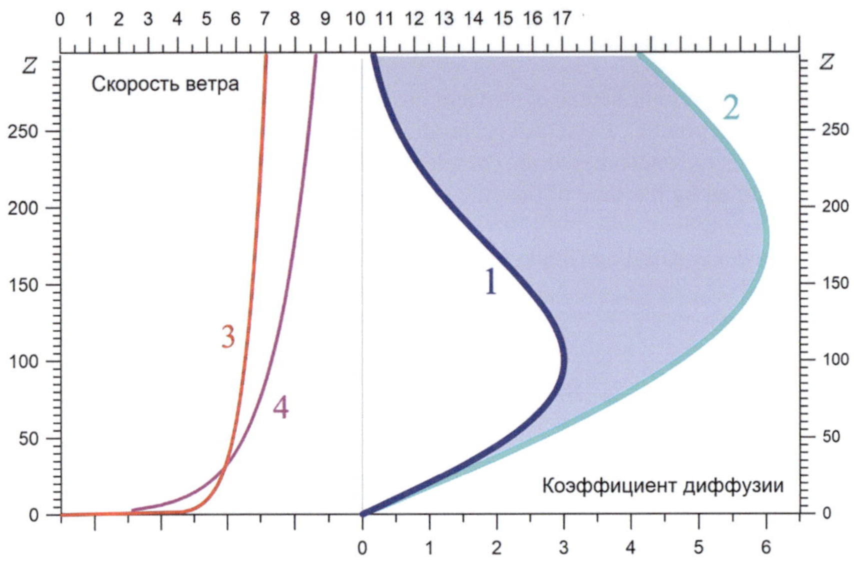


**Fig. 1** Examples of typical distributions of the turbulent diffusion coefficient [$m^2/s$] (curves 1 and 2) and wind speed [m/s] (curves 3 and 4) in the atmospheric boundary layer

characteristic is the diameter distribution density (dispersion composition) $\phi^{(d)}$ with normalization $\int\limits_0^\infty \phi^{(d)}(d^{(a)})dd^{(a)} = 1$.

On the other hand, the numerical solution (4) gives time dependences $\rho^{(d)}_{mod}(t)$ for various sets of free parameters. Matching of modeling data with measurement results can provide calculations of atmospheric characteristics, or obtain restrictions on these parameters. Such matching can be based on minimizing the standard deviation between time series of dust particle concentrations

$$\varepsilon^{(a)} = \sqrt{\frac{1}{N^{(t)}} \sum_{i=1}^{N^{(t)}} \left( AA^{(d)}_{obs}(t_i) - AA^{(d)}_{\text{mod}}(t_i) \right)^2} \tag{6}$$

where $N^{(t)}$ is the number of dust concentration measurements.

Figure 3 shows the diagram of the program using NVIDIA CUDA technology. The process of executing code involves three hardware parts of the computing system: disk storage, central processing unit and graphics accelerator. Disk storage is used to read input data and write computer simulation results. Green rectangles highlight the main file entities. Green dotted arrows indicate the direction of data flows. The first stage of program execution is reading the input data and initializing the system, which is performed on the central processing unit, as well as unloading the results. Control is then transferred to the GPU NVIDIA, where mass flow calculations are performed, boundary conditions are applied, and sensors are operated to measure

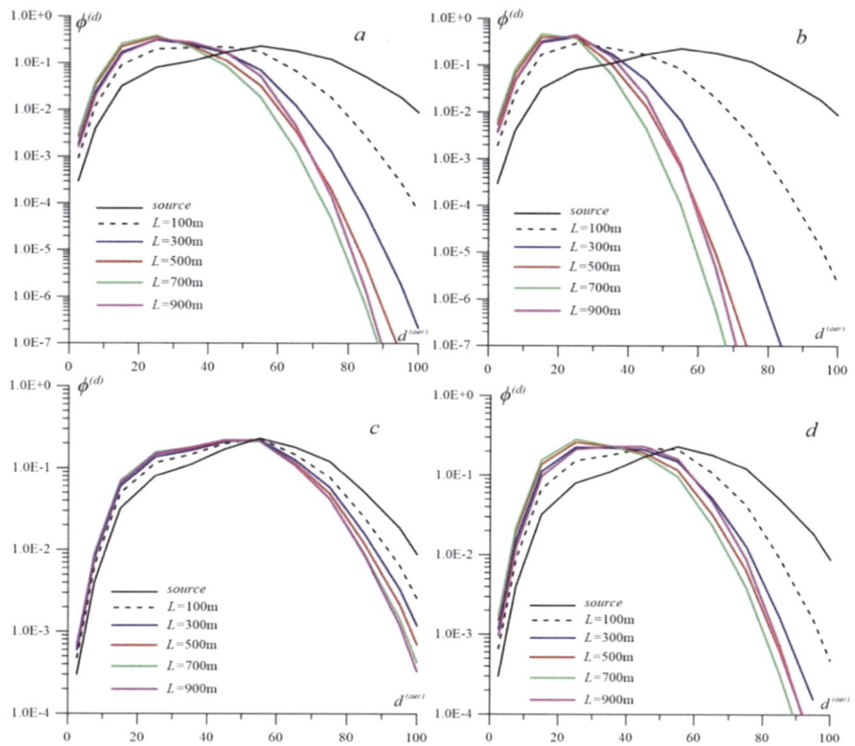


**Fig. 2** Dispersion composition of dust $\phi^{(d)}$ in numerical experiments at different distances from the source for four atmospheric states at $U_{01}^{(w)} = 5$ m/s: **a** $D_{max}^{(turb)} = 1$ m$^2$/s, $z_{max}^{(turb)} = 25$ m, **b** $D_{max}^{(turb)} = 1$ m$^2$/s, $z_{max}^{(turb)} = 100$ m, **c** $D_{max}^{(turb)} = 10$ m$^2$/s, $z_{max}^{(turb)} = 25$ m, **d** $D_{max}^{(turb)} = 10$ m$^2$/s, $z_{max}^{(turb)} = 100$ m

concentrations. Control then returns to the CPU. The main stages of the program algorithm are indicated by blue rectangles. Blue arrows show the sequence of control transfers between stages.

Figure 2 shows changes in the dispersion composition depending on the parameters of the turbulent state of the atmosphere and the distance between the source and the measurement point. The value $\phi^{(d)}$ measured by the sensor is normalized to the total value of dust concentration for all particle sizes: $\phi^{(d)} = \rho/\rho_{tot}$, where $\rho$ is the mass density of dust of a given diameter, averaged over the period of the source, $\rho_{tot}$ is the total mass density of aerosols. There is a relative redistribution of the dispersion composition as one moves away from the source in the direction of the wind. The general trend is a strong decrease in the proportion of large fractions and an increase in the proportion of small particles. There are systematic dependences of $\phi^{(d)}(d^{(a)})$ on the profile $D^{(turb)}(z)$, which can be the basis for reconstructing the parameters of the atmosphere state from the measured $\phi^{(d)}(d^{(a)})$.

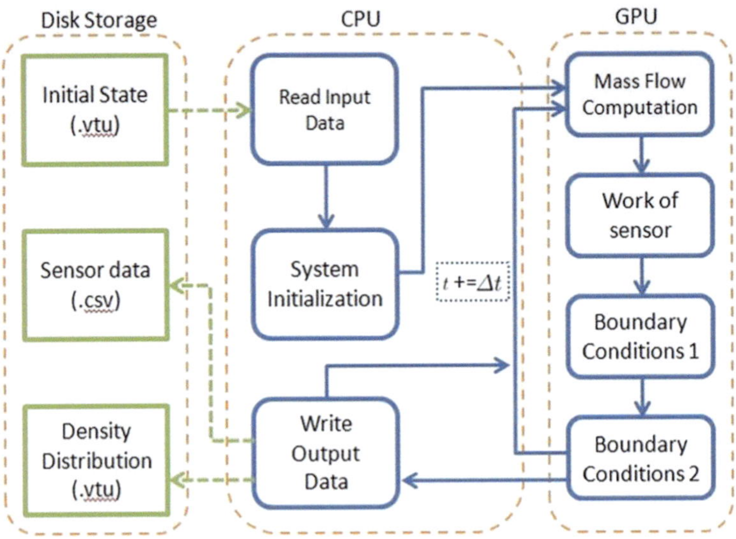


**Fig. 3** Diagram of data flows in the software for direct modeling of pollutants transport

Figure 4 demonstrates examples of calculations of the function $\varepsilon^{(a)}(D_{max}^{(turb)}, z_{max}^{(turb)})$ for different values of aerosol diameter $d^{(a)}$, measured atmospheric diffusion parameters $D_{max,obs}^{(turb)}$, $z_{max,obs}^{(turb)}$, distances from the source to the measurement point $L_{obs}$. Areas of absolute minimum ($\varepsilon^{(a)} = 0$) are marked with a white cross. First of all, the significant conclusion seems to be the "canyon" type of this function. Another feature is the behavior of the function in panels of Fig. 4 *d, e* when one of the slopes is very gentle.

The aerosol source is periodic and is characterized by its dispersion composition, depicted by the solid black line in Fig. 2. At the measurement point, the functions $\rho(t)$ at a distance $L$ reproduce the source mode with a phase shift and a changed amplitude (Fig. 5).

Figure 6 shows the standard deviation (6) for two vertical turbulent diffusion profiles and four different particle diameters. There is a strong dependence of the form of the function $\varepsilon^{(a)}$ on the diameter of the dust particle. The white cross indicates the position of the exact solution ($\varepsilon^{(a)} = 0$). The top row corresponds to models where large particles are more preferable for determining diffusion parameters. The bottom row demonstrates the opposite situation, when large particles do not allow one to reconstruct the vertical profile of $D^{(turb)}$ with satisfactory accuracy.

The results of calculating $\varepsilon^{(a)}$ using the dispersed composition are shown in Fig. 7. This makes it possible on average to improve the conditions for determining the parameters of turbulent viscosity compared to models for one fixed particle size, ince the slope of the well $\varepsilon^{(a)}$ changes noticeably for different diameters $d^{(a)}$.

Minimization of the objective function $\varepsilon^{(a)}$ relative to the atmospheric state parameters is carried out using a multilayer neural network, the architecture of which is

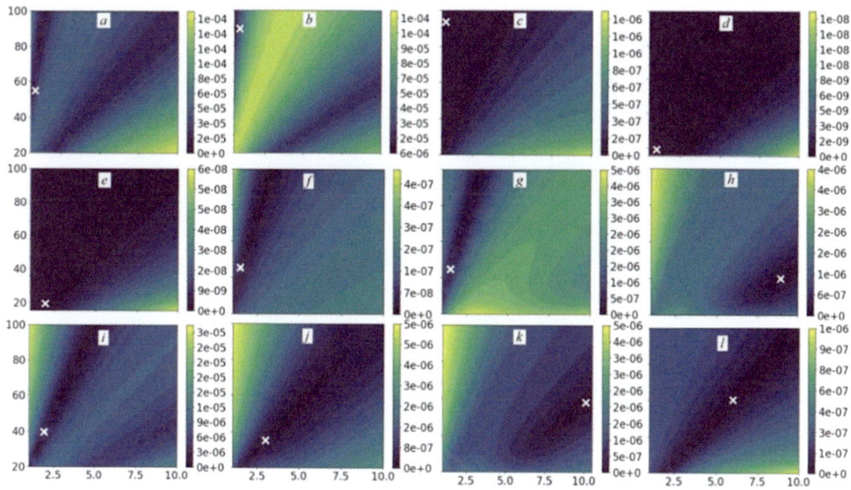

**Fig. 4** Distributions of the function $\varepsilon^{(a)}(D_{max}^{(turb)}, z_{max}^{(turb)})$ for different atmospheric states and particle diameters. White crosses show the positions of exact solutions

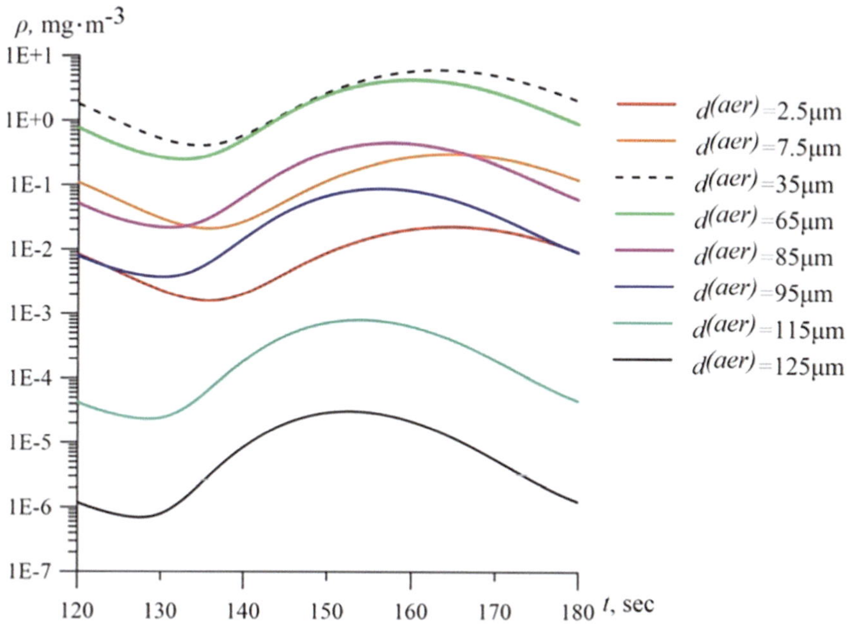


**Fig. 5** Time dependences of the density of particles of a fixed diameter at a distance of $L = 500$ m from the source

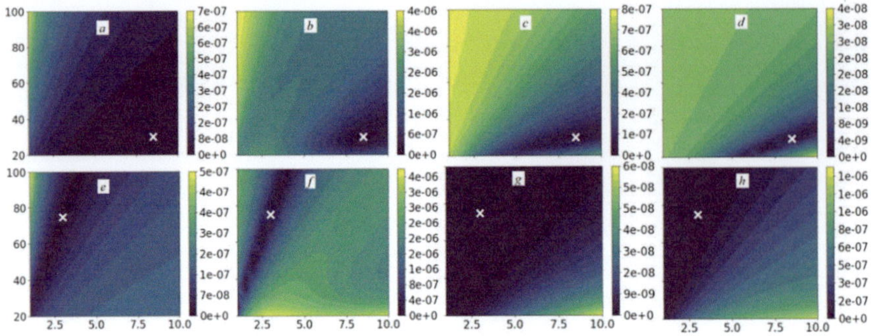


**Fig. 6** Function (6) for two different states of the atmosphere: top—$D = 8\ \text{m}^2/\text{s}$, $z = 35$ m; **a** $d^{(a)} = 2.5\ \mu\text{m}$, **b** 55 μm, **c** 75 μm, **d** 95 μm; bottom—$D = 3\ \text{m}^2/\text{s}$, $z = 75$ m; **e** $d^{(a)} = 2.5\ \mu\text{m}$, **f** 55 μm, **g** 75 μm, **h** 95 μm

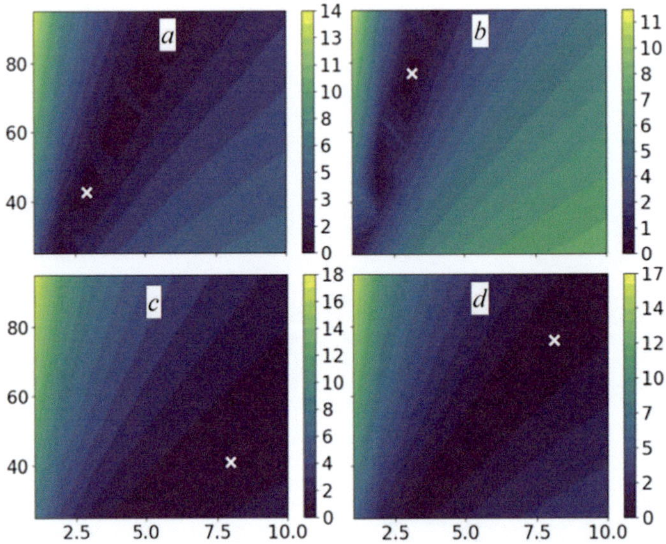


**Fig. 7** Error function for diameter spectra

shown in Fig. 8. The input layer is a matrix of size $K^{(d)} \times K^{(t)}$, The row index determines the particle diameter and the second index is the density measurement number with time step $\tau^{(K)}$. The values of $K^{(d)} = 14$, $K^{(t)} = 10$ are indicated in the diagram in figure. Since mass densities can differ by several orders of magnitude, normalization is necessary. The neural network contains an LSTM layer (Long Short-Term Memory), which is associated with the processing of time series for more efficient inclusion of long memory. We selected $2^{12} = 4096$ neurons in the LSTM layer. The next three layers of the same type are fully connected layers.

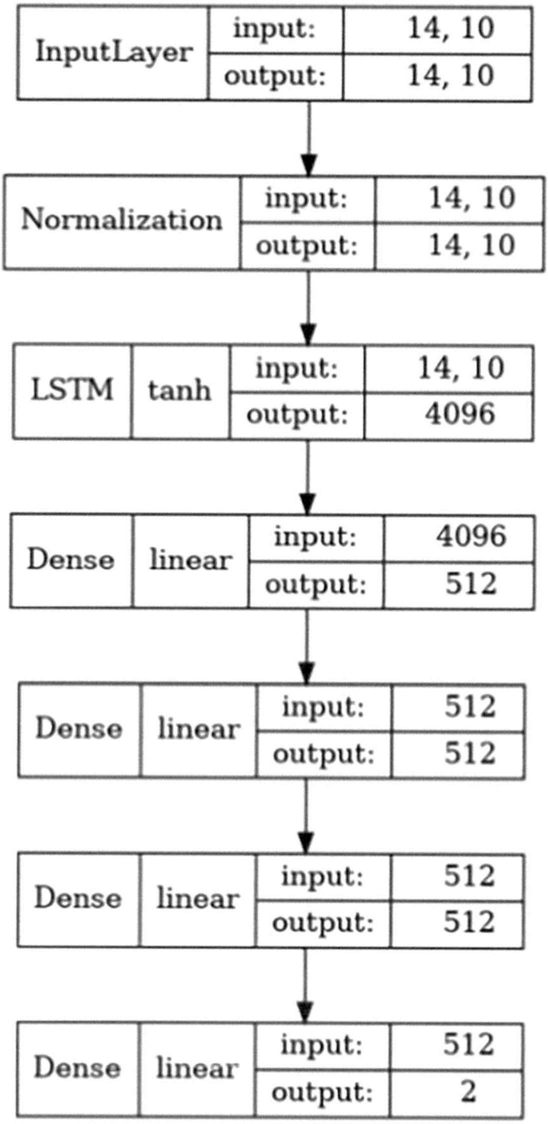


**Fig. 8** Neural network architecture diagram

The NN-LSTM training results for three vertical profiles of the turbulent diffusion coefficient in Fig. 9 show variations in the $\varepsilon$ error versus epoch for different atmospheric states. Satisfactory results are obtained, for example, with ($D_{max}^{turb} = 1$ m$^2$/s, $z_{max}^{turb} = 50$ m). The state with ($D_{max}^{turb} = 1$ m$^2$/s, $z_{max}^{turb} = 25$ m) is unfavorable for determination of atmospheric parameters.

Figure 10 summarizes the analysis of the possibility of restoring the parameters $D_{max}^{turb}, z_{max}^{turb}$, using minimization of differences in the dispersion compositions of dust

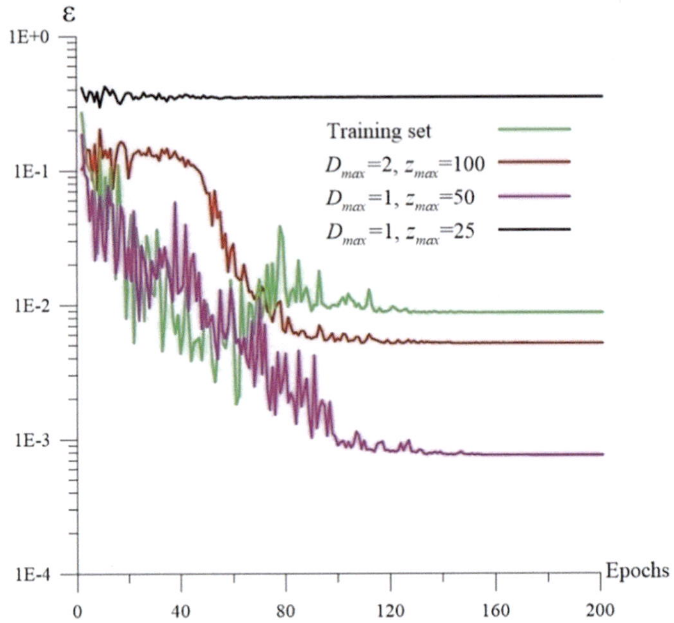


**Fig. 9** Absolute error modules (loss) versus epoch (epoch) when training a neural network for different atmospheric models

particles from a non-stationary source. The neural network outputs both $D^{turb}_{max,mod}$ and $z^{turb}_{max,mod}$, the accuracy of which varies greatly. Therefore, we plot these errors separately in the left and right panels. Experimental values are plotted along the axes. The color of the cells corresponds to the difference between the exact and predicted values of $D^{turb}_{max,mod}$ and $z^{turb}_{max,mod}$. We see a rather complex non-monotonic distribution of errors, which makes it difficult to reconstruct the turbulent properties of the atmosphere.

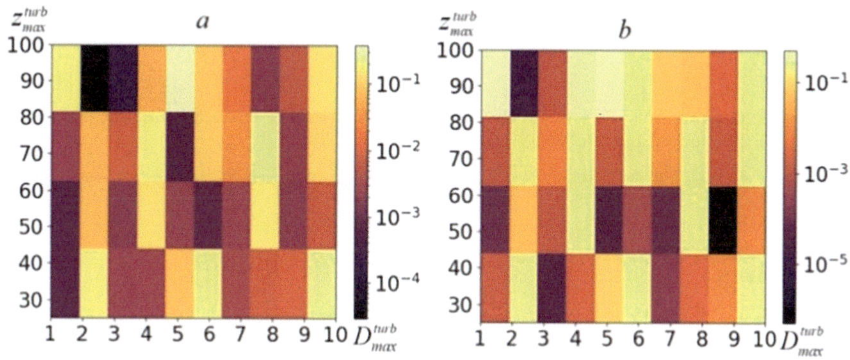


**Fig. 10** Distributions of relative errors of neural network modeling on the parameter plane $D_{max}^{turb}$, $z_{max}^{turb}$ для: $D_{max,mod}^{turb}$ (**a**) and $z_{max,mod}^{turb}$ (**b**)

## 3 Conclusion

Reconstruction of vertical profiles of the turbulent diffusion coefficient based on the analysis of time series of dust concentrations from a non-stationary source is considered. The dynamics of dust depend significantly on the particle size and the distance from the source to the measurement point. This manifests itself in shifts in dust concentration maxima. Changes occur in the distribution of the dispersion composition of aerosols as they move away from the source of pollutants. The proportion of large particles decreases significantly even at short distances from the source, which greatly changes the shape of the dust distribution density along the diameters. This is important for solving the problem of estimating atmospheric turbulence parameters.

Two ways of specifying objective functions are considered. The first approach is based on assessing the degree of similarity of time series of dust concentrations. This allows in many cases to calculate the range of parameters $z_{max}^{turb}$ и $D_{max}^{turb}$, characterizing the real state of the atmosphere. However, due to the peculiarities of the turbulent diffusion coefficient model used, the scatter of parameters can be quite large. The second approach involves analyzing the dispersion composition of dust, which can narrow down the areas of possible solutions.

A method is proposed for determining the parameters of the state of the atmosphere based on machine learning. The neural network contains a recurrent LSTM layer, which improves the processing of time series. The data for the training set uses the results of ~600 computational experiments for particles of different diameters ($d^{(a)} = 2.5 \div 125$ µm) and different states of the atmosphere ($z_{max}^{turb} = 20 \div 100$ m и $D_{max}^{turb} = 1 \div 10$ m$^2$/sec).

The developed neural network shows high accuracy in determining atmospheric parameters for some combinations of turbulent diffusion parameters, but there is a large error (40%) for certain parameter areas. In addition, the use of the neural network reduces the required number of computational experiments by ~10 times compared to direct methods of reconstructing the vertical profile of the diffusion coefficient.

**Acknowledgements** This work supported by the Russian Science Foundation (grant no. 23-71-00016, https://rscf.ru/project/23-71-00016/).

## References

1. Ding, M., Tong, C.: Multi-point Monin-Obukhov similarity of turbulence cospectra in the convective atmospheric boundary layer. Bound. Layer Meteorol. **178**, 185–199 (2021)
2. Glinyanova, I., Asanova, N.: Research of urban atmospheric aerosols of the Lower Volga under conditions of anthropogenic load and active zones of Earth. Environ. Sci. Pollut. Res. **29**, 71380–71399 (2022)
3. Aloyan, A.E., Arutyunyan, V.O., Yermakov, A.N.: Modeling the regional influence of ions on aerosol formation in the atmosphere. Izvestiya Atmos. Ocean. Phys. **58**(3), 246–253 (2022)
4. Kakareka, S.V., Salivonchyk, S.V.: Modeling atmospheric air pollution of an industrial city with fine particulate matter. Russian Meteorol. Hydrol. **46**, 331–340 (2021)
5. Sánchez-Ccoyllo, O.R., Llacza, A., Ayma-Choque, E., Alonso, M., Castesana, P., Andrade, M.F.: Evaluating the impact of vehicular aerosol emissions on particulate matter (PM2.5) formation using modeling study. Atmosphere **13**(11), 1816 (2022)
6. Martins, C.A., Moraes, O.L.L., Acevedo, O.C., et al.: Turbulence intensity parameters over a very complex terrain. Bound. Layer Meteorol. **133**, 35–45 (2009)
7. Grisogono, B., Kraljevic, L., Jericevic, A.: The low-level katabatic jet height versus Monin-Obukhov height. Quart. J. Royal Meteorol. Soc. **133**(139), 2133–2136 (2007)
8. Elansky, N.F., Shilkin, A.V., Ponomarev, N.A., Zakharova, P.V., Kachko, M.D., Poliakov, T.I.: Spatiotemporal variations in the content of pollutants in the Moscow air basin and their emissions. Izvestiya Atmosp. Ocean. Phys. **58**, 80–94 (2022)
9. Frehlich, R.: Doppler lidar measurements of winds and turbulence in the boundary layer. IOP Conf. Ser. Earth Environ. Sci. **1**, 012017 (2008)
10. Frick, P.G.: Turbulence Models and Approaches, 1st edn. Regular and chaotic dynamics, Moscow (2010)
11. Zaichik, L.I., Alipchenkov, V.M.: Statistical Models of Particle Motion in Turbulent Fluid. Fizmatlit, Moscow (2006)
12. Volkov, K.N., Emelyanov, V.N.: Gas Flows with Particles. Fizmatlit, Moscow (2008)
13. Jain, S.K., Agarwal, P.K., Singh, V.P.: Evaporation and other meteorological data. Hydrol. Water Resour. India **57**, 155–192 (2007)
14. Vinnichenko, N.K., Pinus, N.Z., Shmeter, S.M., Shur, G.N.: Turbulence in the Free Atmosphere, 1st edn. Springer, New York (1980)
15. Bulgakov, K.Y., Fokina, K.V.: Modeling the surface layer above sea with parameterization of stratification and the elements of the wind waves influence. Fundam. Prikl. Gidrofiz. **14**(2), 3–16 (2021)
16. Lukin, V.P.: Outer scale of turbulence and its influence on fluctuations of optical waves. Phys. Uspekhi **64**, 280–303 (2021)
17. Davydova, M.A., Zakharova, S.A., Elansky, N.F.: A new approach to the problem of reconstructing the vertical turbulent diffusion coefficient in the atmosperic boundary layer. Doklady Earth Sci. **490**(2), 92–96 (2020)
18. Wu, Y., Li, Q., Li, G., He, B., Dong, L., Lan, H., et al.: Vertical wind speed variation in a metropolitan city in South China. Earth Space Sci. **9**(4), e2021EA002095 (2022)
19. Gualtieri, G., Secci, S.: Methods to extrapolate wind resource to the turbine hub height based on power law: a 1-h wind speed versus Weibull distribution extrapolation comparison. Renew. Energy **43**, 183–200 (2012)

20. Grisogono, B., Oerlemans, J.: Katabatic flow: analytic solution for gradually varying eddy diffusivities. J. Atmosp. Sci. **58**(21), 3349–3354 (2001)

# Secondary Pollution of the Urban Atmosphere by Vehicles of Various Emission Standards and Traffic Natures

**Vadim A. Zaytsev and Anastasia V. Morozova**

**Abstract** Existing methods for calculating the emission rates of primary pollutants from cars are analyzed from the position of the predominance of their various emission standards in the traffic flow. These masses of substances per unit time are used in further calculations of the concentration fields of these compounds obtained during the study of dispersion processes. Emissions are considered at various flow densities and vehicle speeds during continuous movement, as well as in the area of a pedestrian crossing equipped with a traffic light. In this case, the concentrations of nitrogen oxides and reduced carbon compounds, as well as their ratios, change over a fairly wide range. The content of a more toxic secondary pollutant, ozone, is determined using the proposed mathematical model. It is built on the basis of a stiff system of differential equations, reflecting the kinetics of this process and solved by the Rosenbrock method. During numerical experiments, a search is made for the best and worst options in terms of impact on the atmosphere.

**Keywords** Mathematical modeling · Systems of stiff differential equations · Environmental impact assessment · Environmental chemistry · Atmospheric chemistry · Chemical kinetics · Ecology

## 1 Introduction

Due to the complexity and diversity of chemical reactions and accompanying transport processes occurring in the polluted troposphere, when constructing mathematical models that describe their kinetics, one often has to face the problem of a lack of computing resources.

This may occur, on the one hand, due to the impossibility of considering the transformations of a huge number of chemical components, because thousands of organic substances are emitted into the atmosphere [1]. For the convenience of calculations in such cases, which occur, for example, in [2, 3], more compact process

V. A. Zaytsev (✉) · A. V. Morozova
Dmitry Mendeleev University of Chemical Technology of Russia, Moscow, Russia
e-mail: 8zaytsev@mail.ru

 
A. G. Kravets and A. A. Bolshakov (eds.), *Human-Centric Cyber-Society*, Studies in Systems, Decision and Control 639, https://doi.org/10.1007/978-3-032-06787-6_11

mechanisms are proposed that make it possible to reasonably reduce the number of reactions and interacting substances, and, accordingly, differential equations. Since the prevailing processes (their type, quantity and directions), as well as the concentrations of certain substances in the urban, regional and unpolluted atmosphere are significantly different, they were considered in article [4] to create an appropriate mechanism dedicated only to these first two cases. Sometimes, in an earlier period of time, simplifications were achieved by considering the magnitude and nature of deviations from the photostationary state in the $O_3$–NO–$NO_2$ system, which made it possible to replace part of the differential equations with algebraic ones [5].

On the other hand, an attempt to take into account in models, along with chemical transformation, which is presented even in a simplified form, the transfer of pollutants leads to an increase in the number of variables. Concentrations begin to change not only in time, but also in space. From stiff systems of ordinary differential equations we have to move to partial differential equations. The latter circumstance, for the sake of preserving the chemical block of the model, forces, in some cases, to use the approach implemented in the so-called box models. It is based on the assumption that the concentrations of substances inside the box are equalized due to mixing [1, 6].

In publications [7], based on calculations of the dispersion process, it is shown that the content of pollutants near the highway changes relatively little up to a certain distance from it. This circumstance allows us to consider this problem as one-dimensional, in which concentrations depend only on time.

Throughout the development of research in atmospheric chemistry, to implement any of the above-mentioned approaches, data on reaction rate constants has been collected and summarized. Review papers are regularly published, updating and expanding previous estimates on kinetics and photochemistry, which were contained in several articles at once. Thus, work [8] covers 658 reactions and occupies 491 pages, and for all of them the preferred values of rate coefficients at 298 K are given (some of them are used in this chapter), and where possible, temperature dependences. In addition, the available experimental data are summarized with notes detailing the experimental procedures. Subsequently, voluminous papers containing kinetic and photochemical characterizations were regularly published and reviewed by the IUPAC Task Force on the Assessment of Atmospheric Chemical Kinetic Data. They contain a huge amount of material. Directly related to this work on the reactions considered here are [9–11]. The authors are usually members of this target group.

According to articles [1–6], one of the most common secondary pollutants, the formation of which leads to an increase in air toxicity, is tropospheric ozone. It has an extremely negative effect on the human cardiovascular and respiratory systems, causing an increase in mortality [12–17]. Because of the above, a large number of works are regularly devoted to such a very pressing problem as ozone pollution of the lower atmosphere. They are described in reviews [18–20]. In this publication, its concentration in the area of influence of the highway is determined using mathematical modeling.

## 2 Scientific Novelty, Purpose and Objectives of the Research

The airspace near highways with intense traffic flows contains significant amounts of primary pollutants, as well as products of their chemical transformation. To study the kinetics of the processes of these transformations, systems of differential equations are compiled. In this case, the concentrations of primary pollutants are used as initial conditions for solving them. In the author's publication [21] these concentrations are given on the basis of direct measurements. Their various changes, which may occur with an increase in traffic density, are being studied. This occurs in cramped aerodynamic conditions of the urban atmosphere, which impede the transfer of substances from the reaction zone and lead to an increase in their content, which is typical for mass transfer processes [22]. The latter circumstance helps to accelerate chemical processes by increasing the concentrations of interacting substances. In the author's publication [7], they are calculated on the basis of linking the number of transport units on the highway, the emission rates of each primary pollutant and their average concentrations obtained during dispersion.

In this work, the powers of emissions of moving vehicles are used as initial values in the model. As already mentioned above, this approach has already been applied by the author in his work [7]. However, serious differences in the masses of substances emitted by cars per unit time due to the predominance of their different emission standards in the traffic flow were not considered. In addition, variations in different densities of continuous traffic flows with simultaneous changes in vehicle speeds, as well as the influence of the presence of traffic lights in all these cases, have not been studied.

Based on the above, we can formulate the goal of the work as an attempt to connect the number of units of vehicles of various emission standards continuously moving at a certain speed on the highway, the powers of emissions of each primary pollutant with their concentrations after dispersion and the concentration of the resulting chemical transformations of these compounds secondary pollutant—ozone. Similar actions are also carried out on a section of this road, but with a pedestrian crossing equipped with a traffic light. Next, these areas with different patterns of movement are compared in terms of the intensity of primary and secondary pollution [7, 21].

To do this, the following tasks are solved:

1. Calculation of the number of vehicles moving along the highway and on its section with a pedestrian crossing equipped with a traffic light.
2. Determination of the emission rates of pollutants at different speeds, in areas of continuous traffic and near traffic lights.
3. Calculation of concentration fields of primary pollutants (mono- and nitrogen dioxide, carbon monoxide, methane, formaldehyde) near the highway, necessary for studying their further chemical transformation.
4. Calculation using a mathematical model of the concentrations of a secondary pollutant—ozone [7].

## 3 Proposed Mathematical Model and Numerical Experiments

A detailed description of the mathematical model is provided in [7, 21], and in this work, for ease of perception, the most important aspects are shown. Calculation of emissions from moving vehicles can be carried out using methods "according to formula (1):

$$M_{L_i} = \frac{L}{1200} \cdot \sum_{1}^{k} M_{k,i}^{L} \cdot G_k \cdot r_{V_{k,i}}, \tag{1}$$

where $M_{L_i}$—emission of the $i$-th pollutant by moving traffic flow, g/s; $M_{k,i}^{L}$—specific mileage emission of the $i$-th harmful substance by cars of the $k$-th group, determined from the Table 1, g/km; $k$—number of groups of cars; $G_k$—actual maximum traffic intensity, i.e. the number of cars of each of the $k$ groups passing through a fixed section of the selected section of the road per unit of time (20 min) in both directions along all lanes; $r_{V_{k,i}}$—correction factor taking into account the average speed of traffic flow $V_{k,i}$,, km/h on the selected road (or section thereof), determined from Table 2; $L$ is the length of the investigated part of the road, on which there should be no traffic lights (for them the calculation is carried out differently), km".

These methods for calculating the emission rates of primary pollutants from cars are analyzed from the position of the predominance of their various emission standards in the transport flow. Thus, in there is a shift towards the worst—Euro 0, and in [25] the best—Euro 5.

Calculation of vehicle emissions in the traffic light area can also be carried out using methods. To calculate them under red light signals, for a 20-min period of time, for each $i$-th pollutant $M_{T_i}^{P}$ (g), formula (2) is used:

$$M_{T_i}^{P} = \frac{P_c}{60} \cdot \sum_{1}^{N_c} \sum_{1}^{k} \left( M_{T_{i,k}}^{\prime} \cdot G_k \right), \tag{2}$$

"where $P_c$ is the duration of the prohibitory traffic light signal (including yellow) for 20 minutes, s; $N_c$—number of cycles of a prohibiting traffic light signal over a 20-minute period of time; $M_{T_{i,k}}^{\prime}$—specific emission of the $i$-th pollutant by cars of the $k$-th group, which are in the "queue" at the prohibitory traffic light signal, g/min; $G_k$ is the number of cars of the $k$-th group that are in the "queue" in the area of the intersection at the end of each cycle of the prohibitory traffic light signal".

Specific emissions $M_{T_{i,k}}^{\prime}$ (g/min), which are average characteristics for such modes of vehicle movement in the area of a traffic light as braking, idling and acceleration, are presented in Table 3. The values of $R_C$, $N_C$, $G_k$ are determined experimentally.

The calculation of the total one-time emission of the $i$-th pollutant (g/s) for a 20-min period of time near the traffic light $M_{T_i}^{S}$ is performed according to the relation (3). In turn, $M_{L_i}^{A}$ is calculated according to the formula (4):

**Table 1** Specific mileage emissions of pollutants $M_{k,i}^{L}$ (g/km) for different groups of vehicles according to methods

| Number and name of the car group | The normative value of the emission, g/km | | | | | | |
|---|---|---|---|---|---|---|---|
| | CO | $NO_x$ (in terms of $NO_2$) | CH | Soot | $SO_2$ | Formalde-hyde | Benz (a) pyrene |
| I. Passenger cars | 19.0 | 1.8 | 2.1 | – | 0.065 | 0.006 | $1.7 \times 10^{-6}$ |
| | 3.5 | 0.9 | 0.8 | $0.7 \times 10^{-2}$ | $1.5 \times 10^{-2}$ | $3.2 \times 10^{-3}$ | $0.3 \times 10^{-6}$ |
| | 0.9 | 0.33 | 0.26 | $0.55 \times 10^{-2}$ | $0.66 \times 10^{-2}$ | $1.5 \times 10^{-3}$ | $0.18 \times 10^{-6}$ |
| II. Vans and minibuses weighing up to 3.5 t | 69.4 | 2.9 | 11.5 | – | 0.2 | 0.02 | $4.5 \times 10^{-6}$ |
| | 8.4 | 2.1 | 2.4 | $3.8 \times 10^{-2}$ | $2.8 \times 10^{-2}$ | $8.4 \cdot 10^{-3}$ | $0.8 \times 10^{-6}$ |
| | 4.6 | 1.8 | 0.7 | $3.7 \times 10^{-2}$ | $1.4 \times 10^{-2}$ | $2.5 \times 10^{-3}$ | $0.2 \times 10^{-6}$ |
| III. Freight weighing from 3.5 to 12 t | 75.0 | 5.2 | 13.4 | – | 0.22 | 0.022 | $6.3 \times 10^{-6}$ |
| | 6.8 | 6.9 | 5.2 | 0.4 | $5.1 \times 10^{-2}$ | $2.2 \times 10^{-2}$ | $2.1 \times 10^{-6}$ |
| | 5.3 | 6.4 | 1.5 | 0.37 | $2.6 \times 10^{-2}$ | $0.7 \times 10^{-2}$ | $0.6 \times 10^{-6}$ |
| IV. Freight weighing over 12 t | 7.3 | 8.5 | 6.5 | 0.5 | $7.3 \times 10^{-2}$ | $2.5 \times 10^{-2}$ | $2.6 \times 10^{-6}$ |
| | 5.6 | 7.5 | 2.0 | 0.44 | $3.9 \times 10^{-2}$ | $0.8 \times 10^{-2}$ | $0.73 \times 10^{-6}$ |
| V. Buses weighing over 3.5 t | 5.2 | 6.1 | 4.5 | 0.3 | $4.2 \times 10^{-2}$ | $1.8 \times 10^{-2}$ | $1.8 \times 10^{-6}$ |
| | 3.9 | 4.7 | 0.5 | 0.15 | $2.2 \times 10^{-2}$ | $0.22 \times 10^{-2}$ | $0.2 \times 10^{-6}$ |

**Table 2** Values of coefficients $r_{Vk,i}$, taking into account changes in the amount of emitted harmful substances depending on the average speed of movement according to methods

| | The average flow rate of cars, $V$, km/h | | | | | | | | | | | | | | |
|---|---|---|---|---|---|---|---|---|---|---|---|---|---|---|---|
| | 5 | 10 | 15 | 20 | 25 | 30 | 35 | 40 | 45 | 50 | 60 | 70 | 80 | 100 | 110 |
| $r_{Vk,i}$ | - | 1.35 | 1.28 | 1.2 | 1.1 | 1.0 | 0.88 | 0.75 | 0.63 | 0.5 | 0.3 | – | 0.5 | 0.65 | – |
| | 1.4 | 1.35 | 1.3 | 1.2 | 1.1 | 1.0 | 0.9 | 0.75 | 0.65 | 0.5 | 0.3 | 0.4 | 0.5 | 0.65 | 0.75 |
| | 1.4 | 1.35 | 1.3 | 1.2 | 1.1 | 1.0 | 0.9 | 0.75 | 0.6 | 0.5 | 0.3 | 0.4 | 0.5 | 0.65 | 0.75 |
| $r_{Vk,i}$ (NOx) | 1.0 | 1.0 | 1.0 | 1.0 | 1.0 | 1.0 | 1.0 | 1.0 | 1.0 | 1.0 | 1.0 | 1.0 | 1.0 | – | – |
| | 1.0 | 1.0 | 1.0 | 1.0 | 1.0 | 1.0 | 1.0 | 1.0 | 1.0 | 1.0 | 1.0 | 1.0 | 1.0 | 1.0 | 1.2 |
| | 1.0 | 1.0 | 1.0 | 1.0 | 1.0 | 1.0 | 1.0 | 1.0 | 1.0 | 1.0 | 1.0 | 1.0 | 1.0 | 1.0 | 1.2 |

**Table 3** Specific emissions for vehicles located in the traffic light zone $M'_{T_{i,k}}$ according to the methods

| Number and name of the car group | The normative value of the emission, g/min | | | | | | |
|---|---|---|---|---|---|---|---|
| | CO | $NO_x$ (in terms of $NO_2$) | CH | Soot | $SO_2$ | Formalde-hyde | Benz (a) pyrene |
| I. Passenger cars | 3.5 | 0.05 | 0.25 | – | 0.01 | $8 \times 10^{-4}$ | $2 \times 10^{-6}$ |
| | 0.5 | 0.015 | 0.1 | 0.015 | $0.5 \times 10^{-2}$ | $0.4 \times 10^{-3}$ | $0.15 \times 10^{-6}$ |
| | 0.17 | $8.8 \times 10^{-3}$ | 0.033 | 0.011 | $0.17 \times 10^{-2}$ | $0.26 \times 10^{-3}$ | $0.13 \times 10^{-6}$ |
| II. Vans and minibuses weighing up to 3.5 t | 6.3 | 0.075 | 1.0 | – | 0.02 | $1.5 \times 10^{-3}$ | $4 \times 10^{-6}$ |
| | 2.0 | 0.04 | 0.30 | 0.08 | $0.9 \times 10^{-2}$ | $1.4 \times 10^{-3}$ | $0.4 \times 10^{-6}$ |
| | 1.0 | $30 \times 10^{-3}$ | 0.07 | 0.033 | $0.33 \times 10^{-2}$ | $0.3 \times 10^{-3}$ | $0.13 \times 10^{-6}$ |
| III. Freight weighing from 3.5 to 12 t | 18.4 | 0.2 | 2.96 | – | $2.8 \times 10^{-2}$ | $6 \times 10^{-3}$ | $4.4 \times 10^{-6}$ |
| | 2.5 | 0.12 | 0.66 | 0.90 | $1.7 \times 10^{-2}$ | $7.2 \times 10^{-3}$ | $1.1 \times 10^{-6}$ |
| | 1.0 | $9.9 \times 10^{-2}$ | 0.17 | 0.22 | $0.55 \times 10^{-2}$ | $0.76 \times 10^{-3}$ | $0.33 \times 10^{-6}$ |
| IV. Freight weighing over 12 t | 2.7 | 0.14 | 0.83 | 1.10 | $2.4 \times 10^{-2}$ | $9.5 \times 10^{-3}$ | $1.3 \times 10^{-6}$ |
| | 2.0 | $13 \times 10^{-2}$ | 0.26 | 0.45 | $0.66 \times 10^{-2}$ | $1.16 \times 10^{-3}$ | $0.4 \times 10^{-6}$ |
| V. Buses weighing over 3.5 t | 1.9 | 0.10 | 0.57 | 0.67 | $1.5 \times 10^{-2}$ | $4.8 \times 10^{-3}$ | $0.9 \times 10^{-6}$ |
| | 0.9 | $9.9 \times 10^{-2}$ | 0.07 | 0.12 | $0.6 \times 10^{-2}$ | $0.25 \times 10^{-3}$ | $0.1 \times 10^{-6}$ |

$$M^S_{T_i} = \frac{1}{1200}\left(M^P_{T_i} + M^A_{L_i}\right) \quad (3)$$

$$M^A_{L_i} = L^T \cdot \sum_{1}^{N'_c} \sum_{1}^{k} M^L_{k,i} \cdot G^A_k, \quad (4)$$

"where $L^T$ is the distance covered by cars in one direction with permissive traffic lights for 20 minutes, consisting of the length of the queue of cars formed with a prohibiting traffic light and the length of the corresponding transition zone, km; $N'_c$—number of cycles of operation of the permitting traffic light signal within 20 minutes; $G^A_k$—the number of cars of each $k$-th group passing through the traffic light zone in one direction with a signal allowing it".

A section of a highway in a part of the city where there is a ban on the movement of trucks is considered. Thus, the flow includes cars, as well as vans and minibuses, which make up 5% of the total number of cars. On the highway in each direction there are three lanes. Knowing the average length of vehicles and their speeds and considering these values to be uniform on all lanes of the road, the number of vehicles crossing its fixed section in 20 min in both directions is determined [7, 21]. The initial distance between them is 50 m at a speed of 60 km/h. In this case, the correction factor, which takes into account the average speed of the traffic flow, takes on a minimum value. Further, based on the correction coefficient gradually approaching the maximum values, taking into account the average speed of the traffic flow (3 more speed options: 100; 40; 15 km/h), the conditions for conducting numerical experiments were selected. They are carried out at fixed vehicle flow densities (3 density options: the same as with a distance between vehicles of 100 m, 50 m and 25 m at a speed of 60 km/h, which corresponds to 58, 111, and 207 cars passing along all lanes and directions of movement through a fixed section of the highway per minute). The distances between cars change with speed. Numerical experiments at a speed of 100 km/h allow us to evaluate the impact on the atmosphere of an expressway [7].

With the above values, formula (1) calculates the masses of primary pollutants (carbon monoxide, nitrogen monoxide and dioxide, methane and formaldehyde) emitted per unit time. The emission rates of NO and $NO_2$ were calculated taking into account the transformation coefficients of nitrogen oxides in the atmosphere, typical for air near roads, and methane was considered as the main hydrocarbon in the exhaust gases of the studied groups of vehicles.

A standard certified program was used to calculate dispersion. They were carried out at a wind speed of 1 m/s. These are the most unfavorable conditions. Calm as a rule is not considered, because in a number of calculation methods, wind speed is in the denominator and, for obvious reasons, values at wind speeds even close to zero cannot be considered reliable. The wind direction under study is perpendicular to the road, which is essentially an area source of emissions with the origin in its middle. Consequently, concentrations obtained outside it—at a distance of 20 m and further—will be reliable.

The best options from the point of view of the intensity of primary pollution with reduced carbon compounds and formaldehyde for continuous flows of vehicles at constant densities indicated above (in the case of pairwise comparison) are shown in Figs. 1, 5, 9, the worst—in Figs. 2, 6, 10. The figures of the first group show the dependences of the concentrations of emitted substances on the distance from the highway at a vehicle speed of 60 km/h, the second group—15 km/h. For these cases, a change in speed only leads to an increase in the concentrations of CO, $CH_4$ and formaldehyde, while the concentrations of nitrogen oxides do not depend on the speed in a fairly wide range. When moving from flows with lower densities to more dense ones at the same speeds, the concentrations of all compounds increase multiple times. For all these cases, the maximum concentration values for each compound differ from the arithmetic mean by approximately 17%, and the minimum by 20%. Consequently, the content of pollutants near the highway up to a distance of 110 m from it changes relatively little. In Figs. 3, 4, 7, 8, 11, 12 show the dependences of the concentrations of emitted substances on the distance from the highway in the traffic light area at the considered densities and speeds of traffic flows. It is necessary to note the increase in concentrations of CO, $CH_4$ and formaldehyde at a traffic light compared to continuous traffic. This effect is clearly manifested in less dense flows at a speed of 60 km/h and decreases or disappears completely as the flow density increases and the speed approaches 15 km/h. Here, similar differences between the maximum and minimum concentration values with the arithmetic mean are also generally small. Thus, in these cases, concentrations can be considered quasi-constant up to a distance of 110 m. Consequently, these circumstances allow us to consider this problem as one-dimensional, in which concentrations depend only on time.

These average concentrations, taking into account the background, act as initial conditions when solving a system of differential equations that describes the chemical transformations of the substances in question in polluted air. It is described in detail in [7, 21], so now, for ease of perception, the most important aspects are shown. "In the Table 4 shows the chemical reactions that are essential for this process under the above conditions, as well as their rate constants." [7, 21].

Modeling problems of this type are described by the following systems of ordinary differential Eqs. (5):

$$dc_i/d = F_i - R_i, \tag{5}$$

where $c_i$ are the concentrations of each of the starting substances and products of these reactions (the unit of measurement is the number of particles per cm3 of air), τ is time, $F_i$ and $R_i$ are terms that take into account the generation and consumption of substances, respectively. The numerical solution of the stiff system of 17 differential

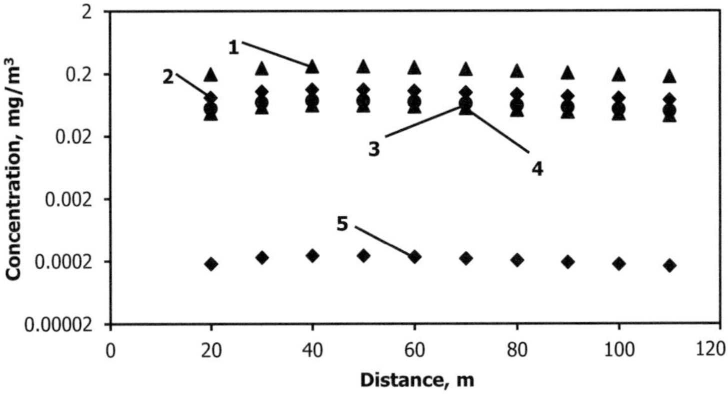


**Fig. 1** Dependence of the concentration of the emitted substance on the distance from the highway at a traffic density of 58 cars/min and a car speed of 60 km/h: 1—CO; 2—$NO_2$; 3—NO; 4—$CH_4$; 5—formaldehyde

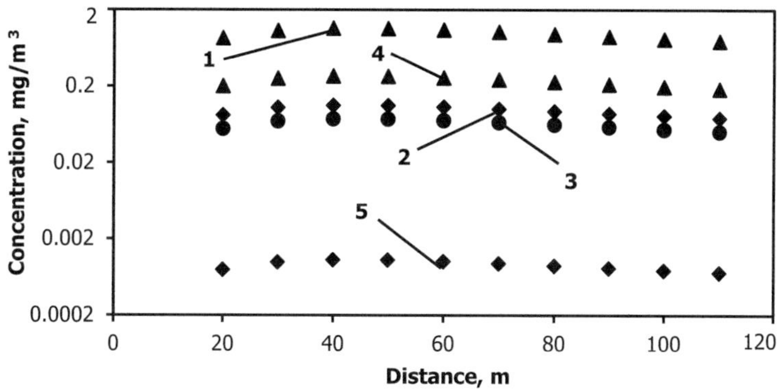


**Fig. 2** Dependence of the concentration of the emitted substance on the distance from the highway at a traffic density of 58 cars/min and a car speed of 15 km/h: 1—CO; 2—$NO_2$; 3—NO; 4—$CH_4$; 5—formaldehyde

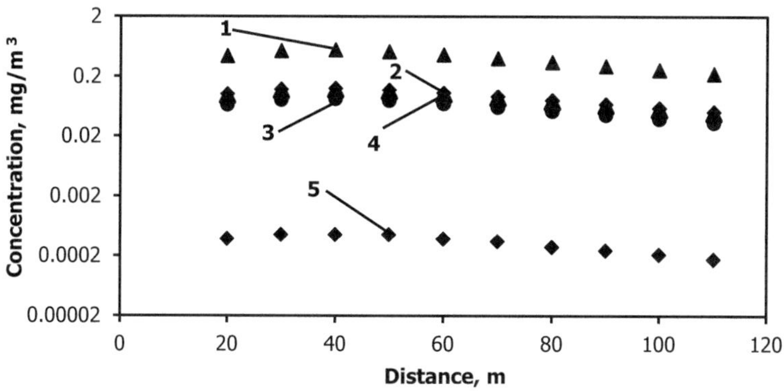


**Fig. 3** Dependence of the concentration of the emitted substance on the distance from the highway in the traffic light area at a traffic density of 58 cars/min and a car speed of 60 km/h: 1—CO; 2—$NO_2$; 3—NO; 4—$CH_4$; 5—formaldehyde

equations derived in this work is carried out by Rosenbrock's method. It is one of the specially developed methods for solving systems of this class [7, 21].

In Fig. 13 presents the results of numerical experiments with a traffic flow density of 111 cars/min for secondary air pollution with ozone on the leeward side. At the considered wind speed, the transfer of substances to a distance of 110 m from the center of the highway will occur in 100 s. The initial ozone concentration is 0.03 mg/m$^3$. The figure shows that the highest values of its concentrations are achieved at a vehicle speed of 15 km/h, followed by dependencies for 40, 100 and 60 km/h.

In this way, the speeds corresponding to the best and worst case scenarios in terms of secondary pollution are selected. In Figs. 14, 15 and 16 show the results of numerical experiments on secondary air pollution at the considered traffic flow

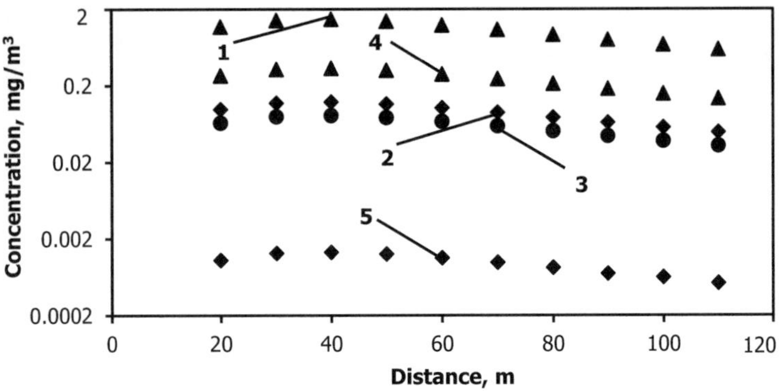


**Fig. 4** Dependence of the concentration of the emitted substance on the distance from the highway in the traffic light area at a traffic density of 58 cars/min and a car speed of 15 km/h: 1—CO; 2—$NO_2$; 3—NO; 4—$CH_4$; 5—formaldehyde

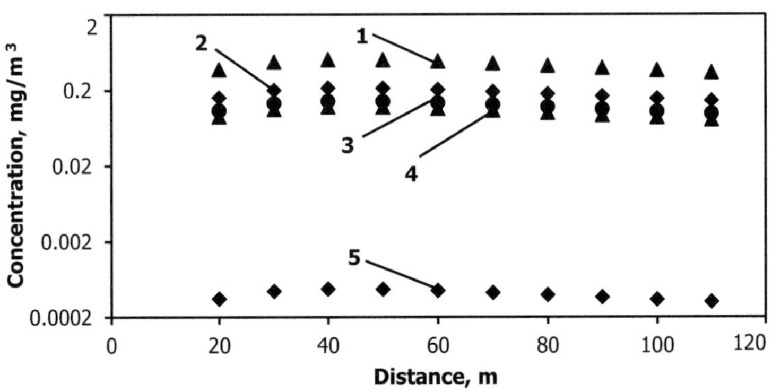


**Fig. 5** Dependence of the concentration of the emitted substance on the distance from the highway at a traffic density of 111 cars/min and a car speed of 60 km/h: 1—CO; 2—$NO_2$; 3—NO; 4—$CH_4$; 5—formaldehyde

densities and these speeds for areas of continuous traffic and in the presence of a traffic light.

It can be seen from the figures that secondary ozone pollution in the presence of traffic lights is only slightly higher than in areas of continuous traffic with the same flow densities and speeds. As the flux density increases, the ozone content increases and the differences between concentrations at different speeds become larger. But these differences are not as great as for a number of primary pollutants.

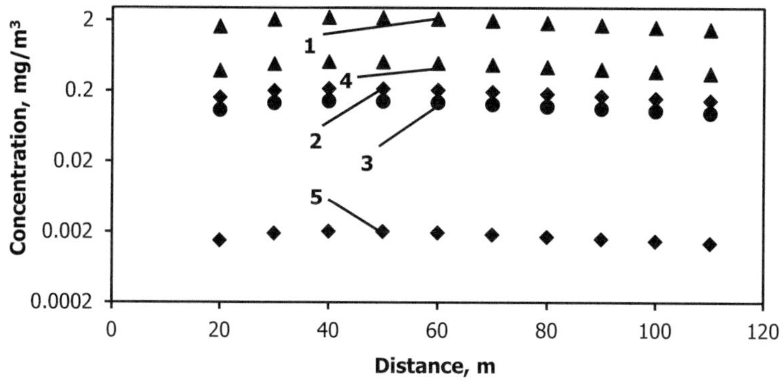


**Fig. 6** Dependence of the concentration of the emitted substance on the distance from the highway at a traffic density of 111 cars/min and a car speed of 15 km/h: 1—CO; 2—$NO_2$; 3—NO; 4—$CH_4$; 5—formaldehyde

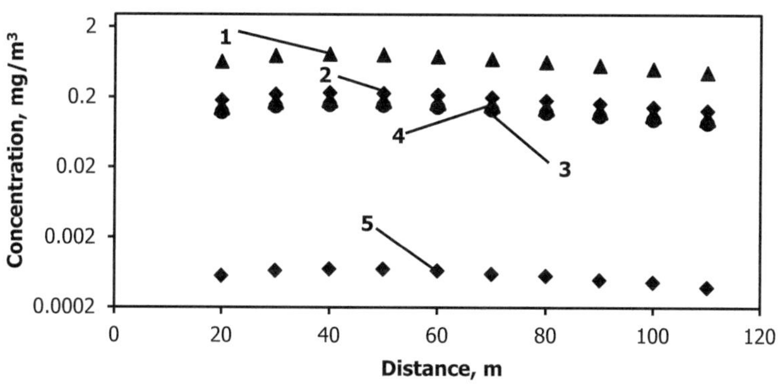


**Fig. 7** Dependence of the concentration of the emitted substance on the distance from the highway in the traffic light area at a traffic density of 111 cars/min and a car speed of 60 km/h: 1—CO; 2—$NO_2$; 3—NO; 4—$CH_4$; 5—formaldehyde

## 4 Conclusions

When moving from flows with lower density to more dense ones at the same speeds, the concentrations of all compounds increase multiple times. This is true for any speed values that can be physically achieved at given flux densities.

It is necessary to note the increase in the concentrations of CO, $CH_4$ and formaldehyde in the most significant cases from this point of view (traffic flow density of 58 cars/min and speed of 60 km/h), respectively, by 78, 63 and 63.5% at the traffic light compared to continuous movement. This effect is most clearly manifested in less dense flows at a speed of 60 km/h and decreases as their density increases and the speed approaches 15 km/h. But this absolutely does not mean that less of these substances are emitted at traffic lights. It's just that in similar continuous flows more

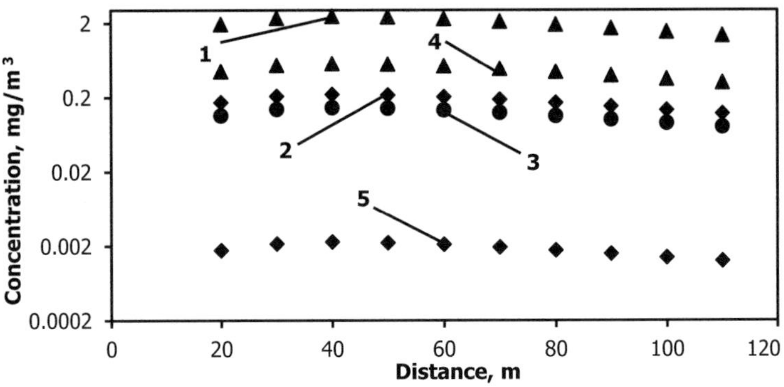


**Fig. 8** Dependence of the concentration of the emitted substance on the distance from the highway in the traffic light area at a traffic density of 111 cars/min and a car speed of 15 km/h: 1—CO; 2—$NO_2$; 3—NO; 4—$CH_4$; 5—formaldehyde

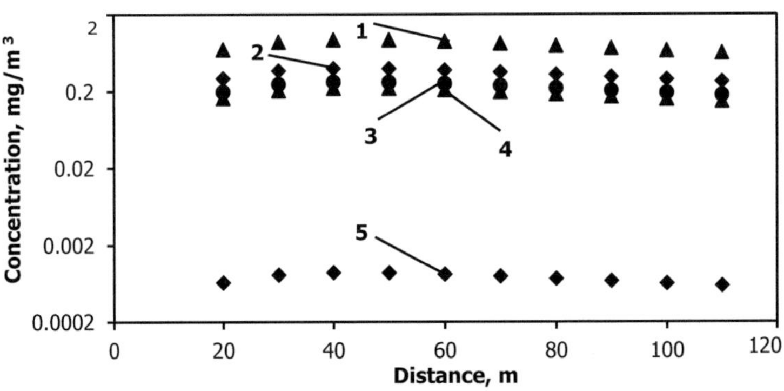


**Fig. 9** Dependence of the concentration of the emitted substance on the distance from the highway at a traffic density of 207 cars/min and a car speed of 60 km/h: 1—CO; 2—$NO_2$; 3—NO; 4—$CH_4$; 5—formaldehyde

of them will be released. For these cases, a change in speed only leads to an increase in the concentrations of CO, $CH_4$ and formaldehyde, while the concentrations of nitrogen oxides do not depend on the speed in a fairly wide range.

During chemical transformations, a quantitative increase in the content of ozone, which is a more toxic secondary pollutant, was noted. In the worst cases considered, i.e. at a traffic density of 207 cars/min and a speed of 15 km/h for continuous traffic and in the presence of a traffic light, by 53% and 55%, respectively. However, the maximum permissible maximum single concentration is not exceeded, although the maximum permissible average daily concentration, which is less suitable for assessments in these circumstances, is exceeded by the same amounts, because it acts as an initial concentration (which quite often occurs in an urban atmosphere).

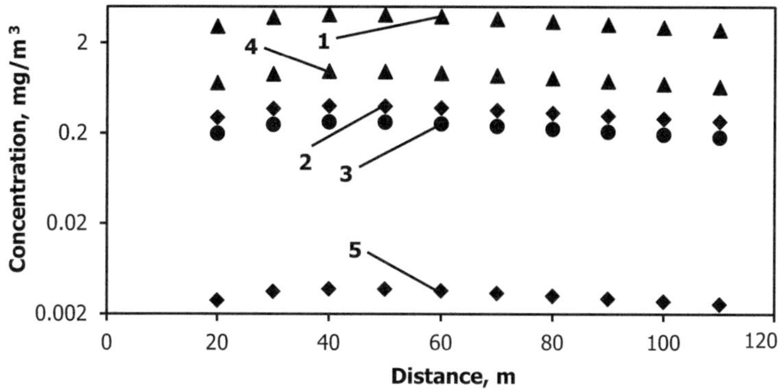


**Fig. 10** Dependence of the concentration of the emitted substance on the distance from the highway at a traffic density of 207 cars/min and a car speed of 15 km/h: 1—CO; 2—$NO_2$; 3—NO; 4—$CH_4$; 5—formaldehyde

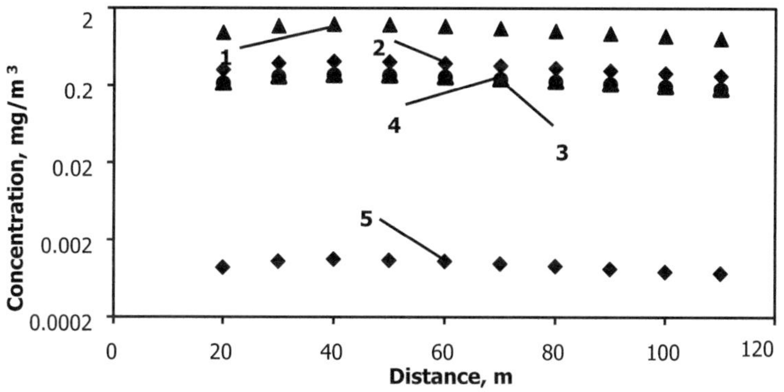


**Fig. 11** Dependence of the concentration of the emitted substance on the distance from the highway in the traffic light area at a traffic density of 207 cars/min and a car speed of 60 km/h: 1—CO; 2—$NO_2$; 3—NO; 4—$CH_4$; 5—formaldehyde

Secondary ozone pollution in the presence of traffic lights is only slightly higher than in areas of continuous traffic with the same flow densities and speeds. As the flux density increases, the ozone content increases, and the differences between concentrations at different speeds become greater. But these differences are not as great as for a number of primary pollutants.

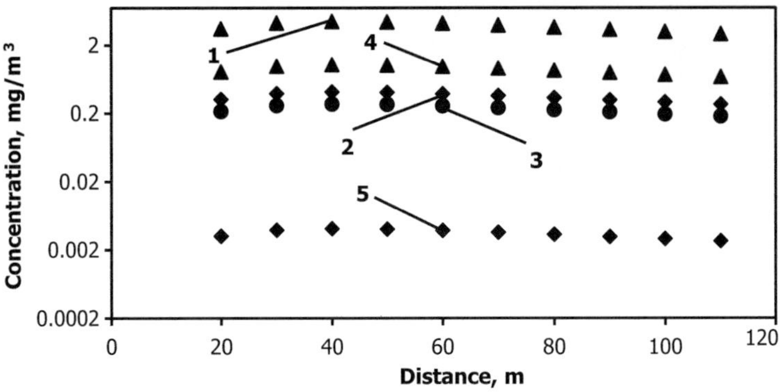


**Fig. 12** Dependence of the concentration of the emitted substance on the distance from the highway in the traffic light area at a traffic density of 207 cars/min and a car speed of 15 km/h: 1—CO; 2—$NO_2$; 3—NO; 4—$CH_4$; 5—formaldehyde

**Table 4** Chemical reactions taken into account in the mathematical model and their rate constants [7, 21]

| Atmospheric reaction | Rate constant |
|---|---|
| $CH_4 + HO^{\cdot} \rightarrow {}^{\cdot}CH_3 + H_2O^{\cdot}$ | $k = 7.9 \times 10^{-15}$ |
| ${}^{\cdot}CH_3 + O_2 + M \rightarrow CH_3O_2^{\cdot} + M^*$ | $k\,[M] = 1.8 \times 10^{-12}$ |
| $CH_3O_2^{\cdot} + NO \rightarrow CH_3O^{\cdot} + NO_2$ | $k = 7.6 \times 10^{-12}$ |
| $CH_3O^{\cdot} + O_2 \rightarrow CH_2 = O + HO_2^{\cdot}$ | $k = 1.3 \times 10^{-15}$ |
| $CH_2 = O + h\nu \rightarrow {}^{\cdot}CH = O + H^{\cdot}$ | $k = 3.7 \times 10^{-5}$ |
| $CH_2 = O + h\nu \rightarrow CO + H_2$ | $k = 4.9 \times 10^{-5}$ |
| $CH_2 = O + HO^{\cdot} \rightarrow {}^{\cdot}CH = O + H_2O$ | $k = 1.0 \times 10^{-11}$ |
| ${}^{\cdot}CH = O + O_2 \rightarrow CO + HO_2^{\cdot}$ | $k = 5.5 \times 10^{-12}$ |
| $CO + HO^{\cdot} \rightarrow CO_2 + H^{\cdot}$ | $k = 3.0 \times 10^{-13}$ |
| $H^{\cdot} + O_2 + M \rightarrow HO_2^{\cdot} + M^*$ | $k\,[M] = 1.4 \times 10^{-12}$ |
| $HO_2^{\cdot} + NO \rightarrow NO_2 + HO^{\cdot}$ | $k = 8.3 \times 10^{-12}$ |
| $NO_2 + h\nu \rightarrow NO + O(^3P)$ | $k = 8.9 \times 10^{-3}$ |
| $O(^3P) + O_2 + M \rightarrow O_3 + M^*$ | $k\,[M] = 1.8 \times 10^{-14}$ |
| $O_3 + NO \rightarrow NO_2 + O_2$ | $k = 1.8 \times 10^{-14}$ |

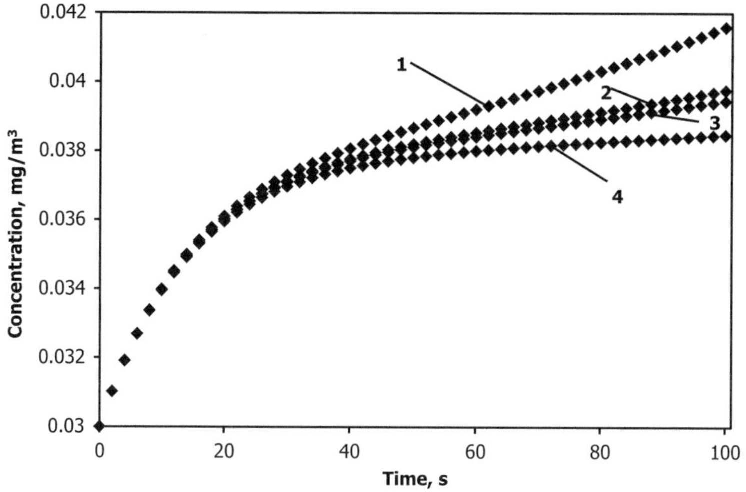


**Fig. 13** Dependence of ozone concentration in the air on time at a traffic flow density of 111 cars/min for a section with continuous traffic at different speeds: 1—15 km/h, 2—40 km/h, 3—100 km/h, 4—60 km/h

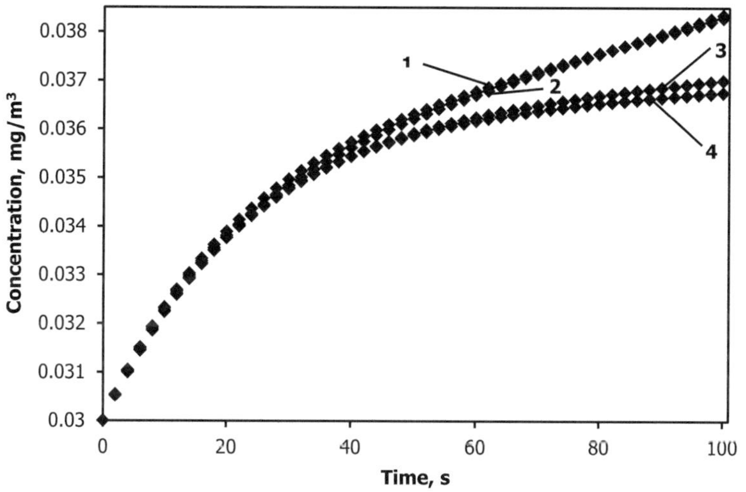


**Fig. 14** Dependence of ozone concentration in the air on time at a traffic density of 58 cars/min: 1 and 3—traffic lights at a speed of 15 and 60 km/h, 2 and 4—continuous movement at a speed of 15 and 60 km/h

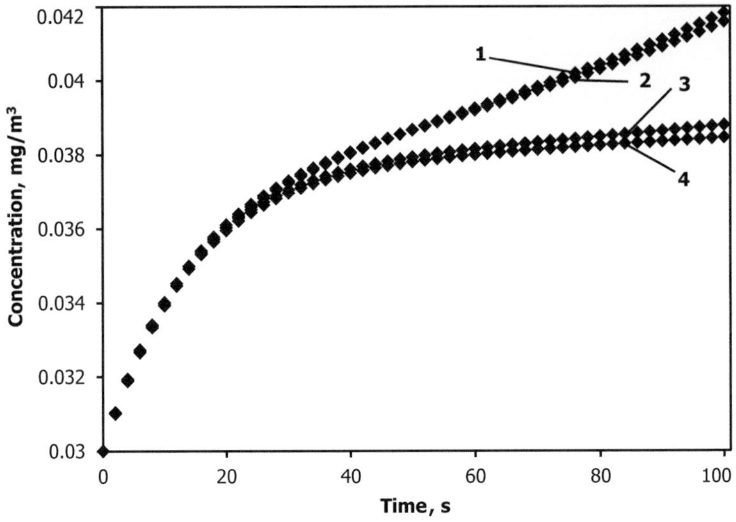


**Fig. 15** Dependence of ozone concentration in the air on time at a traffic density of 111 cars/min: 1 and 3—traffic lights at a speed of 15 and 60 km/h, 2 and 4—continuous movement at a speed of 15 and 60 km/h

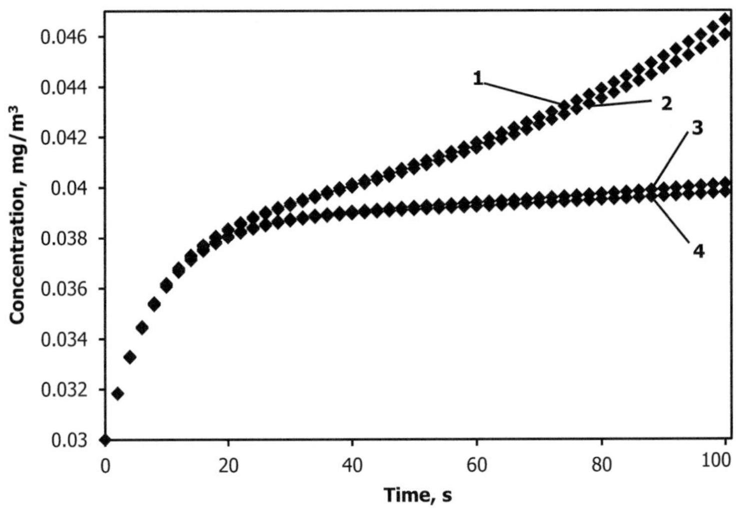


**Fig. 16** Dependence of ozone concentration in the air on time at a traffic density of 207 cars/min: 1 and 3—traffic lights at a speed of 15 and 60 km/h, 2 and 4—continuous movement at a speed of 15 and 60 km/h

**Acknowledgements** The research was supported by the Russian Foundation for Basic Research (grant 16-08-01252).

## References


1. Stockwell, W.R., Lawson, C.V., Saunders, E., Goliff, W.S.: A review of tropospheric atmospheric chemistry and gas-phase chemical mechanisms for air quality modeling. Atmosphere **3**, 1–32 (2012). https://doi.org/10.3390/atmos3010001
2. Bates, K.H., Jacob, D.J., Li, K., Ivatt, P.D., Evans, M.J., Yan, Y., Lin, J.: Development and evaluation of a new compact mechanism for aromatic oxidation in atmospheric models. Atmos. Chem. Phys. **21**, 18351–18374 (2021). https://doi.org/10.5194/acp-21-18351-2021
3. Bates, K.H., Jacob, D.J.: A new model mechanism for atmospheric oxidation of isoprene: global effects on oxidants, nitrogen oxides, organic products, and secondary organic aerosol. Atmos. Chem. Phys. **19**, 9613–9640 (2019). https://doi.org/10.5194/acp-19-9613-2019
4. Stockwell, W.R., Kirchner, F., Kuhn, M., Seefeld, S.: A new mechanism for regional atmospheric chemistry modeling. J. Geophys. Res. **102**, 25847–25879 (1997)
5. Calvert, J.G., Stockwell, W.R.: Deviations from the $O_3$–NO–$NO_2$ photostationary state in tropospheric chemistry. Can. J. Chem. **61**, 983–992 (1983)
6. Karol', I.L., Kiselev, A.A.: Photochemical models of the atmosphere and their application in ozonosphere and climate studies: a review. Izv. Atmos. Ocean. Phys. **42**(1), 1–31 (2006). https://doi.org/10.1134/S0001433806010014
7. Zaytsev, V.A.: Secondary pollution of the urban atmosphere by motor transport at various emission capacities. In: Kravets, A.G., Bolshakov, A.A., Shcherbakov, M. (eds.) Society 5.0: Human-Centered Society Challenges and Solutions. Seria "Studies in Systems, Decision and Control", vol. 416, pp. 253–263. Springer, Switzerland, Cham (2022). https://doi.org/10.1007/978-3-030-95112-2_21
8. Atkinson, R., Baulch, D.L., Cox, R.A., et al.: Evaluated kinetic, photochemical and heterogeneous data for atmospheric chemistry: supplement V. J. Phys. Chem. Ref. Data **26**(3), 521–1011 (1997)
9. Atkinson, R., Baulch, D.L., Cox, R.A., et al.: Evaluated kinetic and photochemical data for atmospheric chemistry: volume I—gas phase reactions of $O_x$, $HO_x$, $NO_x$ and $SO_x$ species. Atmos. Chem. Phys. **4**, 1461–1738 (2004). https://doi.org/10.5194/acp-4-1461-2004
10. Atkinson, R., Baulch, D.L., Cox, R.A., Crowley, J.N., Hampson, R.F., Hynes, R.G., Jenkin, M.E., Rossi, M.J., Troe, J., IUPAC Subcommittee: Evaluated kinetic and photochemical data for atmospheric chemistry: volume II—gas phase reactions of organic species. Atmos. Chem. Phys. **6**, 3625–4055 (2006). https://doi.org/10.5194/acp-6-3625-2006
11. Mellouki, A., Ammann, M., Cox, R.A., et al.: Evaluated kinetic and photochemical data for atmospheric chemistry: volume VIII – gas phase reactions of organic species with four, or more, carbon atoms ($\geq$C4). Atmos. Chem. Phys. **21**, 4797–4808 (2021). https://doi.org/10.5194/acp-21-4797-2021
12. Gryparis, A., Forsberg, B., Katsouyanni, K., Analitis, A., Touloumi, G., Schwartz, J.: Acute effects of ozone on mortality from the "Air pollution and health: a European project approach." Amer. J. Respir. Crit. Care Med. **170**, 1080–1087 (2004)
13. Bates, D.V.: Ambient ozone and mortality. Epidemiology **16**(4), 427–429 (2005)
14. Longphre, M., Zhang, L.-Y., Harkema, J.R., Kleeberger, S.R.: Ozone-induced pulmonary inflammation and epithelial proliferation are partially mediated by PAF. J. Appl. Physiol. **86**(1), 341–349 (1999)
15. Fang, Y., Naik, V., Horowitz, L.W., Mauzerall, D.L.: Air pollution and associated human mortality: the role of air pollutant emissions, climate change and methane concentration increases from the preindustrial period to present. Atmos. Chem. Phys. **13**, 1377–1394 (2013)

16. Bell, M., Dominici, F.: Effect modification by community characteristics on the short-term effects of ozone exposure and mortality in 98 US communities. Am. J. Epidemiol. **167**(8), 986–997 (2008). https://doi.org/10.1093/aje/kwm396
17. Hollingsworth, J.W., Kleeberger, S.R., Foster, W.M.: Ozone and pulmonary innate immunity. Proc. Amer. Thorac. Soc. **4**, 240–246 (2007)
18. Elansky, N.F.: Russian studies of atmospheric ozone in 2007–2011. Izvestiya. Atmos. Oceanic Phys. **48**(3), 281–298 (2012). https://doi.org/10.1134/S0001433812030024
19. Elansky, N.F.: Russian studies of atmospheric ozone in 2011–2014. Izvestiya. Atmos. Oceanic Phys. **52**(2), 132–146 (2016). https://doi.org/10.1134/S0001433816020031
20. Elansky, N.F.: Russian studies of atmospheric ozone and its precursors in 2015–2018. Izvestiya. Atmospheric Oceanic Phys. **56**(2), 141–155 (2020). https://doi.org/10.1134/S0001433820020048
21. Zaytsev, V.A.: Transformation of reduced compounds of carbon and nitrogen oxides in the cramped aerodynamic conditions. In: Kravets, A.G., Bolshakov, A.A., Shcherbakov, M. (eds.). Society 5.0: Cyberspace for Advanced Human-Centered Society. Seria "Studies in Systems, Decision and Control", vol. 333, pp. 157–167. Springer, Switzerland, Cham (2021). https://doi.org/10.1007/978-3-030-63563-3_13
22. Pokusaev, B.G., Zaitsev, A.A., Zaitsev, V.A.: Transfer processes under slug flow conditions in three-phase media. Theor. Found. Chem. Eng. **33**(6), 539–549 (1999)

# Healthcare 5.0 and Inclusive Technologies

# Detection of Breast Cancer Risk Using Machine Learning Methods Based on Features Selected According to Diagnostic Models

**Kirill Dyomin, Ilya Germashev, and Ibrahima Niang**

**Abstract** A machine learning model for assessing the development of breast cancer based on medical history and other indicators, which do not require special medical measuring devices, has been presented. Various machine learning models have been compared. Models based on decision trees and random forests have proven to be the most effective. The result was a random forest model that correctly classifies 95% of healthy individuals and 69% of the at-risk group. The generalization ability of the model was tested on a different dataset.

**Keywords** Breast cancer · Cancer screening · Machine learning · Artificial intelligence · Random forest

## 1 Introduction

Modern medicine is constantly searching for new methods of diagnosing, preventing, and treating various diseases. Despite significant advancements in medicine, there are still many pathologies for which existing methods of diagnosis and selection of appropriate treatment are insufficiently effective or inaccessible to a wide range of individuals. For example, some pathologies, such as breast cancer, still remain a leading cause of death and disability among the population. According to the WHO data for 2022, breast cancer was the most common type of cancer among women in 157 out of 185 countries [1]. In such cases, the development and implementation of innovative methods of diagnosis, prevention, and treatment are necessary to improve

K. Dyomin (✉) · I. Germashev
Volgograd State University, Volgograd, Russia
e-mail: diominkirill@yandex.ru

I. Germashev
e-mail: germashev@volsu.ru

I. Niang
Cheikh Anta Diop University, Dakar, Senegal
e-mail: ibrahima1.niang@ucad.edu.sn

 
A. G. Kravets and A. A. Bolshakov (eds.), *Human-Centric Cyber-Society*, Studies in Systems, Decision and Control 639, https://doi.org/10.1007/978-3-032-06787-6_12

therapy effectiveness and prognosis of diseases. Currently, diagnostic approaches associated with the application of machine learning and artificial intelligence methods for diagnosis, including breast cancer, are actively being developed. For instance, in the study [2], a convolutional neural network is used for automatic classification of breast cancer based on images obtained through computer tomography. The dataset contained 2620 breast images, and the developed model has a sensitivity of 80.95% and a specificity of 97.63%.

In another study [3], machine learning methods were applied for the classification of different types of cancer based on digital mammography images. The artificial intelligence system demonstrated a sensitivity of 96.6% and a false-negative rate of 3.4% in detecting various types of breast cancer.

Additionally, there are works related to the application of machine learning methods for identifying the risk of developing breast cancer. In the study [4], natural language processing was used to identify breast cancer from medical descriptions of patients' conditions. The developed model has a specificity of 80%, sensitivity of 82%, and efficiency of 86%.

In the study [5], logistic regression was applied to determine the risk of developing breast cancer using microwave radiometry data. The results showed that the developed model has a sensitivity of 0.89 and a specificity of 0.81.

Machine learning algorithms, such as generalized linear models, random forests, gradient boosting algorithms, and deep learning models, were used in the study to identify complex patterns in the embryonic DNA lineage that correlate with breast cancer incidence [6]. The analysis of patterns was conducted by measuring changes in the length of the embryonic DNA lineage on a chromosomal scale. These measurements represent a sum of many insertions, divisions, and copy number changes in the chromosome. As a result, a software product was developed using the gradient boosting algorithm, with a ROC AUC of 0.83 for binary classification.

In another study, in addition to genetic profile information, lifestyle and reproductive factors were used in a logistic regression model to identify the risk of breast cancer [7]. The model yielded a ROC AUC of 0.73, and conclusions were drawn that this model could serve as a starting point for breast cancer screening.

In studies [8, 9] a dataset with the results of examinations conducted by medical institutions in Slovenia was used to diagnose the risk of developing breast cancer based on radiometric indications. In addition to temperature data, the dataset also includes medical history and other indicators not related to temperature measurement, which were not used in these works. The study considered indicators that were not used in [8, 9] and assessed their impact on the preliminary diagnosis of the risk of developing breast cancer.

## 2 Data Analysis and Task Formulation

The first step in the work is to analyze the available dataset and gather information on relevant indicators for the subsequent classification of breast cancer development risk. Table 1 presents the indicators of the dataset that are not related to temperature measurements.

The degree of disease risk is assessed on a 6-point scale, described in Table 2.

To assign a numerical value to the degree of risk, a scale [10] is used. According to this scale, values th0-th2 indicate that the risk of disease is practically absent, while values th3-th5 indicate being in the risk group.

To select from the available indicators, it is advisable to analyze breast cancer diagnostic models used by doctors. Based on this analysis, the indicators of the

**Table 1** Assignment of dataset indicators

| Indicator | Meaning, unit of measurement |
|---|---|
| birthday_date | Age, years |
| breast_diameter | Diameter of the breast, cm |
| figure | Body shape, O.x.—very thin; X.—thin; N.—normal; P.—overweight; O.p.—obese |
| cycle | Duration of the cycle, days |
| cycle_day | Cycle day, number |
| hormone | Hormonal dependency, yes/no |
| pain_location | Complaints of pain, yes/no |
| preg_count | Number of pregnancies, pcs |
| birth_count | Number of children born, pcs |
| first_pregnancy_age | Age at the time of the 1st childbirth, years |
| first_pregnancy_result | The 1st pregnancy ended in childbirth, yes/no |
| breast_operations | Presence of surgery on the breast, yes/no |

**Table 2** Degree of disease risk

| Value | Description |
|---|---|
| Th0 | There are almost no thermal changes |
| Th1 | Decreased thermal activity of tissues |
| Th2 | Minor thermal changes |
| Th3 | Increased thermal activity without local foci and without high thermoassimetry |
| Th4 | High level of thermal activity with the presence of focal thermoasymmetry without local temperature elevation |
| Th5 | High level of thermal activity with the presence of focal thermoasymmetry with local temperature elevation |

dataset that correspond to the features relied upon by doctors in diagnosis should be selected.

There are various breast cancer diagnostic models that can help understand which features and factors influence the manifestation of the disease and the condition of patients. According to the Gale model [11], important risk indicators for the development of invasive breast cancer are:

1. Age.
2. Race.
3. Number of biopsies.
4. Age at menarche.
5. Age at first childbirth.
6. Number of first-degree relatives with breast cancer.

According to the Claus model [12], important risk assessment indicators for cancer development are:

1. Age.
2. Number of first-degree relatives with breast cancer.
3. Number of second-degree relatives with breast cancer.
4. Age of onset of breast cancer manifestation.
5. Medical history.

The IBIS model (Tyra-Cuzick model) [13] considers the following risk factors:

1. Age.
2. Body mass index.
3. Age at menarche.
4. Age at first childbirth.
5. Menopause stage.
6. Hormone therapy usage.
7. BRCA1 or BRCA2 gene mutation.
8. Breast biopsy.
9. Ovarian cancer presence.
10. Breast density.
11. Ashkenazi heritage.
12. Number of first-degree relatives with breast cancer.
13. Number of second-degree relatives with breast cancer.

In addition to the above-mentioned models, there are studies where similar linguistic descriptions are taken as input indicators. For example, in [14], risk factors for breast cancer development were identified as race (higher incidence among African American women compared to white women), obesity, and young age at first childbirth. Another study used genetic profiles and data on lifestyle and reproductive factors to identify breast cancer risk [15].

The most frequent feature present in all three models was age. The second most frequent feature was age at first childbirth, which was also associated with the indicator '1st pregnancy ended with delivery' in the Gale and IBIS models. The indicator 'Hormonal dependency' was also present twice, in the IBIS and Claus models (in the Claus model, it was classified as medical history). Indicators such as 'Pain complaints', 'History of breast surgery', 'Number of pregnancies', and 'Number of children born' were classified under the Claus model as features of 'Medical history'. 'Build' and 'Breast diameter' were only present in the IBIS model. Indicators such as 'Day of cycle' and 'Cycle duration' were not used in the aforementioned models.

Based on the analysis of the studied models, all parameters from Table 1 were taken, except for the duration and cycle day data, as input for assessing the risk of breast cancer. Figure 1 shows the block diagram of the distribution of indicators across models. The data were also scaled to conform to a normal distribution with a mean of 0 and a variance of 1.

Classification utilizes two combined classes from Table 2: th0-th2—healthy, th3-th5—risk group. Thus, a dataset has been processed, containing 5858 records of healthy patients and 491 records of patients in the risk group. A labeled dataset was obtained.

The main task is to use the labeled dataset and build a classifier that, based on the above indicators, determines whether a patient belongs to the risk group or not.

## 3 Solution Method and Computational Experiment

To solve this problem, it was decided to use the following machine learning methods for identifying the risk group: artificial neural network, logistic regression, linear regression, decision tree, and random forest, as these methods have shown high diagnostic accuracy in similar studies [16–18]. The Python programming language and the Scikit-learn machine learning library were used as the platform for developing these models. This open-source machine learning library supports both supervised and unsupervised learning. It also provides various tools for model selection, data preprocessing, model evaluation, and more.

Neural networks, logistic regression, and linear regression did not yield the expected results due to almost zero identification of the «risk group» class. The most effective solutions turned out to be models based on decision trees and random forests. Figures 2 and 3 show the classification results for the decision tree and random forest models, respectively, for the training dataset in the form of a confusion matrix. Default values were used for the hyperparameters of the machine learning models.

As a basis for the classification system, it was decided to use a model based on decision trees and random forests, with subsequent hyperparameter optimization and selection of the best model. Through computational experiments, the best hyperparameters for binary classification were identified.

Below are the hyperparameters used for the decision tree:

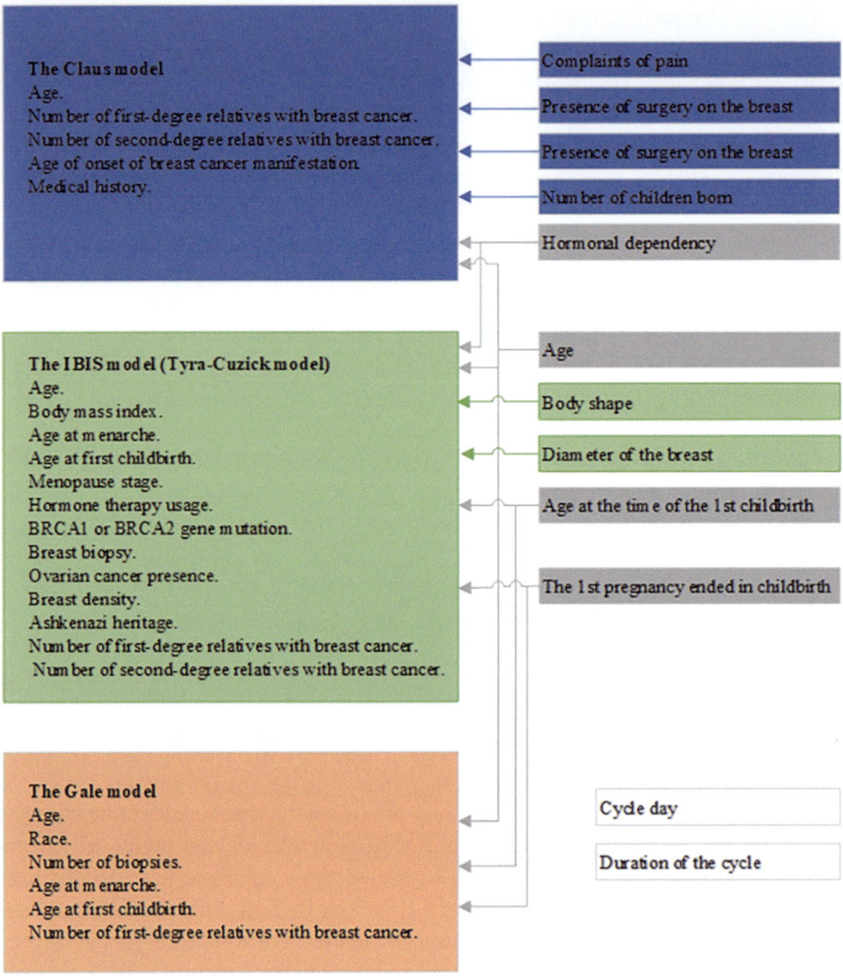


**Fig. 1** Distribution of patient dataset indicators by breast cancer risk degree indicators in diagnostic models (indicators of the dataset corresponding to a specific diagnostic model are shown in the same color; if the dataset indicator corresponds to multiple diagnostic models, it is shown in gray, if it does not belong to any models, then it is shown in white)

Splitting criterion—«Gini», which is determined by the formula 1:

$$H(Q_m) = \sum_k p_{mk}(1 - p_{mk}) \tag{1}$$

where k is the class at node m.

Maximum depth of the tree—unlimited.

Maximum number of features to consider when splitting—0.8.

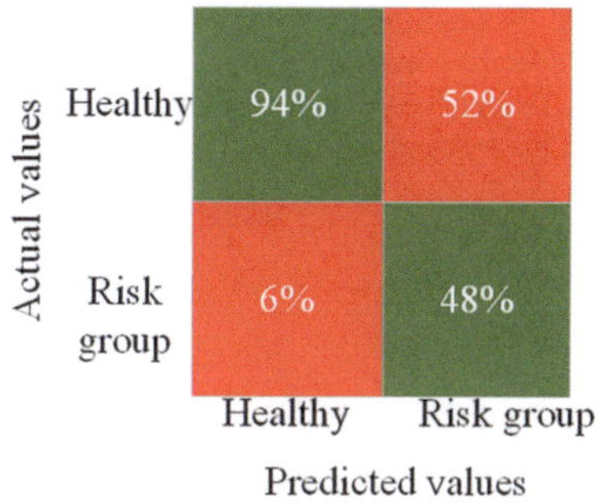


**Fig. 2** Confusion matrix of the decision tree on the training dataset

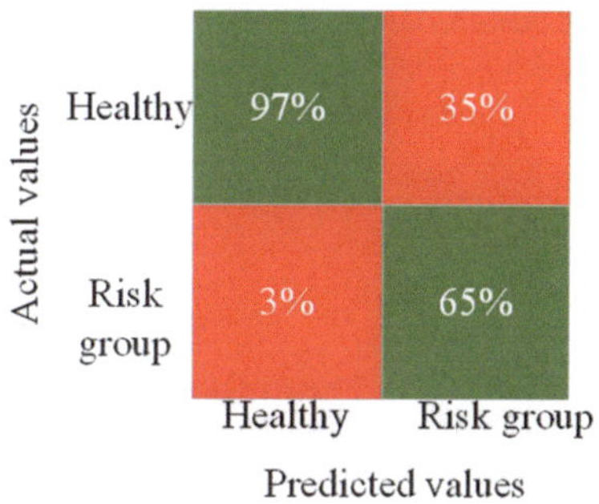


**Fig. 3** Confusion matrix of the random forest on the training dataset

Strategy used for selecting the split at each node—random.

Below are the hyperparameters used for the random forest:

Number of trees used in the model—50.

Size of random subsets of features to consider when splitting a node—5.

Maximum depth of the trees—24.

Minimum number of samples required to split an internal node—1.

The labeled dataset was split in a 90%/10% ratio, ensuring that the labels in each part were represented in the same proportion as the entire dataset. 90% of the records were used as the training dataset, while 10% were used as the testing dataset. The testing dataset was not involved in the training of the classifier and only served for testing the resulting model. To classify each example in the training dataset, cross-validation with 5 equal and non-overlapping folds was used. Figure 4 shows the averaged classification results of the decision tree during cross-validation for the training dataset in the form of a confusion matrix. Figure 5 shows the classification results of the decision tree on the testing dataset in the form of a confusion matrix.

Figure 6 shows the averaged classification results of the random forest during cross-validation for the training dataset in the form of a confusion matrix. Figure 7 shows the classification results of the random forest on the testing dataset in the form of a confusion matrix.

Thus, the decision tree correctly classified 98% of healthy patients and 51% of patients in the risk group, while the random forest correctly classified 95% of healthy

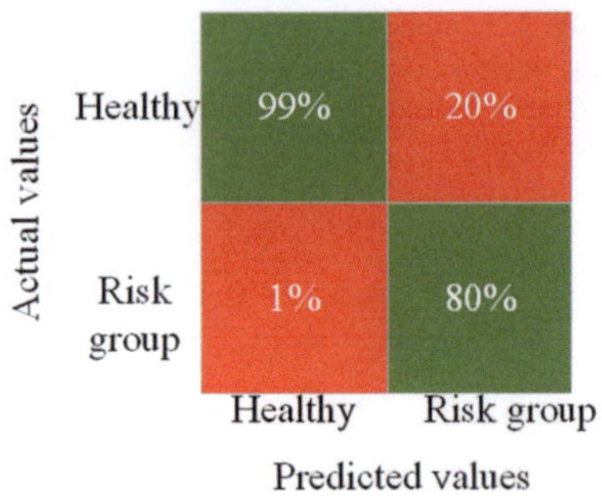


**Fig. 4** Confusion matrix of the decision tree on the during cross-validation for the training dataset

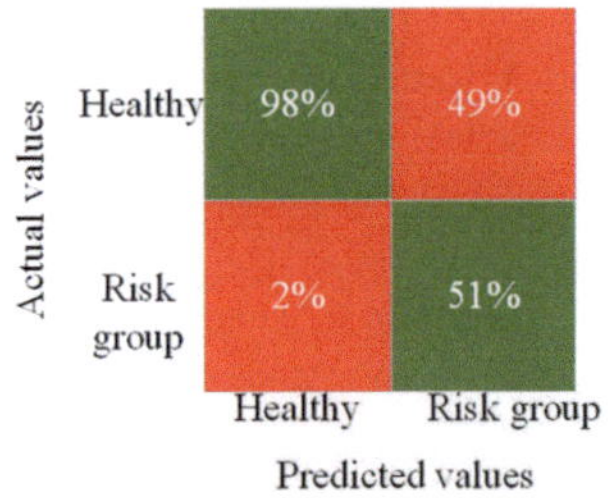


**Fig. 5** Confusion matrix of the decision tree on the test dataset

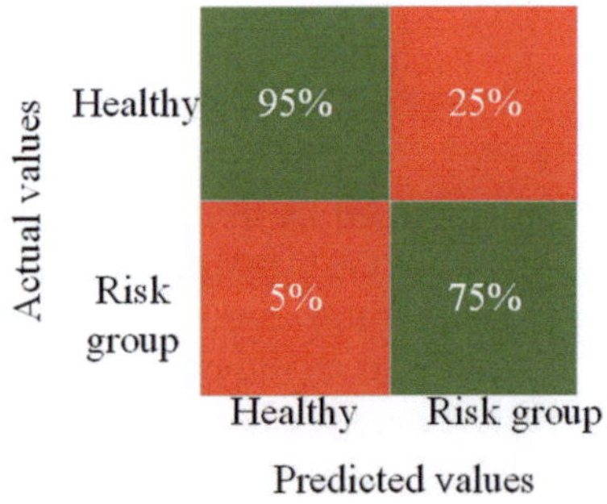


**Fig. 6** Confusion matrix of the random forest on the during cross-validation for training dataset

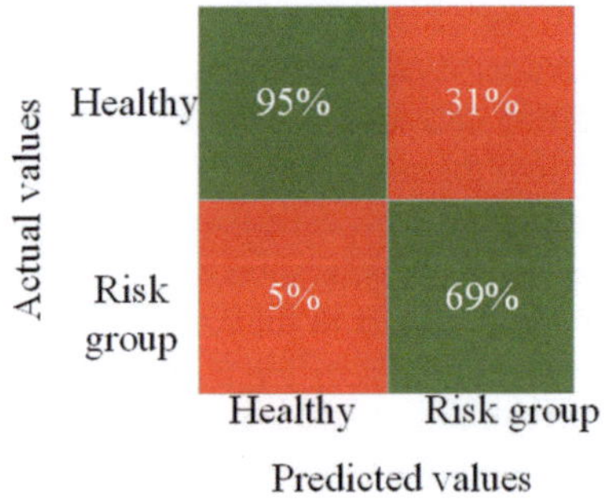


**Fig. 7** Confusion matrix of the random forest on the test dataset

**Table 3** Correlation matrix

| Indicator | Importance (%) |
|---|---|
| Age | 27 |
| Diameter of the breast, cm | 20.3 |
| Body shape | 7.7 |
| Hormonal dependency | 3 |
| Number of pregnancies | 7.5 |
| Number of children born | 5.4 |
| Complaints of pain | 2.9 |
| Age at the time of the 1st childbirth | 23 |
| The 1st pregnancy ended in childbirth | 2 |
| Presence of surgery on the breast | 1.2 |

patients and 69% of patients in the risk group. The random forest model proved to be the best classifier. Additionally, a correlation matrix was obtained to assess the influence of input variables on the output, as presented in Table 3.

Based on the obtained data, the most significant indicator is the patient's age, which corresponds to the analyzed models and other studies [19, 20]. The diameter of the breast and the age at first childbirth were also significant parameters, which are supported by models and other research [21, 22].

## 4 Assessing Generalizability of Model

For the purpose of assessing generalizability, another dataset with the results of examinations conducted by medical institutions in Russia was taken, which includes 596 records of healthy patients and 83 records of patients at risk. This dataset contains similar patient indicators as presented in Table 1. To test the generalizability of the random forest model, the entire dataset was subjected to classification. Figure 8 shows the classification results of the trained model on the dataset of patients from Russia in the form of a confusion matrix.

In comparison with the dataset from Slovenia, the prediction percentage for the «Risk group» class sharply decreased, but at the same time, the percentage for the «Healthy» class decreased to acceptable values. This may indicate that the identified features are not common to different groups of people, and it is necessary to expand the features for building a model capable of classifying the risk of breast cancer development in various groups of people (ethnicity, geographical location, ecology, etc.). Nevertheless, based on the obtained results, it can be concluded that data that does not require special devices for measuring indicators can be used for identifying the risk of breast cancer development.

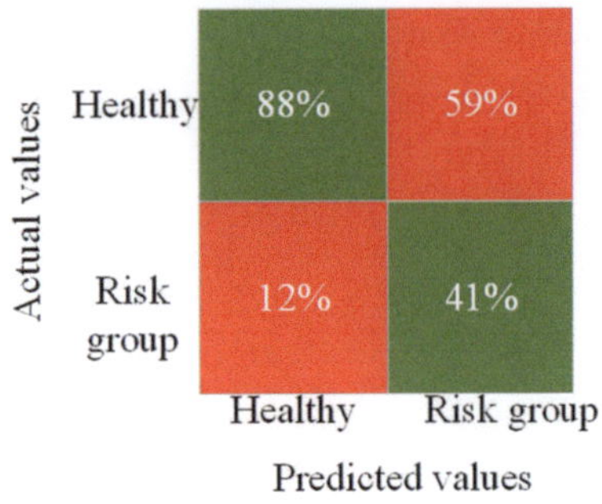


**Fig. 8** Confusion matrix on the dataset of patients from Russia

## 5 Conclusion

The obtained results can serve as valuable information for the development of more effective screening and prevention methods for this disease using machine learning models. Overall, the results confirm the importance of indicators such as patient age, breast diameter, and age at first childbirth in the context of breast cancer risk. The study also revealed the need for additional research on factors influencing the formation of the dataset, which have a significant impact on the machine learning model (ethnicity, geographical location, ecology, etc.). Further research is planned to incorporate additional features into the obtained model that can provide more accurate results considering these factors, as well as to validate the model using other machine learning methods.

## References


1. World Health Organization.: Breast cancer. Retrieved from: https://www.who.int/news-room/fact-sheets/detail/breast-cancer (2024)
2. Wang, Y., Yang, F., Zhang, J., et al.: Application of artificial intelligence based on deep learning in breast cancer screening and imaging diagnosis. Neural Comput. Applic. **33**, 9637–9647 (2021). https://doi.org/10.1007/s00521-021-05728-x
3. Raafat, M., Mansour, S., Kamal, R., et al.: Does artificial intelligence aid in the detection of different types of breast cancer? Egypt J Radiol Nucl Med **53**, 182 (2022). https://doi.org/10.1186/s43055-022-00868-z
4. Naydenko, A.V., Germashev, I.V., Popov, I.E., Losev, A.G.: Data extraction from the medical description patients state in the breast canser identification task. Math. Methods Technol. Eng. **4**, 81–84 (In Russ)
5. Losev, A.G., Svetlov, A.V.: Artificial intelligence algorithms in diagnosis of breast cancer. In: Inshakova, E.I., Inshakova, A.O. (eds.) New Technology for Inclusive and Sustainable Growth. Smart Innovation, Systems and Technologies, vol. 287. Springer, Singapore (2022). https://doi.org/10.1007/978-981-16-9804-0_15
6. Ko, C., Toh, C., Brody, J.P.: Genetic risk scores for breast cancer based on machine learning analysis of chromosomal-scale length variation [abstract]. Clin. Cancer Res. **27**(5_Suppl), PR-09

7. Zou, S., Lin, Y., Yu, X., et al.: Genetic and lifestyle factors for breast cancer risk assessment in Southeast China. Cancer Med. **12**(14), 15504–15514 (2023). https://doi.org/10.1002/cam4.6198
8. Germashev, I., Dubovskaya, V., Losev, A.: Substantiation of the result of the work of the cyber-physical system for breast cancer diagnostics. In: Kravets, A.G., Bolshakov, A.A., Shcherbakov, M. (eds) Society 5.0: Human-Centered Society Challenges and Solutions. Studies in Systems, Decision and Control, vol. 416. Springer, Cham (2022). https://doi.org/10.1007/978-3-030-95112-2_1
9. Polyakov, M.V., Popov, I.E., Losev, A.G., Khoperskov, A.V.: Application of computer simulation results and machine learning in analysis of microwave radiothermometry data. Math. Phys. Comput. Simul. **24**(2), 27–37 (2021)
10. Rozhkova, N.I., Smirnova, N.A., Nazarov, A.A.: Factors influencing the effectiveness of radiothermometric measurements of the breast using the diagnostic complex RTM-01-RES. Tumors of the female reproductive system (2007) (In Russ)
11. Gail, M.H., Brinton, L.A., Byar, D.P., et al.: Projecting individualized probabilities of developing breast cancer for white females who are being examined annually. J. Natl. Cancer Inst. **81**(24), 1879–1886 (1989). https://doi.org/10.1093/jnci/81.24.1879
12. Claus, E.B., Risch, N., Thompson, W.D.: Autosomal dominant inheritance of early-onset breast cancer. Implications for risk prediction. Cancer **73**(3), 643–651 (1994). https://doi.org/10.1002/1097-0142(19940201)73:3%3c643::aid-cncr2820730323%3e3.0.co;2-5
13. Tyrer, J., Duffy, S.W., Cuzick, J.: A breast cancer prediction model incorporating familial and personal risk factors. Stat. Med. **23**(7), 1111–1130 (2004). https://doi.org/10.1002/sim.1668
14. Levine, P.H., Veneroso, C.: The epidemiology of inflammatory breast cancer. Semin. Oncol. **35**(1), 11–16 (2008). https://doi.org/10.1053/j.seminoncol.2007.11.018
15. Zou, S., Lin, Y., Yu, X., et al.: Genetic and lifestyle factors for breast cancer risk assessment in Southeast China. Cancer Med. **12**, 15504–15514 (2023). https://doi.org/10.1002/cam4.6198
16. Magna, A.A.R., Allende-Cid, H., Taramasco, C., Becerra, C., Figueroa, R.L.: Application of machine learning and word embeddings in the classification of cancer diagnosis using patient anamnesis. IEEE Access **8**, 106198–106213. https://doi.org/10.1109/ACCESS.2020.3000075
17. Mortazavi, S.A.R., Tahmasebi, S., Parsaei, H., et al.: Machine learning models for predicting breast cancer risk in women exposed to blue light from digital screens. J. Biomed. Phys. Eng. **12**(6), 637–644 (2022). https://doi.org/10.31661/jbpe.v0i0.2105-1341
18. Afrash, M.R., Bayani, A., Shanbehzadeh, M., Bahadori, M., Kazemi-Arpanahi, H.: Developing the breast cancer risk prediction system using hybrid machine learning algorithms. J. Educ. Health Promot. **11**, 272 (2022). https://doi.org/10.4103/jehp.jehp_42_22
19. Alghamdi, I.G., Hussain I.I., Alghamdi M.S., El-Sheemy M.A. The incidence rate of female breast cancer in Saudi Arabia: an observational descriptive epidemiological analysis of data from Saudi Cancer Registry 2001–2008. Breast Cancer: Targets Ther. **5**, 103–109 (2013)
20. Srur-Rivero, N., Cartin-Brenes, M.: Breast Cancer Characteristics and Survival in a Hispanic Population of Costa Rica. Breast cancer (Auckl). **8**, 103–108 (2014)
21. Kotsopoulos, J., Gronwald, J., Lynch, H.T., et al.: Age at first full-term birth and breast cancer risk in BRCA1 and BRCA2 mutation carriers. Breast Cancer Res. Treat. **171**(2), 421–426 (2018). https://doi.org/10.1007/s10549-018-4822-y
22. Yaghjyan, L., Darville, L.N.F., Cline, J., et al.: Associations of established breast cancer risk factors with urinary estrogens in postmenopausal women. Cancer Causes Control **33**(2), 279–291 (2022). https://doi.org/10.1007/s10552-021-01528-9

# Malignant Neoplasms Risk Determination Based on Images Analysis of Internal and External Breast Thermograms Asymmetry

**Amdy Moustapha Drame, Ilya Germashev, Elena Markushevskaya, and Ibrahima Niang**

**Abstract** Currently, breast cancer is one of the main problems of medicine, from which a significant number of women around the world constantly die. Therefore, its detection becomes important, that is, the detection of malignant breast tumors in the early stages is the main task. In recent years, the use of thermal imaging techniques to identify and differentiate the stages of breast cancer has aroused exceptional interest. Microwave (RMT) and infrared (IR) radio-thermometric imaging techniques are very useful for preclinical assessment because these methods are relatively fast and can be reused over a longer period without the risk of exposure. Some studies show that microwave radio thermometry, when used with well-defined protocols, can detect the first symptoms of cancer several years earlier than mammography. This method allows for diagnosis in women of any age and allows for periodic screening programs in countries with any level of resources. Thus, this method can be considered as a good alternative to mammography. This chapter proposes a model for the identification of a malignant breast tumor based on thermogram data obtained using microwave radio-thermometry.

**Keywords** Asymmetry · Breast cancer · Tumors · Deep learning · Thermogram · Convolutional transformation · Classification

A. M. Drame (✉) · I. Germashev · E. Markushevskaya
Volgograd State University, Volgograd, Russia
e-mail: knia-221_681944@volsu.ru

I. Germashev
e-mail: germashev@volsu.ru

E. Markushevskaya
e-mail: markushevskaya@volsu.ru

I. Niang
Cheikh Anta Diop University, Dakar, Senegal
e-mail: ibrahima1.niang@ucad.edu.sn



A. G. Kravets and A. A. Bolshakov (eds.), *Human-Centric Cyber-Society*, Studies in Systems, Decision and Control 639, https://doi.org/10.1007/978-3-032-06787-6_13

# 1 Introduction

In several regions of the world, especially in India, breast cancer remains the most common disease in 2023, with one woman dying every 13 min [1]. It is estimated that one in eight women (12.5%) in British Columbia develops breast cancer during their lifetime, which leads to the death of about one in 24 women, and the country itself ranks second in terms of mortality among women [2, 3]. Therefore, it is necessary to find ways to diagnose the disease early to effectively prevent and overcome its progression.

Compared with other methods of examination, thermal imaging of the breast has several advantages due to the possibility of deploying organized screening systems. Its peculiarity lies in the fact that it allows the detection of certain types of malignant tumors at an early stage, which significantly increases the chances of a cure for this pathology, which is an urgent problem in medicine [4]. This information is the basis for the search and justification of new methods of breast research.

By replacing traditionally used models with homogeneous parameters in a multilayer approximation (usually limited to four types of tissues-skin, muscles, mammary gland, tumors), a geometric structure of tissues with heterogeneous characteristics close to the real one was used in modeling. The developed models also considered filamentous connective tissues [5], lobules of the breast, nipple, excretory ducts, and adipose tissue. It should be noted that many inputs physical parameters seriously complicate the modeling process, but at the same time significantly increases the relevance of the model. In real conditions, the discrepancy between the parameter values is very large. Also, in a study based on a thorough analysis of medical data and the use of machine learning algorithms, an algorithm for checking the simulation model was proposed [2].

On the other hand, in recent years, the possibility of using machine learning algorithms in the development and diagnostic solution based on the substantiation of microwave radio thermometry data has been actively studied [2, 6]. It was soon discovered that an attempt to make a diagnosis using artificial intelligence methods based solely on temperature data did not provide the required sensitivity and specificity. Spatial thermometric characteristics have been proposed in various variants [6]. Their construction is based on the established descriptive hypothesis about the behavior of temperature fields in patients of different diagnostic classes.

In patients with pathology, there is an increased value of thermal asymmetry between the same points of the breast; an increase in the temperature gap between individual points of the affected breast; an increase in the temperature of the nipples in the affected breast compared with the average breast temperature, in particular taking into account age-related temperature changes; during these recent studies, The characteristics of quantitative thermometric factors are described.

In modern clinical conditions, many traditional and modern non-invasive methods of cancer detection are being studied, namely the use of microwave radio thermometry [6–8]. In the Russian Federation, the successes achieved in this field of diagnostics are associated with the development of devices that allow combining the measurement of the internal (deep) temperature of tissues based on their natural thermal radiation in the microwave range and the measurement of skin temperature based on microwave radiation, infrared thermal radiation.

A feature of our work is the use of thermo-asymmetry. It's about finding the difference in images in 2–3 different ways. The first way is to measure the difference between the internal and external thermograms of the same patient to see the peculiarity of the RGB color difference in each area of the breast. The second approach to thermo-asymmetry is to find and study the difference between points symmetrical with respect to a line parallel to the spine separating the two mammary glands. This comparison will highlight the anomaly if there is a clear difference between two symmetrical regions. The third method of asymmetry examines the double difference, that is, the difference in symmetrical image points resulting from the difference between the internal and external thermograms of the same patient.

The purpose of this study is to study the presence of malignant neoplasms in the mammary glands using thermograms obtained using microwave radio-thermometry. To achieve this goal, a set of asymmetry calculation data is created based on the first approach, which is used to build a cyberphysical classification system.

# 2 Methods of Breast Cancer Diagnosis Based on Thermograms and Neural Networks

## *2.1 The Basic Principles and Features of the Microwave Radiothermometry Method*

Microwave radio thermometry is based on measuring the intensity of electromagnetic radiation from the patient's internal tissues in the microwave frequency range. There is a proportionality between the intensity of this radiation and the temperature of the tissues. Thus, the measured value of the internal temperature of the tissues is displayed on the monitor screen [9]. A temperature anomaly or temperature change may be caused by an increased metabolism of cancer cells, on which early cancer diagnosis is based. According to research, in most cases, the structural changes detected by traditional methods of breast examination: ultrasound, mammography, palpation, are replaced by changes in tissue temperature; hence his interest in early diagnosis of diseases.

It was in the middle of the nineteenth century that radio thermometry was first used to diagnose breast cancer [10]. However, despite the obvious advantages of the method, for a long time its effectiveness remains noticeable in medical practice and deserves the attention of scientific research. This is due, in particular, to the complexity of the design of measuring equipment and the quality of measurements. Since the creation of the RTM-01-RES microwave radio thermometer in the Russian Federation in 1997, interest in this technique has increased significantly. Currently, the method of radio-thermometric diagnostics is used in practice in several countries of the world, such as the USA, European countries, Australia, and many other countries. It is based on the sequential measurement of tissue temperature in certain areas of the body, depending on the intensity of their thermal radiation in the radio range (RTM) and in the infrared range (IR). The measurement results are displayed on the computer screen in the form of thermograms and temperature fields. Analyzing the information received, the doctor makes a diagnosis [10, 11]. Previous studies have shown that the release of specific heat in a tumor is directly proportional to the rate of its growth, that is, fast-growing tumors are "warmer" and, therefore, more noticeable on thermograms. Thus, microwave radio thermometry has a unique ability to detect mainly fast-growing tumors. Therefore, the use of microwave radio thermometry in complex diagnostics can lead to the natural selection of breast cancer patients with rapid tumor growth.

## 2.2 A Brief Overview of Breast Cancer Thermography

The purpose of research on early detection of breast cancer was to replace old methods (ultrasound, mammography, etc.). They are based on infrared thermal imaging and microwave radio thermometry. The basic principle of diagnosis using infrared tomography is that unregulated cell growth leads to a higher metabolic rate and requires more blood flow than the surrounding tissues. The extra heat released is transferred to the tissues surrounding the tumor, causing a temperature spike on the surface. Infrared imaging detects the tumor at the reached peak temperature. Thermography was first proposed for the diagnosis of breast cancer in 1956 [9]. Then the scientists continued their research until 1982, when this method became an additional tool for mammography in the United States [9]. Thus, the research continued until Gauteri and Gamagami concluded that infrared imaging can be useful for early detection of breast cancer and fast-growing neoplasms, despite the limitations of its methods due to the lack of control of thermographic materials by specialists in this field [12]. The first decade of the twenty-first century, coinciding with the improvement of thermographic materials (cameras), led to the development of methods for detecting breast tumors, such as dynamic infrared thermography. This method is aimed at detecting the thermal contrast between cancerous and non-cancerous areas when warming the breast after exposure to cold stress. The principle is based on the fact that temperature parameters in breast cancer remain unchanged, while healthy areas are affected by cold stress [12].

Researchers such as Ohashi and Uchida studied 728 cases of breast cancer using thermograms from infrared images in a stable mode and found an accuracy of 82% [13]. By solving thermal equations in different coordinate systems and examining different breast models, the researchers showed that the location and size of the tumor have a primary effect on the temperature distribution over the breast surface. The phenomenon of temperature contrast on the surface of the chest may be intensified because of cold stress. There are several studies using a genetic algorithm (ga) and an artificial neural network (RNA) or regression method to estimate the size, location, and heat of tumor metabolism. For example, a neural network was created to map the thermograms obtained, and then AG was used to determine tumor parameters such as size, location, and metabolic heat release rate.

### 2.3 Application of Various Machine Learning Methods for Breast Cancer Diagnosis

The vast majority of artificial intelligence algorithms used to detect breast cancer are mainly related to classification. The purpose of these algorithms is to distinguish a healthy breast from a breast with the help of malignant tumors. Data sets generated using thermographic images of healthy and diseased breasts can be used for their training. Currently, machine learning is one of the most widely used technologies [14, 15]. It is a technology focused on pattern recognition using algorithms that can be developed to examine data and establish relationships with statistical and mathematical calculations. Such applications include the support vector machine (SVM), the relevance vector machine (RVM), the naive Bayesian method and the decision tree (DT) (Table 1).

**Table 1** Machine learning technologies

| The authors of the study | The method used in the study | Population of thermograms | Accuracy factor (%) | Specificity (%) | Sensitivity (%) |
|---|---|---|---|---|---|
| Pramanik et al. [13] | Block variance (BV), a pending artificial neural network (FANN) with a decent gradient learning rule | A dataset of 100 thermal breast images in front view mode from DMR (database for Mastology research) | 90 | 95 | 85 |
| Gayatri Sumathi [16] | Relevance Vector Machine (RVM) | 300 data sets | 97 | 98 | 98 |
| Acharya et al. [17] | Support Vector Machine (SVM) | 50 IR thermograms | 85 | 85.71 | 90.48 |
| Tan et al. [18] | Gaussian Mixture Model (GMM) | 90 patients; 6,000 sets of temperature data | 90.6 | 94.8 | 78 |
| Nguyen et al. [19] | A machine learning method based on a random forest classifier and a feature selection method | Wisconsin Database | 99.82 | 99.83 | 99.72 |

## 2.4 Exploring Approaches to Asymmetry

Since the thermal symmetry of a healthy human body is so great, any change from the original symmetry can be considered abnormal. Therefore, thermograms are used to detect these deviations. Any deviation from the original symmetry can be considered abnormal. Therefore, thermograms are used to detect these deviations.

The skin, one of the largest organs of the body, is equipped with a network of vessels, which are accompanied by dense nerve fibers. It serves as the body's thermoregulator, controlling blood flow within a few millimeters of the body surface. The system is both anatomically and physiologically symmetrical.

For this reason, localized and asymmetric temperature changes on the surface of the body interested doctors. The recent development of sophisticated thermographic measuring devices capable of displaying body surface temperature has increased this interest. However, the thermography method has been criticized for possible bias resulting from the interpretation of a color thermogram. To minimize this subjectivity, it is proposed to use temperature differences calculated by thermographic devices between homogeneous areas of the body. The indicators of the degree of thermal asymmetry are following:

- **The asymmetry between the internal and external image of the same person**. Interested in studying the temperature fields represented by RGB colors, we perform image processing, which results in a difference between internal and external, as shown in Fig. 1 and 2.

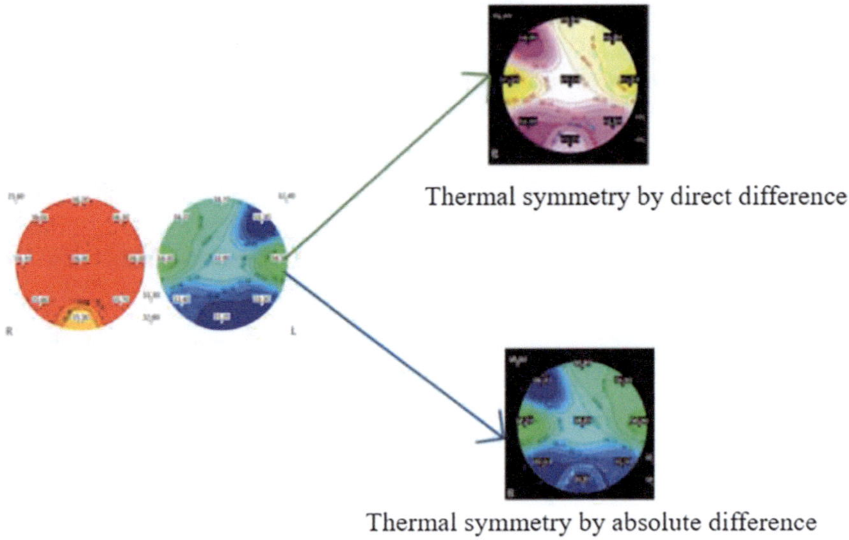


**Fig. 1** Thermal symmetry of symmetric points

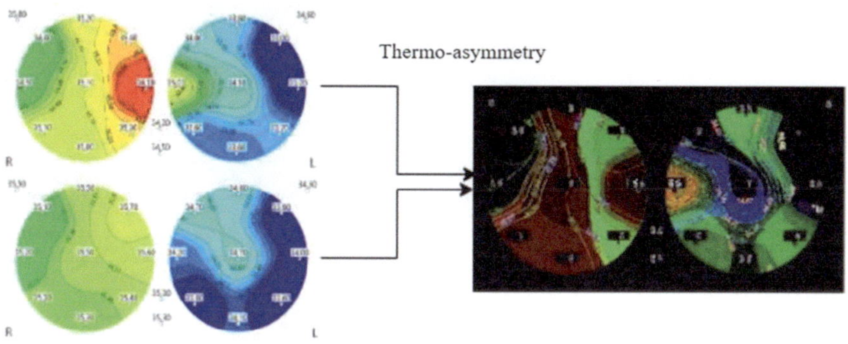


**Fig. 2** Thermal symmetry at the same point

- **Asymmetry between points that are symmetrical vertically**. It's about seeing the differences that exist between 2 points symmetrical with respect to a vertical line parallel to the spine for the same person and for the same image (internal or external image). This asymmetry will be the subject of our next study.
- **Diagonal asymmetry.** In this part, we are interested in comparing the image of the left gland from below with the image of the right gland from above and vice versa. We will consider all the features obtained using asymmetry. Similarly, double asymmetry may be of research interest, that is, to apply the asymmetry between the left and right mammary glands in the image obtained between the asymmetry of internal and external thermograms.

## 3 The Problem Statement

Thermography is used in various fields of medicine, such as cancer detection, pain management and diabetes. Many methods have been proposed for the computer analysis of these thermal images, such as descriptive matches, statistical methods, and asymmetric thermal methods.

The modern method of diagnosing breast cancer using microwave radio thermometry is based on thermograms [20, 21]. In our case, we have a data set of thermograms obtained from the surface of the mammary glands, and a data set of thermograms obtained with an accuracy of several millimeters inside the glands. The asymmetry of temperature fields between the internal sample and the external sample from the same patient was performed.

In thermograms obtained from a closer distance, the resulting thermal image has a maximum active area, which is protected by the chest and is considered an active area, while the frame of the thermal image is considered a passive area of the thermal image (Fig. 3).

Our initial data is taken from the RTM-01-RES device. Our dataset contains 6525 thermograms, including 4960 healthy patients and 1565 at-risk patients. Our

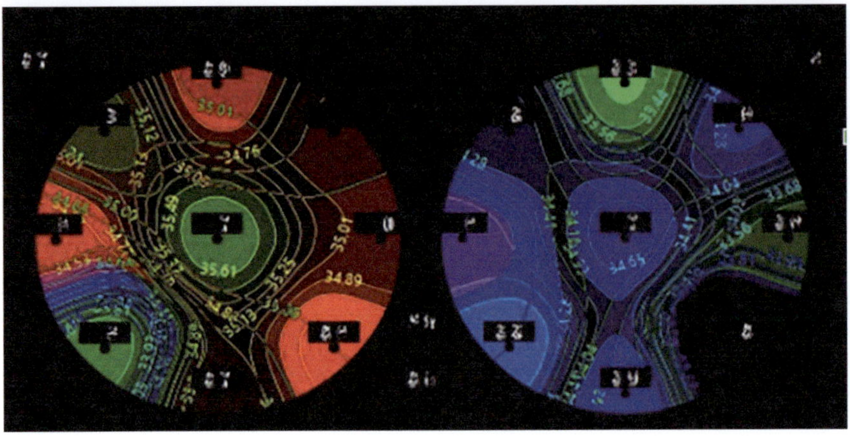

**Fig. 3** Data sample

asymmetry operation is based on the difference in the color of the thermal field inside and outside each point or area of the gland. The black areas on our thermograms are not of interest to our study, as they are the result of a difference in the color of the same field. In our case, the zones of anomalies correspond, in particular, to the localization of malignant neoplasms [6, 22]. Since it is obvious that this imaging method is available to medical professionals to identify areas of abnormalities, our model will follow the same principle when building a classification model.

The initial data also contains information about the diagnosis made by the doctor. Based on this diagnosis, labels were applied to each thermogram: 0-the patient is healthy, 1-the patient is at risk.

## 4 The Proposed Classification Method

First, using convolutional image transformation, each thermogram resulting from asymmetry is transformed into a feature vector. These functions then become input data for our artificial neural network. The general scheme of the classifier is as follows (Fig. 4).

### *4.1 Convolutional Image Conversion*

This template loads a 500 $\times$ 375 original image (thermogram) and converts it to a 224 $\times$ 224 image. The resulting image is then split into RGB components in the form of a 224 $\times$ 224 $\times$ 3 array. The VGG16 model uses 64 characteristics for classification. Each function uses its own transformation from a 224 $\times$ 224 $\times$ 3 array to a 224

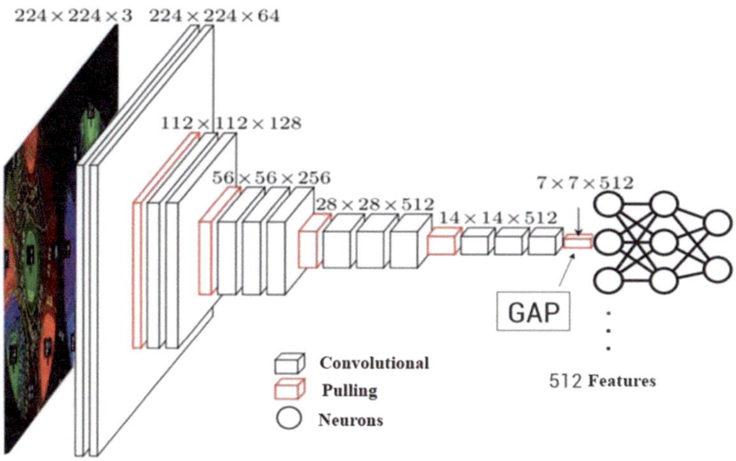


**Fig. 4** The proposed model structure

× 224 array, resulting in a 224 × 224 × 64 array. These array transformations continue in the following order of the resulting arrays: 112 × 112 × 128, 56 × 56 × 256, 28 × 28 × 512, 14 × 14 × 512, 7 × 7 × 512. To obtain the feature vector, the VGG16 image convolution template is used, implemented in the Python keras library. This model was trained by the authors [23] and is widely used to solve various image classification problems [24]. The resulting array is then applied to the global intermediate pool transformation implemented in the Python Keras library, which converts a 7 × 7 matrix to a number. As a result, we get a vector of 512 integers, which is used as features to form a classifier, as which an artificial neural network was chosen convolutional image transformation (Fig. 5).

## *4.2 Artificial Neural Network Training*

As an artificial neural network, the perceptron, consisting of 3 layers-input, output and hidden layers, was taken from 512.1 and 64 neurons, respectively. The training took place over 25 epochs. During the training, the parameters of the artificial neural network were changed, such as the loss function, the method of minimizing the loss function. As a result, the best values of these parameters were empirically selected.

To implement this training procedure, the data set is divided into three parts: a training sample (70%), an evaluation sample (15%) and a test sample (15%). The evaluation sample was used to evaluate intermediate results during training, the test sample was used to verify the operation of an already fully trained neural network (these data were not involved in training). During the training, the following data were obtained (Fig. 6) and the values of the loss function (Fig. 7).

The ratio of true positive results to the total number of predicted positive results:

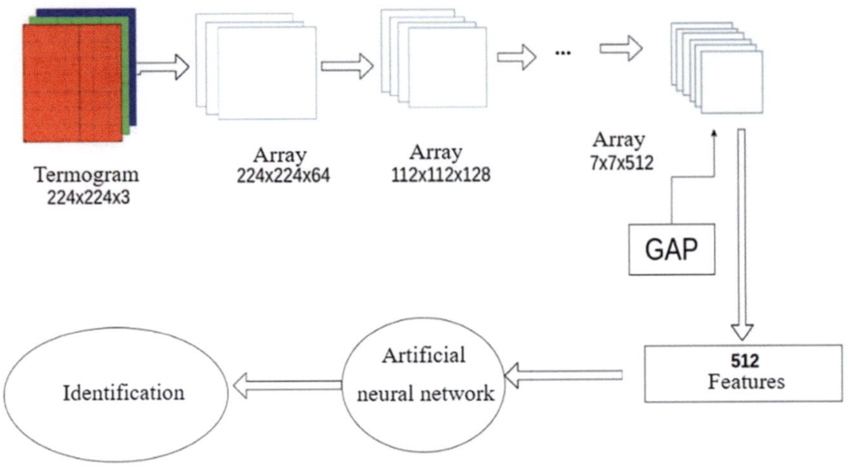


**Fig. 5** The stages of the proposed method—from extraction to classification

**Fig. 6** Accuracy in training and validation

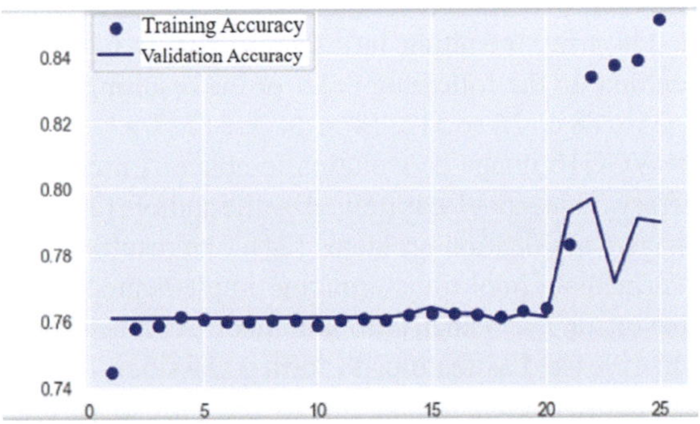


**Fig. 7** Data loss function during training and validation of tests

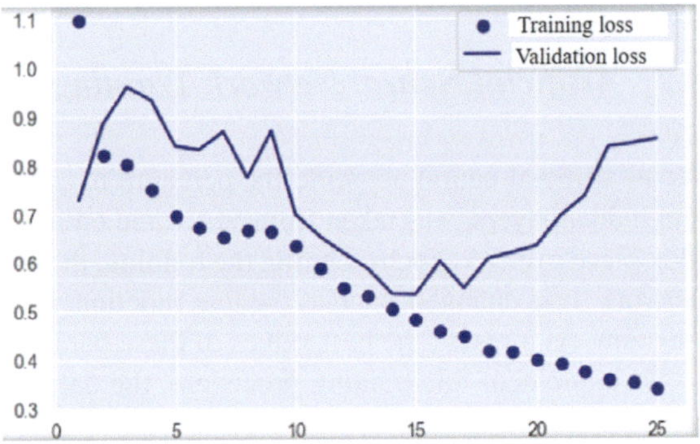

**Table 2** Metric values for testing and comparing developed models

| Class | Precision (P) | Recall (R) | f1-score (f1-s) |
|---|---|---|---|
| 0 | 0.85 | 0.88 | 0.86 |
| 1 | 0.57 | 0.50 | 0.53 |
| Accuracy (A) | 0.79 | | |
| The average value (avg) | 0.71 | 0.69 | 0.70 |
| Weighted average (weighted avg) | 0.78 | 0.79 | 0.78 |

$$\text{precision P} = \frac{Tp}{Tp + Fp} \tag{1}$$

The number of correct forecasts divided by the total number of forecasts:

$$\text{accuracy A} = \frac{Tp + Tn}{Tp + Tn + Fp + Fn} \tag{2}$$

The ratio of true positive results to all positive ones in the basic truth:

$$\text{completeness (recall)R} = \frac{Tp}{Tp + Fn} \tag{3}$$

$$\text{specificity} = \frac{Tn}{Tn + Fp} \tag{4}$$

and the average harmonic precision and completeness:

$$\text{(f1 - score) f1 - s} = 2\frac{R * P}{R + P} \tag{5}$$

where Tp true positive results are the number of health patients correctly predicted as “0”, Tn true negative results are the number of patients at risk correctly predicted as “1”, Fp false positive results are the number of patients at risk but who were mistakenly predicted as “0”, and Fn false negative results are the number of patients who are healthy, but who were mistakenly predicted to be “1”. The results found are presented in Table 2.

### *4.3 Training an Artificial Neural Network Based on a Clustered Dataset*

In this part of the study, a method is used based on the division of data into different groups with similar characteristics. Common clustering algorithms are K-means, hierarchical, and declasterized. In our case, K-means are used. This is a method that allows you to divide n images into k clusters so that each data point belongs to only one group.

Mathematically, this algorithm is described in detail below:

K-means [25]. splits a set of data points X = (x_1,x_2,...,x_n) into several clusters (k). Then the partition is performed according to the similarity criterion, which is usually the sum of the quadratic error defined in (6).

$$J = \sum_{i=1}^{k} \sum x_j \in Xx_j - m_i^n \tag{6}$$

where $m_i$—the centroid of cluster i, which is collectively represented as

$$M = (m_1, m_2, \ldots, m_k) \tag{7}$$

all relevant clusters,

$$C = (c_1, c_2, \ldots, c_k) \tag{8}$$

The algorithm K-MEANS It was applied to 6525 images (4960 healthy, 1565 risky). Three groups (K = 3) were obtained, which allowed us to obtain three groups of images.

The results obtained using cluster 1.

Using this dataset of the first subgroup from cluster analysis, divided into training, validation and testing described above, an artificial neural network was built, the results of which are presented in the Table 3.

The results obtained using cluster 2.

As for subgroup 2, it has the largest number of images. The results are shown in the Table 4.

Results obtained using cluster 3.

Compared to each of the clusters, cluster 3 gives the best result (also the best results for all data, point 4.2), which we summarize in the following Table 5.

**Table 3** Metric values selected for testing in cluster 1

| Class | Precision (P) | Recall (R) | f1-Score (f1-s) |
|---|---|---|---|
| 0 | 0.89 | 0.92 | 0.90 |
| 1 | 0.68 | 0.59 | 0.63 |
| Accuracy (A) | 0.85 | | |
| The average value (avg) | 0.78 | 0.75 | 0.77 |
| Weighted average (weighted avg) | 0.84 | 0.85 | 0.84 |

**Table 4** Metric values selected for testing in cluster 2

| Class | Precision (P) | Recall (R) | f1-Score (f1-s) |
|---|---|---|---|
| 0 | 0.96 | 0.97 | 0.97 |
| 1 | 0.66 | 0.60 | 0.63 |
| Accuracy (A) | 0.94 | | |
| The average value (avg) | 0.81 | 0.79 | 0.80 |
| Weighted average (weighted avg) | 0.94 | 0.94 | 0.94 |

**Table 5** Metric values selected for testing in cluster 3

| Class | Precision (P) | Recall (R) | f1-Score (f1-s) |
|---|---|---|---|
| 0 | 0.86 | 0.90 | 0.88 |
| 1 | 0.51 | 0.43 | 0.47 |
| Accuracy (A) | 0.80 | | |
| The average value (avg) | 0.69 | 0.66 | 0.67 |
| Weighted average (weighted avg) | 0.79 | 0.80 | 0.80 |

## 5 Conclusion

Breast cancer is a serious health problem for women all over the world. Breast cancer patients have a high mortality rate. Mortality rates can be reduced to a greater extent through early detection through appropriate and effective screening. Thermography has become a serious competitor to other screening methods such as mammography. Breast cancer is a serious health problem for women all over the world. Breast cancer patients have a high mortality rate.

This chapter presents a deep learning model for the classification of malignant breast tumors. The proposed model consists in the fact that the data are thermograms constructed using the RTM-RES-01 device, and image processing begins with an asymmetry between internal and external space. These images are then processed by extracting characteristics from the images to create a simpler artificial neural network architecture. This extraction was performed using a VG16 convolutional neural network specialized in this field. The results obtained are satisfactory, and the accuracy of the accepted classification model ranges from 0.72 to 0.93. In the future, it is planned to start with the asymmetry of both mammary glands in the same patient.

## References


1. Abstracts from the 22nd Annual Meeting of the North American Neuromodulation Society, Neuromodulation: Technology at the Neural Interface, vol. 22, no. 3, pp, e40-e295 (2019), ISSN 1094–7159. https://doi.org/10.1111/ner.12950
2. Polyakov, M.V., Popov, I.E., Losev, A.G., Khoperskov, A.V.: Application of Computer simulation results and machine learning in the analysis of microwave radiothermometry data. Math. Phys. Comput. Simul. **24**(2), 27–37 (2021). https://doi.org/10.15688/mpcm.jvolsu.2021
3. Kakileti, S.T., Manjunath, G., Madhu, H., Ramprakash, H.V.: Advances in breast thermography.. IntechOpen **2017**, 91–107 (2017)
4. Lee, K.-S., Jung, S.-K., Ryu, J.-J., Shin, S.-W., Choi, J.: Evaluation of transfer learning with deep convolutional neural networks for screening osteoporosis in dental panoramic radiographs. J. Clin. Med. **9**(2), 392 (2020). https://doi.org/10.3390/jcm9020392
5. Levshinskii, V., Galazis, C., Losev, A., Zamechnik, T., Kharybina, T., Vesnin, S., Goryanin I. Using AI and passive medical radiometry for diagnostics (MWR) of venous diseases. Comput. Methods Programs Biomed. **215**, 106611 (2022), ISSN 0169-2607. https://doi.org/10.1016/j.cmpb.2021.106611
6. Germashev, I., Dubovskaya, V., Losev, A.: Hierarchical fuzzy inference of adequacy of highly informative diagnostic signs of breast cancer. Stud. Syst., Decis. Control. **437**, 31–41 (2023)
7. Germashev, I.V., Dubovskaya, V.I., Losev, A.G., Popov, I.E.: Fuzzy inference of the effectiveness factors of the computational model for the diagnosis of breast cancer. In: 3rd International Conference on Control Systems, Mathematical Modelling, Automation and Energy Efficiency, SUMMA 2021; Lipetsk; Russian Federation; 10 November 2021–12 November 2021, 2021, pp. 528–533 (2021). https://doi.org/10.1109/SUMMA53307.2021.9632217
8. Bhowmik, A., Repaka, R., Mulaveesala, R., Mishra, S.C.: Suitability of frequency modulated thermal wave imaging for skin cancer detection—a theoretical prediction. J. Thermal Biol. **51** (2015)
9. Pramanik, S., Bhattacharje, D., Nasipuri, M.: Texture analysis of breast thermogram for differentiation of malignant and benign breast. In: International Conference on Advances in Computing. Communications and Informatics, ICACCI, pp. 8–14 (2016)
10. Germashev, I., Dubovskaya, V., Losev, A.: Substantiation of the result of the work of the cyber-physical system for breast cancer diagnostics. Cyber-Phys. Systems. Model. Intell. Control. Stud. Syst., Decis. Control **416**, 3–14 (2021)
11. Vidyukov, V.I., Mustafin, Ch.K., Kerimov, R.A., Fisher, L.N.: Differential diagnosis of breast tumors on the basis of radio thermometric findings. Tumors Female Reprod. Syst. **2016**, 26–31 (2016)
12. Gayathi, B.M., Sumathi, C.P.: Comparative study of relevance vector machine with various machine learning techniques used for detecting breast cancer. In: IEEE International Conference on Computational Intelligence and Computing Research (ICCIC), pp. 1–5 (2016).https://doi.org/10.1109/ICCIC.2016.7919576
13. Acharya, U.R., Ng, E.Y.K., Tan, J.-H., Sree, S.V.: Thermography based breast cancer detection using texture features and support vector machine J. Med. Syst. **36**(3), 1503–1510
14. Hakim, A., Awale, R.N., Imaging, T.: An emerging modality for breast cancer detection: a comprehensive review. J. Med. Syst. **44**, 136–142 (2020). https://doi.org/10.1007/s10916-020-01581
15. Mittal, H., Pandey, A.C., Saraswat, M.: A comprehensive survey of image segmentation: clustering methods, performance parameters, and benchmark datasets. Multimed Tools Appl. **81**(2022), 35001–35026 (2022). https://doi.org/10.1007/s11042-021-10594-9
16. Tan, J.M.Y., Ng, E.Y.K., Acharya, U.R., Keith, L.G., Holmes, J.: Comparative study on the use of analytical software to identify the different stages of breast cancer using discrete temperature data. J. Med. Syst. **33**(2), 141–153 (2008)
17. Germashev, I.V., Dubovskaya, V.I., Losev, A.G.: Substantiation of the result of the work of the cyber-physical system for breast cancer diagnostics. Stud. Syst. Dec. Control **416**, 3–13 (2022)

18. Yadav, S.S., Jadhav, S.M.: Thermal infrared imaging based breast cancer diagnosis using machine learning techniques. Multimed Tools Appl. **81**(2022), 13139–13157 (2022). https://doi.org/10.1007/s11042-020-09600-3
19. Karthiga, R., Narasimhan, K.: Medical imaging technique using curvelet transform and machine learning for the automated diagnosis of breast cancer from thermal image. Pattern. Anal. Applic. **24**(2021), 981–991 (2021). https://doi.org/10.1007/s10044-021-00963-3
20. Simonyan, K., Zisserman, A.: Very deep convolutional networks for large-scale image recognition. arXiv preprint arXiv:1409.1556 (2014)
21. Husaini, A., Habaebi, M.A.S., Gunawan, M.H., Islam, M.R., Elsheikh, E.A.A., Suliman, F.M.: Thermal-based early breast cancer detection using inception V3, inception V4 and modified inception MV4. Neural Comput. Appl. **34**(2022), 333–348 (2022). https://doi.org/10.1007/s00521-021-06372-1
22. Mashekova, A., Zhao, Y., Ng, E.Y.K., Zarikas, V., Fok, S.C., Mukhmetov, O.: Early detection of the breast cancer using infrared technology—a comprehensive review. Thermal Sci. Eng. Prog. **27**, 101142 (2022). https://doi.org/10.1016/j.tsep.2021.101142
23. Kakileti, S.T., Madhu, H.J., Manjunath, G., Wee, L., Dekker, A., Sampangi, S.: Personalized risk prediction for breast cancer pre-screening using artificial intelligence and thermal radiomics. Artif. Intell. Med. **2020**, 105–114 (2020). https://doi.org/10.1016/j.artmed.2020.101854
24. Sanchez-Cauce, R., Perez-Martin, J., Luque, M.: Multi-input convolutional neural network for breast cancer detection using thermal images and clinical data. Comput. Methods Programs Biomed. **204**, 106045 (2021). https://doi.org/10.1016/j.cmpb.2021.106045

# Advanced Framework ISIDA-Based Approach to Brain Tumor Classification from MRI Images

**Daniil Skaldin, Ilya Sokolov, and Alla G. Kravets**

**Abstract** Recent advancements in artificial intelligence (AI) have led to significant improvements in medical image analysis, especially for early detection and classification of various pathologies. This chapter presents a comprehensive approach to developing a system for automated brain tumor classification using MRI images. Our system, named ISIDA (Intelligent System for Identifying and Diagnosing Anomalies), leverages a convolutional neural network (CNN) for image-based detection and classification, as well as a large language model (LLM) to generate patient-specific diagnostic reports and treatment recommendations. Experimental results on two datasets demonstrate high classification accuracy, even in the presence of noisy images. Additionally, the automated report generation component translates raw classification outputs into meaningful, human-readable text, serving as a useful decision-support tool for medical professionals. This unified approach significantly reduces the workload of healthcare providers, improves diagnostic speed, and offers a scalable solution for medical institutions seeking to integrate AI into their diagnostic workflows.

**Keywords** Brain tumor classification · MRI · Convolutional neural networks · Large language models · Diagnostic automation · Medical image analysis

---

D. Skaldin (✉) · I. Sokolov · A. G. Kravets
Dubna State University, Dubna, Russian Federation
e-mail: danya.skaldin@mail.ru

I. Sokolov
e-mail: ilya20834@gmail.com

A. G. Kravets
e-mail: allagkravets@yandex.ru

A. G. Kravets
Volgograd State Technical University, Volgograd, Russian Federation



A. G. Kravets and A. A. Bolshakov (eds.), *Human-Centric Cyber-Society*, Studies in Systems, Decision and Control 639, https://doi.org/10.1007/978-3-032-06787-6_14

# 1 Introduction

The proliferation of medical imaging technology has resulted in a substantial increase in the volume of patient data that healthcare providers must analyze daily. Manual inspection and diagnosis are time-consuming and prone to human error, which can delay critical treatments. In the case of brain tumors, early and accurate detection is vital for improving patient outcomes and reducing mortality rates.

To address this challenge, AI-based methods for computer-aided diagnosis (CAD) are rapidly emerging, combining powerful image-processing algorithms with advanced machine-learning techniques. These methods enable robust detection and classification of pathologies in medical scans with high precision. However, many existing solutions focus on a single task (e.g., classification only) and do not provide the physician with more extensive guidance, such as automated textual reports or recommendations for further testing.

In this work, we propose an integrated solution named ISIDA, which merges:

1. A convolutional neural network (CNN) for classification of brain tumors from MRI images into four categories: meningioma, glioma, pituitary, or normal.
2. A large language model (LLM) for generating structured medical reports and initial recommendations based on the classification results.

Through this synergy of image-based classification and natural language generation, the system aims to significantly reduce diagnostic time while also delivering a degree of interpretability and guidance that purely numerical outputs lack.

# 2 Related Work

## *2.1 Patent Analytics Approaches (Background Context)*

In the broader AI research landscape, multiple methods for handling large text corpora and classification tasks exist. For instance, in patent analytics—where massive volumes of documents must be examined—techniques often revolve around text mining, morphological analysis, and machine learning [1, 2]. Although the problem domain differs (medical imaging vs. patent text), the core technical challenges of processing large, heterogeneous datasets, applying advanced classification algorithms, and providing structured outputs share many similarities. Lessons learned in patent analytics—such as cross-thematic clustering [3] and formal metrics [4]—can inform the development of robust classification pipelines in healthcare.

### 2.2 Medical Image Analysis

Medical image analysis has greatly benefited from the emergence of deep learning, particularly convolutional neural networks. Prominent examples include following.

Litjens et al. [5] conducted an extensive review of deep learning techniques in medical imaging, underscoring the importance of specialized architectures for tasks like segmentation and classification.

Ronneberger et al. [6] introduced the U-Net architecture for biomedical image segmentation, showcasing how encoder-decoder structures can capture complex features while preserving localization details.

Transfer Learning methods have also proven effective, using models pre-trained on large non-medical datasets (e.g., ImageNet) to bootstrap performance in smaller medical datasets [7].

### 2.3 Textual Report Generation with LLMs

Recently, large language models such as GPT and Llama have made it possible to transform structured or semi-structured data into human-readable reports. These models excel at:

- Summarizing the salient features of input data.
- Generating concise or extended text incorporating domain-specific terminology.
- Adapting to specialized prompts to ensure context-relevant output [8, 9].

By integrating an LLM with a CNN-based classification pipeline, the system provides not only numeric or categorical outputs (e.g., "glioma" or "meningioma") but also descriptive commentary, medical considerations, and disclaimers.

## 3 System Description

### 3.1 Overall Architecture

Our system, ISIDA, comprises two key AI modules connected via a web-based interface (Fig. 1):

1. TumorClassifier (CNN): Performs multi-class classification of MRI images.
2. LLM-based Report Generator: Produces a structured medical report, including diagnostic conclusions and initial recommendations.

The application uses Django as its main framework, ensuring a clear separation between front-end (user interface) and back-end (classification, data handling).

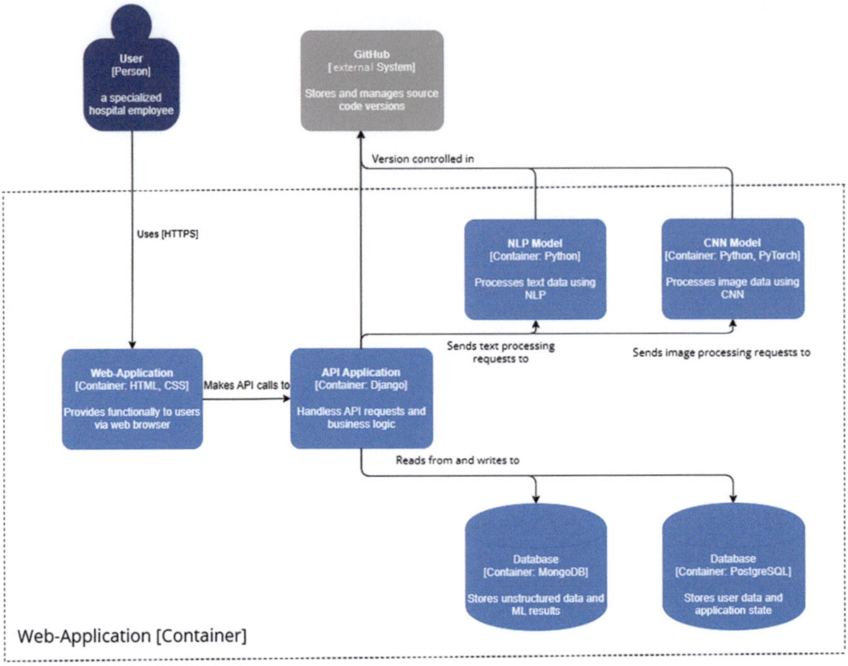


**Fig. 1** The architecture of the C4 notation project

PostgreSQL stores user credentials, patient records, and metadata (e.g., timestamps).

MongoDB manages semi-structured or unstructured data (e.g., generated reports).

# 4 Convolutional Neural Network: TumorClassifier

## *4.1 Network Structure*

The TumorClassifier CNN has the following layout.

1. Input layer: Accepts 256 × 256 color images (3 channels).
2. Convolutional/Pooling Blocks:
   - Conv1: 32 filters of size 3 × 3, padding = 1, followed by ReLU and MaxPool2d (2 × 2).
   - Conv2: 64 filters of size 3 × 3, padding = 1, followed by ReLU and MaxPool2d (2 × 2).
   - Conv3: 128 filters of size 3 × 3, padding = 1, followed by ReLU and MaxPool2d (2 × 2).

3. Fully Connected Layers:
   - FC1: Flattens the output from the convolutional blocks; outputs 128 neurons.
   - FC2: Outputs 4 neurons corresponding to the categories (meningioma, glioma, pituitary, normal).

Softmax/CrossEntropy computes probabilities for each class, returning the highest-likelihood category.

## *4.2 Training Procedures*

We trained the CNN on two primary datasets from Kaggle:

1. Brain Tumors 256 × 256 [10]
2. Crystal Clean: Brain Tumors MRI Dataset [11]

Both datasets contain multi-class tumor images (plus normal scans), ensuring diverse real-world conditions. Data augmentation (noise injection, brightness changes, flips) improves robustness.

- Optimizer: Adam, learning rate = 0.001
- Batch size: 32
- Epochs: 20

Validation accuracy on noisy images reached ~ 92.50%, while clean images yielded up to 100% accuracy. These results underscore the model's robustness and potential clinical applicability.

Model Training Metrics (Loss and Accuracy per Epoch) (table 1).

# 5 LLM-Based Report Generation

## *5.1 Model Description*

We employ a pre-trained large language model, Llama 3.1:8B-Instruct-Q8_0, adapted with prompt engineering techniques:

- Input: A structured summary of classification results (e.g., "Detected meningioma tumor" or "No tumor detected"), additional patient metadata, and disclaimers or rules for output formatting.
- Hidden Layers: 8-billion-parameter transformer architecture, featuring self-attention modules and layer normalization.
- Output: A coherent medical report, summarizing classification findings and suggesting a list of next steps for follow-up.

**Table 1** The training metrics (loss and accuracy) per epoch

| Epoch | Loss | Accuracy |
|---|---|---|
| 1/20 | 1.2502 | 0.4448 |
| 2/20 | 0.8846 | 0.6388 |
| 3/20 | 0.6382 | 0.7531 |
| 4/20 | 0.4347 | 0.8432 |
| 5/20 | 0.2703 | 0.9046 |
| 6/20 | 0.1742 | 0.9358 |
| 7/20 | 0.1079 | 0.9624 |
| 8/20 | 0.0611 | 0.976 |
| 9/20 | 0.0574 | 0.9804 |
| 10/20 | 0.072 | 0.979 |
| 11/20 | 0.0275 | 0.9915 |
| 12/20 | 0.0157 | 0.9958 |
| 13/20 | 0.021 | 0.9947 |
| 14/20 | 0.0266 | 0.9984 |
| 15/20 | 0.0191 | 0.9968 |
| 16/20 | 0.0091 | 0.9984 |
| 17/20 | 0.0078 | 0.9988 |
| 18/20 | 0.0047 | 0.9988 |
| 19/20 | 0.0069 | 0.9992 |
| 20/20 | 0.0039 | 0.9996 |

## 5.2 *Report Content*

The generated reports contain:

1. Diagnosis Statement: E.g., "The patient MRI scan indicates signs of glioma, corresponding to 85% confidence."
2. Clinical Description: Brief explanation of what this tumor type typically entails.
3. Recommended Actions: Suggestions such as consulting a neurosurgeon, additional MRI with contrast, or biopsy for confirmation.
4. Disclaimer: A mandatory section clarifying that the generated text serves as preliminary guidance rather than an official medical diagnosis.

This section presents the key user interfaces of the developed application designed for storing, displaying, and analyzing medical brain MRI images. The interface allows medical professionals to view a list of patients, access their scan history, and examine detailed reports for each analysis. Special attention has been given to intuitive navigation, clear visualization of MRI scans, and structured presentation of clinical data, including preliminary diagnoses and treatment recommendations. Figures 2, 3 illustrate the main components of the application.

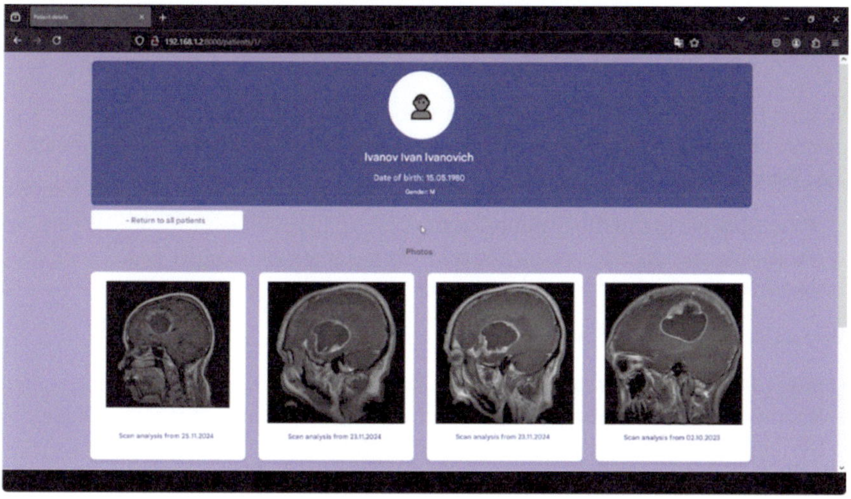


**Fig. 2** Patient profile page with scan list interface

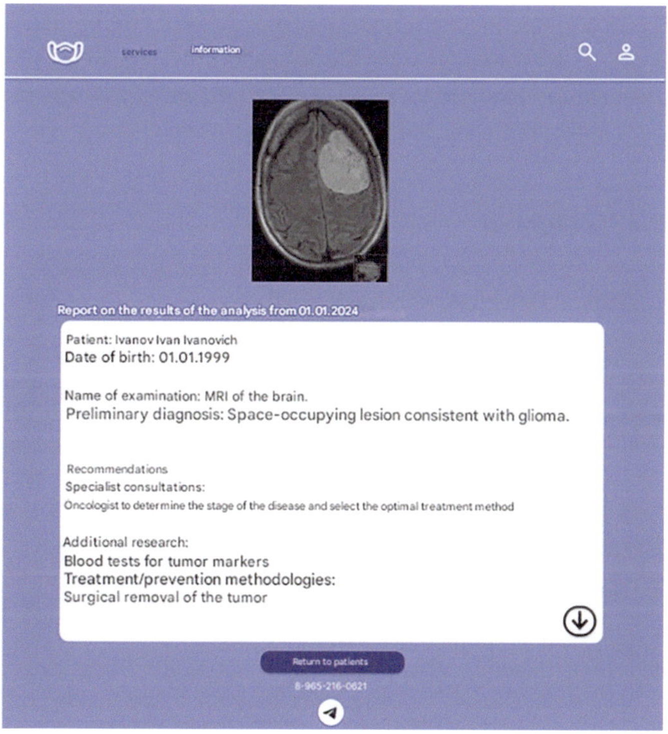


**Fig. 3** MRI scan analysis report with preliminary diagnosis and treatment recommendations

# 6 Experiments and Results

We integrated the CNN and LLM into a web application. The user (doctor) logs in, selects a patient from a database, chooses an MRI scan, and initiates the classification. The system then:

1. Processes the scan with TumorClassifier.
2. Passes the classification result and relevant metadata to the LLM-based generator.
3. Displays the final report in the browser.

We conducted user acceptance tests to assess:

- Ease of use (UI/UX feedback).
- Report comprehensibility (medical professionals examined sample outputs).
- System responsiveness (time from image upload to final report).

Quantitative Evaluation:

- Classification Accuracy: Up to 92.50% on augmented/noisy data; 100% on clean data.
- Inference Time: CNN classification ~ 1 to 2 s per 256 $\times$ 256 image on a GPU. LLM-based report generation ~ 2 to 5 s, depending on complexity.
- User Feedback: Test participants reported that the automatically generated reports were "clear, well-structured, and beneficial as a second opinion."

## *6.1 Example Outputs*

**Scenario A (Tumor Detected—Glioma)**

- CNN Output: Glioma (probability 92%)
- Generated Text (Excerpt): "Based on the MRI findings, there is a high likelihood of glioma. This type of tumor arises from glial cells and may present with headaches, seizures, or neurological deficits. We recommend follow-up imaging with contrast-enhanced MRI and referral to a neuro-oncologist for potential biopsy and staging. Disclaimer: This report is for informational purposes only and should not replace clinical judgment or official medical evaluations."

**Scenario B (No Tumor Detected)**

- CNN Output: Normal (probability 98%)
- Generated Text (Excerpt): "No significant abnormalities were detected in the patient's MRI scan. We advise continued observation and regular check-ups to ensure early detection of any future issues. Disclaimer: This report is preliminary and does not replace professional medical assessment."

## 7 Discussion

By unifying CNN-based classification and LLM-based reporting, ISIDA offers a holistic approach to automated medical diagnostics:

- Reduced Workload: Medical practitioners can expedite routine reviews, focusing on complex cases requiring expert judgment.
- Consistency: Automated methods ensure consistent classifications and systematic reporting, reducing subjective variations.
- Scalability: The system can be integrated into hospital networks, cloud environments, or portable devices, offering flexible deployment for institutions of varying sizes.

Limitations include:

1. Data Availability: The system's performance may degrade if the target population or imaging equipment is significantly different from the training sets.
2. Regulatory Concerns: Automated clinical decision-making tools must adhere to strict legal frameworks, demanding rigorous validation and certification before widespread adoption.
3. Interpretability: While CNNs provide classification accuracy, their black-box nature may prompt questions about how final decisions are made.

Future research will explore:

- Extending tumor categories beyond the four current classes (e.g., metastatic brain tumors, rare tumor subtypes).
- Integration with PACS (Picture Archiving and Communication System) to streamline data flow between the system and clinical imaging databases.
- Real-time data augmentation (e.g., contrast shifting, rotation) to further improve robustness.
- Active learning methodologies, enabling the model to learn from new cases and continuously adapt.

## 8 Conclusion

This chapter presented an advanced framework, ISIDA, that combines a CNN-based brain tumor classification with a large language model for automatic report generation. The approach significantly improves diagnostic efficiency, offering real-time classification and concise textual recommendations that can assist clinicians in making faster, more informed decisions.

Our experiments underscore the potential of such an integrated system:

- High classification accuracy across noisy and clean MRI images.

- Fully automated text generation that is coherent and domain-specific, guiding the subsequent diagnostic steps.
- Scalable deployment in healthcare environments, potentially accelerating the adoption of AI-driven solutions in routine clinical practice.

We anticipate that further enhancements, including broader dataset coverage and refined medical language processing, will position AI-based automated diagnostic platforms as an indispensable tool for modern medicine, ultimately leading to better patient outcomes.

**Acknowledgements** This research was supported by the University "Dubna," which provided computational resources and institutional support. We appreciate the open-source community and Kaggle datasets that facilitated model training and testing. We also thank the medical professionals who contributed their time and expertise during the user acceptance testing phase.

## References

1. Kravets, A.: On approach for the development of patents analysis formal metrics. In: Proceedings of International Conference on Robotics and Automation (2023)
2. Mironenko, A.G., Kravets, A.G.: Automated methods of patent array analysis. In 7th International Conference on Information, Intelligence, Systems & Applications (IISA). IEEE (2016)
3. Korobkin, D.M., Fomenkov, S.S., Kravets, A.G., Kolesnikov, S.G.: Prior art candidate search on base of statistical and semantic patent analysis. In: Proceedings of the International Conference on Computer Graphics, Visualization, Computer Vision and Image Processing, pp 231–238 (2017)
4. Kravets, A.G., Legenchenko, M.S.: Formal metrics for automated assessment of inventions. Casp. J.: Manag. High Technol. **3**(39), 8–19 (2017)
5. Litjens, G., et al.: A survey on deep learning in medical image analysis. Med. Image Anal. **42**, 60–88 (2017)
6. Ronneberger, O., et al.: U-Net: Convolutional networks for biomedical image segmentation. MICCAI **9351**, 234–241 (2015)
7. Tajbakhsh, N., et al.: Convolutional neural networks for medical image analysis: Full training or fine tuning? IEEE Trans. Med. Imaging **35**(5), 1299–1312 (2016)
8. Brown, T.B., et al.: Language models are few-shot learners. Adv. Neural. Inf. Process. Syst. **33**, 1877–1901 (2020)
9. Raffel, C., et al.: Exploring the limits of transfer learning with a unified text-to-text transformer. J. Mach. Learn. Res. **21**(140), 1–67 (2020)
10. Brain tumors 256×256 // Kaggle. [Online]. Available: https://www.kaggle.com/datasets/thomasdubail/brain-tumors-256x256.Accessed: 26 Nov 2024
11. Crystal Clean: Brain Tumors MRI Dataset // Kaggle. [Online]. Available: https://www.kaggle.com/datasets/mohammadhossein77/brain-tumors-dataset. Accessed: 26 Nov 2024

# A Server Subsystem for Diagnosing Depression Disorders Platform

**Alla G. Kravets, Anna A. Smirnova, Alexander S. Miheev, and Inna V. Strukova**

**Abstract** The chapter presents an overview of the server subsystem of the PsyScan platform, designed for self-diagnosis of depressive states and to assist doctors in decision-making. The focus is on the description of the architecture and functional capabilities of the server component, which ensure the collection, processing, and storage of user data. The subsystem leverages modern machine learning approaches and natural language processing techniques to analyze information from surveys, patient records, and other sources. Special attention is given to data protection measures and user privacy, which are critical in medical applications. The chapter provides testing results that demonstrate the subsystem's efficiency and reliability, as well as discusses prospects for its further development within the platform for depression diagnosis.

**Keywords** Depression · Medical diagnostics · Data analysis · Software · Server subsystem · Amazon web services · JSON web token · Elastic compute cloud · .NET 6 · Entity framework

## 1 Introduction

In recent years, the task of diagnosing depression has become increasingly relevant due to the rising number of cases of this mental disorder worldwide. According to the World Health Organization, depression is one of the leading causes of disability and a significant factor affecting the quality of life for millions of people [1]. Psychotropic medications often lead to weight gain, resulting in discomfort and discontinuation of treatment. More importantly, this increases the risk of obesity-related diseases such as diabetes, hypertension, and ischemic heart disease [2]. Traditional methods of

A. G. Kravets · A. A. Smirnova (✉) · A. S. Miheev · I. V. Strukova
Volgograd State Technical University, Volgograd, Russia
e-mail: yesjessmay@mail.ru

A. G. Kravets
Dubna State University, Dubna, Russia



A. G. Kravets and A. A. Bolshakov (eds.), *Human-Centric Cyber-Society*, Studies in Systems, Decision and Control 639, https://doi.org/10.1007/978-3-032-06787-6_15

diagnosing depression, based on clinical surveys and consultations, require significant time and human resources. In light of this, there is a need to develop new approaches that can provide faster, more accurate, and scalable diagnostics.

Modern technologies, including artificial intelligence and big data, open new prospects in the field of medical diagnostics [3]. In particular, server subsystems are becoming an integral part of platforms designed for analyzing and interpreting medical data [4]. They ensure high performance, reliability, and security in processing information, which is critically important for accurate and timely diagnosis of depression [5].

The diagnostic platform is intended for conducting preliminary self-diagnosis of depressive states in individuals who have not yet sought help from a psychotherapist, as well as for supporting clinical decision-making during initial consultations [6]. As part of the project to implement a psychiatric diagnostic system based on heterogeneous information analysis, PsiScan [7–9] proposed an innovative method for automated depression detection based on patients' textual data.

This chapter describes the key components of the server subsystem, presents methods for data collection and processing, and addresses issues of information security and confidentiality. Special attention is given to the integration of the server subsystem with other elements of the platform.

## 1.1 Analysis of the Existing Depression Diagnosis Process

The existing process of diagnosing depression includes the following stages (Fig. 1).

1. Understanding the need to consult a specialist. This stage is crucial when individuals recognize possible deviations from the norm and understand that the help of a psychotherapist may be beneficial. It requires self-awareness, self-analysis, and the courage to overcome the stigma associated with seeking psychiatric help.

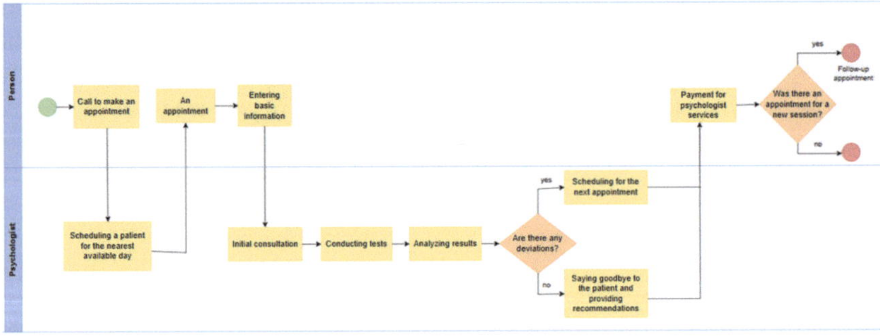


**Fig. 1** The existing process of depression diagnosing

2. Making an appointment with a specialist. It is important to find a competent specialist and schedule an initial consultation at a convenient time and day. This helps align the schedules of the patient and the specialist.
3. Filling out a personal information form. The psychotherapist provides a form to fill out information about oneself, including workplace, position, education, relationship status, and other relevant data. This information helps the specialist form a more complete picture of the patient.
4. Conducting a basic test. After filling out the form, a test is conducted to identify possible issues the patient may have. This test aids in the initial analysis of the mental state and may include tasks, questionnaires, or scales to assess various aspects of mental health. The test results will help the specialist determine the next steps and the need for additional consultations and procedures.

### *1.2 Comparison of Existing Methods of Diagnosing*

We will compare existing methods for analyzing psychiatric conditions based on the following criteria (see Table 1), which include:

- reliability and scientific accuracy;
- ethical aspects and data protection [10];
- cost;
- accessibility;
- degree of automation.

## 2 Architecture of the Server Subsystem of the Psychiatric Diagnosis Platform

The architecture of the Server Subsystem for the Depression Diagnosis Platform (SS DDP) was developed using C# 9 and.NET 6 [11, 12]; Entity Framework tools based on the Code First methodology for database management. The Entity Framework (EF) [13] software allows interaction with data using objects and queries written in C#. JWT (JSON Web Token) [14] mechanisms ensure secure transmission of information between the client and server in JSON format.

The server subsystem was deployed using the AWS (Amazon Web Services) [15] cloud platform, including EC2 and S3 services. EC2 mechanisms [16, 17] provide virtual machines to run applications and handle workloads, offering complete control over virtual servers. Amazon S3 tools provide cloud data storage organized in “buckets” for uploading and managing data.

The use of AWS, EC2, and S3 provides developers with flexibility, reliability, and effective means for scaling applications in the cloud, allowing them to focus on implementing functional capabilities rather than managing infrastructure [18].

The DDP consists of three levels (Fig. 2).

**Table 1** Results of the comparison of existing methods

| Method | Consultations with a psychotherapist in a hospital | Online consultations | Self-tests |
|---|---|---|---|
| Reliability | Depends on the qualifications of the specialist | Depends on the qualifications of the specialist | Absence of subjectivity; deviations can be identified without the assistance of a qualified specialist |
| Ethical aspects, data protection | According to the ethical code of psychotherapists, the specialist protects the patient's confidential information | According to the ethical code of psychotherapists, the specialist protects the patient's confidential information | Under the responsibility of the tested individual |
| Cost | From 1500 to 6000 rubles | From 2000 to 4000 rubles | Free |
| Availability | Poor availability | Averaged availability, as it is necessary to schedule an appointment | High availability, only internet access is required |
| Degree of automation | Not automated | Remote registration | High degree of automation, if tests are conducted on a website with automatic evaluation |

1. The user level provides interaction with the application's frontend through HTTP requests. It is closely related to the domain level.
2. The domain level is the central element of the system, describing the database entities and interfaces for interacting with it. It includes services for indirect interaction with the database.
3. The data access level establishes a connection to the database and performs read and write operations using the interaction descriptions from the domain level.

The designated separation into levels enhances the implementation of functional capabilities and application logic, providing structure and simplified interaction between components [19].

## 2.1 Features of the Data Storage Subsystem Implementation

During the implementation of the data storage subsystem of the depression diagnosis platform (DDP), the following entities were defined (see Fig. 3): User; Text Analysis; Audio Analysis; Preliminary Testing. Each of these entities has its own role and

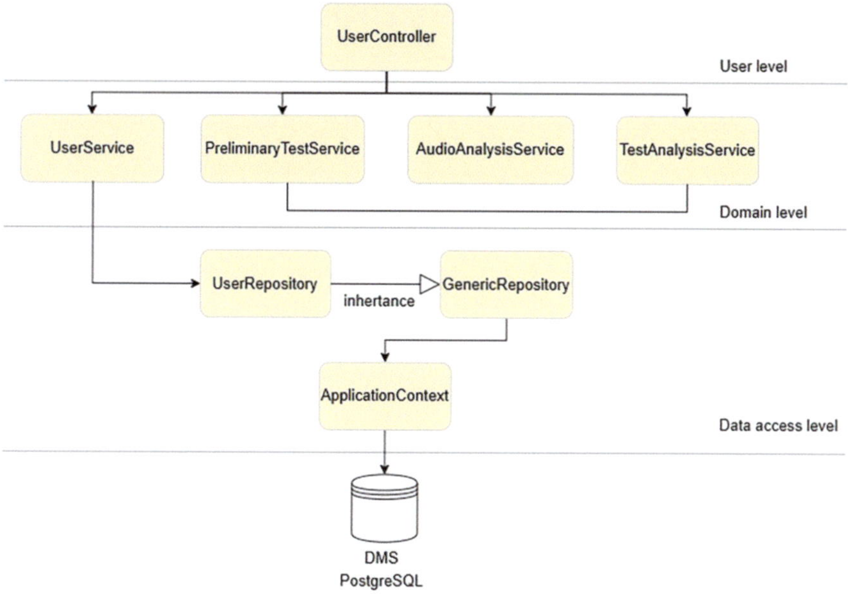


**Fig. 2** Technical architecture of the depression diagnosis platform (DDP)

interacts with other parts of the server subsystem, allowing for the creation of a storage structure [20].

To create tables in the storage of the Psychiatric Diagnosis Platform (see Fig. 4), Entity Framework was used with the Code First methodology. This means that models

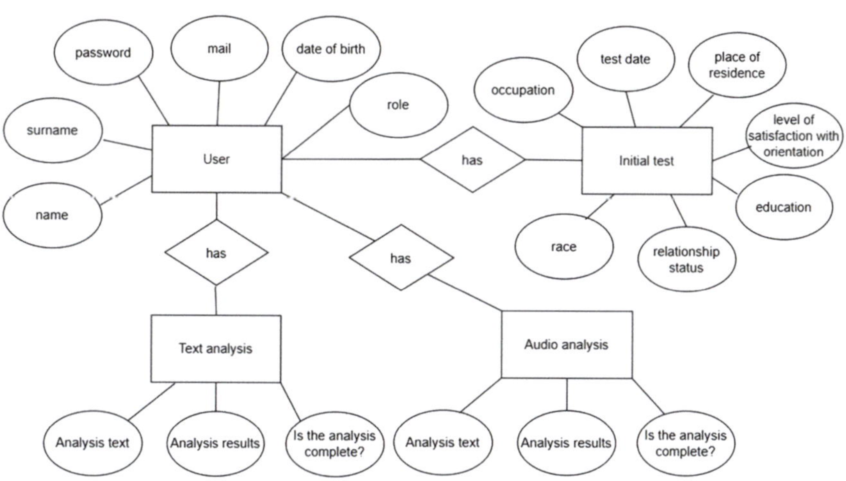


**Fig. 3** ER-diagram of the DDP storage subsystem

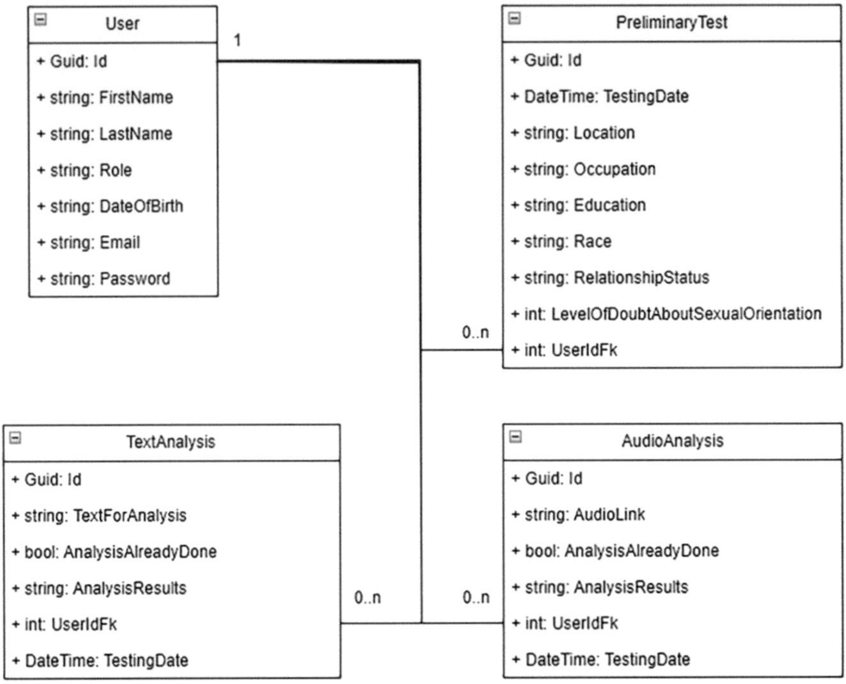


**Fig. 4** Data storage system structure

and entity classes were defined using a programming language, and then the storage was automatically created and configured based on these models. This approach simplifies the process of creating and updating the database, ensures consistency between the code and the storage structure, and facilitates working with data [21].

The use of Entity Framework and the Code First methodology in database creation allows developers to conveniently define and interact with various entities. This approach ensures flexibility and extensibility of the module when adding new features and modifying the database structure.

## 3 Development of the Server Subsystem for the Psychiatric Diagnosis Platform

The first step in developing the server subsystem for the Psychiatric Diagnosis Platform was the creation of the domain layer, which includes data models. The GenericRepository pattern was used to provide a common interface for the repository, which ensured basic functionality for adding, deleting, editing, and reading data, thereby improving maintainability and simplifying development. Also present at the domain level are DTO objects (Data Transfer Objects), which transmit only the necessary information between the user and domain layers, protecting the data and ensuring the security of the application [22].

Services at the domain level facilitate interaction between the user interface and the data storage, providing functional capabilities for processing requests and performing data operations. An interface for generating JSON Web Tokens (JWT) has been used for registration and authorization procedures, ensuring secure data transmission between the client and the server.

### *3.1 Deployment of the Psychiatric Diagnosis Platform (PDP) in AWS*

To operate the PDP, it needs to be deployed on suitable infrastructure. Two AWS services were considered: EC2 and Lambda.

The EC2 (Elastic Compute Cloud) service allows renting virtual servers (instances) with flexible configurations and various operating system and resource options;

The Lambda service runs code in response to events or triggers and automatically scales.

Based on the availability of necessary functionalities and ease of use, EC2 was chosen for deploying the subsystem. To deploy the application on EC2, a virtual machine with Windows 10 was created,.NET 6 was installed and configured, and the source code was transferred (Fig. 5). The project is launched via the console.

Two virtual machines are running Linux, while the second one is using Windows 10. Each virtual machine is configured for the subsequent deployment of the Psychiatric Diagnosis Platform (PDP), and they have the libraries required by the server module and the analysis module installed. To test their capabilities, test programs were run that utilize the same set of libraries as the application modules.

Thus, the choice of EC2 provided a reliable and flexible infrastructure for the operation of the PDP.

| Name | Instance ID | Instance state | Instance type | Status check | Alarm status | Availability Zone |
|---|---|---|---|---|---|---|
| Test | i-09c3d3e071f23990d | Running | t2.micro | 2/2 checks passed | No alarms | eu-west-1b |
| – | i-009fa949244717c41 | Running | t2.micro | 2/2 checks passed | No alarms | eu-west-1b |
| dost | i-06f138be2847b4e1d | Running | t2.micro | 2/2 checks passed | No alarms | eu-west-1b |

**Fig. 5** Configured and running virtual machines

## 3.2 Features of the Implementation of the Server Subsystem

To describe the interaction of the PDP, a case of text information analysis is considered (Fig. 6).

The server subsystem does not have a user interface in the conventional sense. Instead, interaction with the server subsystem is carried out through controller classes that process HTTP requests. Let's examine the operation of the server subsystem using the case of text information analysis as an example.

The frontend module uses a POST request provided by the Psychiatric Diagnosis Platform (PDP) to save the text for analysis. The object received from the frontend subsystem is a Data Transfer Object (DTO). This object is sent to the text analysis subsystem, where the DTO object is transformed into the corresponding database entity. The transformed object is then passed to the repository for storage in the database. After successful saving, the result is returned to the service, and the text object is sent to the S3 storage.

The analysis subsystem receives the text and determines which trained model to apply—either for audio analysis or text analysis. After processing the input information via an HTTP request, the analysis results are stored.

Thus, in the described structure of the platform, the server subsystem interacts with the frontend subsystem and the analysis subsystem for the storage and analysis

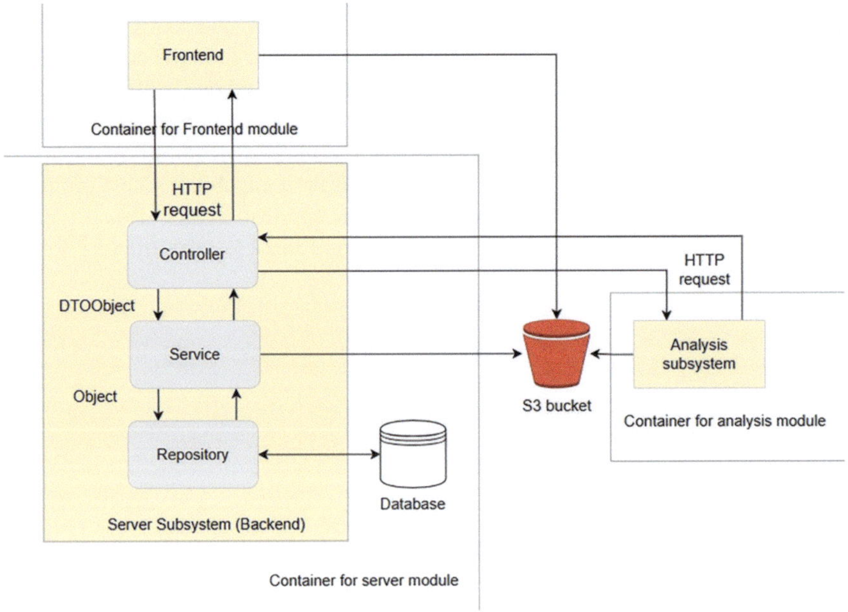


**Fig. 6** Structure of interaction between modules of the psychiatric diagnosis platform (case of text information analysis)

of text. This is achieved through controller classes, analysis services, and repositories, ensuring efficient and reliable interaction between the components.

### *3.3 Fault Tolerance Testing*

Fault tolerance testing included sending invalid data for login and checking the operation during the shutdown of the analysis subsystem. Four tests were conducted:

(1) Entering an already registered user; the subsystem returns an HTTP response with a status of 400 without a JWT token.
(2) Entering an incorrect email when logging into the account; the subsystem returns an HTTP response with a status of 400 without a JWT token.
(3) Entering an incorrect password when logging into the account; the subsystem returns an HTTP response with a status of 400 without a JWT token.
(4) Disabling the analysis subsystem; the server subsystem continues to operate, but records are not analyzed. After restoration, the subsystem operates normally.

## 4 Conclusion

As a result of this work, the server subsystem of the psychiatric diagnosis platform was designed and implemented. The goal was to ensure interaction between the frontend subsystem and the analysis subsystem, as well as to store user information and protect it from malicious attacks.

The results of the testing showed that the developed Psychiatric Diagnosis Platform (PDP) successfully addresses the set tasks, providing interaction between the modules of the server subsystem and safeguarding user data from hacking, facilitated by Amazon services, namely EC2 and S3 buckets.

The area of application for the developed platform PsyScan [8] is the process of conducting initial psychiatric diagnosis. This approach helps save time for both specialists and their patients, and allows patients to quickly understand whether they need to consult a specialist.

The main direction for further development of the system is to expand functionality and implement new analysis models. Integration of additional psychological tests, as well as analysis of a person's condition through video recordings, is planned. This will enhance the overall quality of analysis results.

## References

1. World Health Organization. Depression and other common mental disorders: global information. WHO (2017). https://www.who.int/publications/i/item/depression-and-other-common-mental-disorders-global-health-estimates
2. Serretti, A., Mandelli, L.: Antidepressants and body weight: a comprehensive review and meta-analysis. J. Clin. Psychiatry **71**(10), 1259–1272 (2010). https://doi.org/10.4088/JCP.09r05346blu
3. Ibragimov, A.R.: Application of artificial intelligence in medicine. Mod. Med. **3**, 45–52 (2020). (in Russian)
4. Ivanov, I.V., Petrov, P.S.: Big Data Technologies in Healthcare, 300 p. Medical Publishing House (2019) (in Russian)
5. Russell, S., Norvig, P.: Artificial Intelligence: A Modern Approach, vol. 175, pp. 935–937. Prentice Hall (2020). https://doi.org/10.1016/j.artint.2011.01.005
6. Kravets, A., Poplavskaya, O., Lempert, L.: The development of medical diagnostics module for psychotherapeutic practice. Commun. Comput. Inf. Sci. **754**, 872–883 (2017). https://doi.org/10.1007/978-3-319-65551-2_63
7. Petraevsky, V.A., Kravets, A.G., Galynov, M.S.: Certificate of state registration of the computer program No. 2022685105 Russian Federation. The program for psychiatric diagnostics PsyScan: No. 2022666930, dated 21.12.2022
8. Petraevsky, V.A., Kravets, A.G., Eremenko, I.M.: Certificate of state registration of computer program No. 2024618252 dated April 10, 2024. Russian Federation. Server subsystem of the program for psychiatric diagnostics PsyScan; VolGTU (2024)
9. Petraevsky, V.A.: Method of intellectual analysis of text information for psychiatric diagnostics. Bulletin of Astrakhan State Technical University. Ser.: Manag. Comput. Eng. Inform. **2**, 95–104 (2024). https://doi.org/10.24143/2072-9502-2024-2-95-104. (in Russian)
10. Afanasyeva, I.V., Pimanov, V.A:. Confidentiality as a principle of activity of a practicing psychologist: the current state of legal regulation and prospects [electronic resource]. Psychol Law **13**(2), 143–152 (2023). https://doi.org/10.17759/psylaw.2023130211 (in Russian).
11. Microsoft. C# 9.0. Microsoft Docs (2021). https://docs.microsoft.com/en-us/dotnet/csharp/whats-new/csharp-9
12. Microsoft..NET 6. Microsoft Docs (2021). https://docs.microsoft.com/en-us/dotnet/core/whats-new/dotnet-6
13. Entity Framework Documentation. Microsoft Docs (2021). https://docs.microsoft.com/en-us/ef/
14. Jones, M.B., Bradley, J., Sakimura, N.: JSON Web Token (JWT). IETF (2015). https://tools.ietf.org/html/rfc7519
15. Amazon Web Services. S3. AWS Documentation (2021). https://docs.aws.amazon.com/s3/
16. Amazon Web Services.: EC2. AWS Documentation (2021). https://docs.aws.amazon.com/ec2/
17. Fowler, M.: Patterns of Enterprise Application Architecture. Addison-Wesley Professional (2002)
18. Fielding, R.T.: Architectural styles and the design of network-based software architectures. University of California, Irvine (2000). https://www.ics.uci.edu/~fielding/pubs/dissertation/fielding_dissertation.pdf
19. Lin, J., Tegmark, M.: Why does deep and cheap learning work so well? J. Stat. Phys. **168**, 1223–1247 (2017). https://doi.org/10.1007/s10955-017-1836-5
20. Kotsiantis, S.B., Zaharakis, I., Pintelas, P.: Supervised machine learning: a review of classification techniques. Emerg. Artif. Intell. Appl. Comput. Eng. **31**, 249–268 (2007). https://doi.org/10.1007/s10462-007-9052-3
21. Goodfellow, I., Bengio, Y., Courville, A.: Deep Learning. MIT Press (2016)
22. Zhou, J., et al.: Graph Neural Networks: A Review of Methods and Applications. AI Open (2020)

# Research of Neural Network Models of Sign Language Recognition in the Presence of Negative Factors

**V. A. Gordeev and T. I. Buldakova**

**Abstract** The chapter considers the problem of automatic recognition of sign images for computer vision systems. It is shown that the main purpose of such systems is to extract from images information useful for further use in various applications. In particular, this chapter discusses issues related to sign language recognition and identifies potential negative factors affecting the accuracy of sign recognition. It is noted that for computer vision, hands are a difficult object to perceive during sign language communication. The accuracy and efficiency of sign language recognition algorithms are also affected by low image quality, variety of poses and orientations, non-uniform backgrounds and others. The preparation of initial data, creation of training and test dataset is described. For this purpose, a custom data set is created, and integration and preparation of external data is performed. An example of program code and generated own images is presented. A research of popular neural network methods for recognizing sign language in the presence of negative factors is performed and an evaluation of their effectiveness in the presence of negative factors is obtained. It is noted that knowledge of negative factors and their effect on the recognition quality will help to improve the efficiency of the recognition system. The recommendations for improving the quality of sign recognition are given.

**Keywords** Signs · Images · Recognition methods · Negative factors · Neural networks

V. A. Gordeev (✉) · T. I. Buldakova
Bauman Moscow State Technical University, Moscow, Russia
e-mail: gordeevva@student.bmstu.ru

T. I. Buldakova
e-mail: buldakova@bmstu.ru



A. G. Kravets and A. A. Bolshakov (eds.), *Human-Centric Cyber-Society*, Studies in Systems, Decision and Control 639, https://doi.org/10.1007/978-3-032-06787-6_16

## 1 Introduction

There are currently several hundred million people in the world with hearing loss and this number is growing annually. In particular, in Russia there are more than 13 million people with hearing impairment, including more than 1 million children. To communicate, these people use special sign languages which differ from country to country and from dialect to dialect. These differences make it difficult for this social group to use voice-activated digital assistants (e.g. on smartphones, computers and other devices). Therefore, the development of computer systems that automatically recognize sign language from video clips is relevant [1–4].

The computer vision system can be integrated into devices that support sign language recognition, such as cameras, smartphones or laptops. When a user makes a sign in front of the camera, the system will process the video stream in real time and recognize the signs converting them into text or speech.

That systems capable of effectively recognizing and interpreting signs are noted to have a wide range of applications not only in the field of technology to support people with hearing and speech impairments but also for education or computer games. For example, they can be integrated into educational applications to help with learning sign language and used in virtual and augmented reality to create game scenarios.

In the business sphere sign language recognition technologies can serve as a tool to improve communication in professional environments as well as to increase efficiency in areas where standard forms of communication may be limited. Various methods of sign language recognition have been proposed to realize such possibilities [5–9].

Russian sign language is a full-fledged language that uses signs created by hand movements combined with mimics and body posture. The elements of sign language are hand shape; palm orientation; movement; place of articulation and facial expression. A system for recognizing this language can be useful for improving the quality of life of people with disabilities. It can help them communicate with others more comfortably as well as improve accessibility to medical care and education.

The aim of the work is to analyze the effectiveness of neural network methods for sign recognition in the presence of negative factors affecting the correctness of recognition.

## 2 Primary Negative Factors Affecting Sign Recognition

Computer vision is a rapidly growing area of scientific and applied research that helps computers analyze and understand visual information such as images and video. Computer vision systems mimic human vision but work much faster, using complex algorithms, large amounts of data and cameras.

Computer vision allows systems to interpret visual information and make decisions based on it. It uses advanced techniques to process and analyze images and video, allowing computers to identify objects and react accordingly. At the same time, new technologies based on machine learning enable the creation of robust solutions for complex vision recognition tasks [10–12].

In general, the main steps of computer vision systems are:

1. Collection of visual data, including data collection and pre-processing (to ensure data homogeneity and improve image quality for further analysis);
2. Image processing and analysis (feature extraction and pattern recognition);
3. Machine learning which enhances the ability of systems to interpret and interact with visual data.

Thus, the main goal of computer vision systems is to extract from images information about objects in the external world useful for further use within an application. In particular, this chapter discusses issues related to sign language recognition. Therefore, it is necessary to consider possible negative factors affecting sign recognition.

For computer vision hands are a difficult object to perceive when communicating in sign language. Hands are actively moving and can overlap each other or change shape from open palm to fist, fingers can be crossed. There are no active points on the hands such as the eyes and mouth on the face, which interferes with the natural perception of movements. In order for the system to see a moving palm it must be placed at a given distance from the camera. This allows the algorithm to filter out objects from the image that do not need to be analyzed.

The presence of negative factors can have a significant impact on the accuracy and efficiency of sign language recognition algorithms. These include poor image quality, variety of poses and orientations, non-uniform backgrounds and others. For example, images with insufficient resolution, compression artifacts or poor lighting conditions can make it difficult to extract key features of sign. The presence of complex and heterogeneous backgrounds in images can lead to loss of context and make it difficult for algorithms to extract hands and signs.

Most modern methods of pattern recognition, which in this case are signs, use intelligent technologies [13–16]. Therefore, we are going to perform a study of popular neural network methods for recognizing sign language in the presence of negative factors. We consider the SSD (Single Shot Multibox Detector) method [17], the use of a convolutional neural network (CNN) with an embedded data pipeline PyTorch [18], pre-trained neural network models VGG16 and ResNet-50 [19, 20].

## 3 Creating a Training and Test Dataset

The most responsible step is visual data collection because the quality of the collected data largely determines the efficiency of recognition.

In this case, a program code in the Python programming language using the OpenCV library has been developed to generate a set of images which is presented

```
import cv2
import os
import time
import uuid

IMAGES_PATH = 'Tensorflow/workspace/images/collectedimages'

labels = ['А', 'Б', 'В', 'Г', 'Д', 'Е', 'Ж', 'З', 'И', 'К', 'Л', 'М', 'Н',
'О', 'П', 'Р', 'С', 'Т', 'У', 'Ф', 'Х', 'Ц', 'Ч', 'Ш', 'Щ', 'Э', 'Ю', 'Я']
number_imgs = 30

for label in labels:
    folder_path = os.path.join(IMAGES_PATH, label)
    if not os.path.exists(folder_path):
        try:
            os.makedirs(folder_path)
        except OSError as e:
            print('Error creating directory')
    else:
        print('Directory exists')

    cap = cv2.VideoCapture(0)
    print('Collecting images for {}'.format(label))
    time.sleep(5)
    for imgnum in range(number_imgs):
        ret, frame = cap.read()
        imagename = os.path.join(IMAGES_PATH, label,
label+'.'+'{}.jpg'.format(str(uuid.uuid1())))
        cv2.imwrite(imagename, frame)
        cv2.imshow('frame', frame)
        time.sleep(2)

        if cv2.waitKey(1) & 0xFF == ord('q'):
            break
    cap.release()
```

**Fig. 1** Code for collecting personal images

in Fig. 1. The program code allows collecting images for each label representing sign language.

Each image is saved with a unique file identifier created from the uuid identification timestamp. The code allows to collect a sufficient number of good quality personal images for further training and testing of the model. Negative images are also obtained to evaluate the quality of their recognition (Fig. 2).

The external data set is available as two CSV files, «sign_mnist_train.csv» and «sign_mnist_test.csv». One of the files is for training and the other is for testing. Each line in the CSV file is a training sample with the 0th index having labels from 0 to 28 and the rest of the line contains the values of 4096 pixels of a 64 × 64 image. The value of each pixel is in the range [0, 255].

An example of these files is illustrated in Fig. 3.

These datasets were then augmented with their own set of edited images.

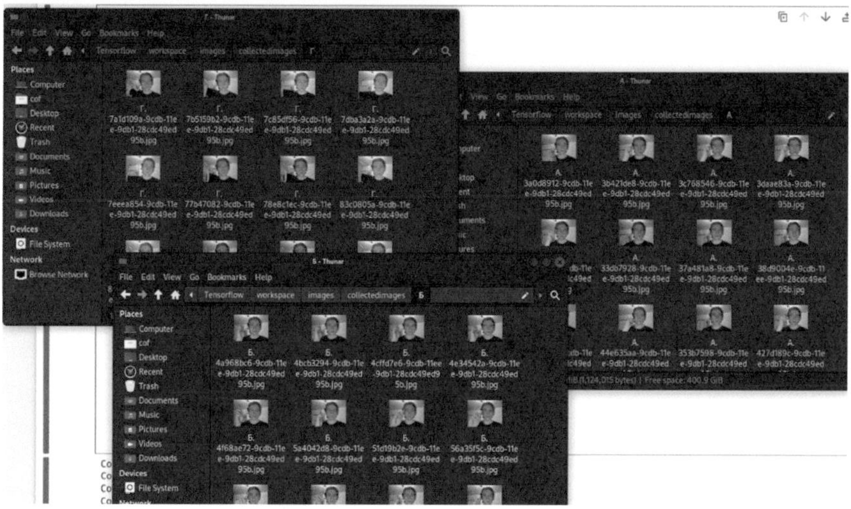


**Fig. 2** Example of received images

## 4 Sign Recognition Methods Under Investigation

Popular neural network models were used as the ones under investigation (for sign language recognition).

Object Detection is a field of computer vision and machine learning that focuses on the development of algorithms and models for automatic detection and localization of objects in images or video streams. In general, a recognition algorithm includes object localization and object classification.

SSD method is a technique for real-time object detection with high speed [17]. This method is popular for gesture or sign detection tasks where high speed real-time video stream processing is required.

Once the model is set up and configured, a command is generated to start the training process. In our case, 20,000 training steps were used. TensorFlow script was used to train the model. After executing this command, the model will start training using the pre-trained weights and data labeled with LabelImg.

The commands for training the model are shown in Fig. 4.

In order to evaluate the effectiveness of the model of sign language recognition testing on a test set of images with different negative factors was carried out. The test results are presented in Table 1.

The table shows that the factors that most strongly interfered with sign recognition were insufficient illumination, low photo resolution and palm rotation along the rotation axis.

As the second method a convolutional neural network CNN (Convolutional Neural Network) using the PyTorch framework was investigated [18]. The neural network contains three convolutional layers followed by three fully connected layers. The

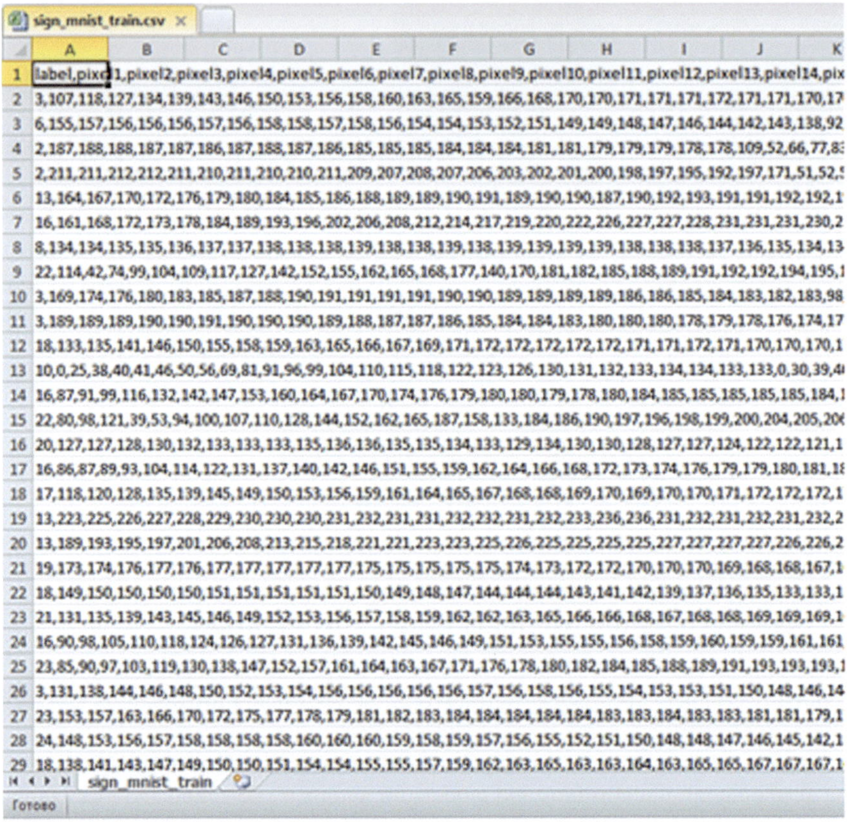


**Fig. 3** Line cuts of the file "sign_mnist_train.csv"

```
print("""python {}/research/object_detection/model_main_tf2.py --model_dir={}/{}
--pipeline_config_path={}/{}/pipeline.config                                    --
num_train_steps=20000""".format(APIMODEL_PATH,
MODEL_PATH,CUSTOM_MODEL_NAME,MODEL_PATH,CUSTOM_MODEL_NAME))

python Tensorflow/models/research/object_detection/model_main_tf2.py --
model_dir=Tensorflow/workspace/models/my_ssd_mobnet --
pipeline_config_path=Tensorflow/workspace/models/my_ssd_mobnet/pipeline.config -
-num_train_steps=20000
```

**Fig. 4** Program code for model training

input of the CNN is an image and the output specifies its class (in our case it is a symbol corresponding to a sign). First, the PyTorch Dataset interface is set up which allows the PyTorch embedded data pipeline to be loaded and used. The dataset is initialized using the SignLanguageMNIST class. Then it is wrapped in a DataLoader. This allows the dataset to be iterated for use. As a result, data preprocessing included data incrementation to improve the robustness of the model (Fig. 5).

**Table 1** Testing the object detection model with SSDs

| Negative factor | Recognition quality (%) |
|---|---|
| No | 92 |
| Insufficient lighting | 83 |
| Excessive lighting | 92 |
| Low resolution | 78 |
| Palm orientation pointing more towards the camera | 90 |
| Palm rotated on the axis of rotation | 84 |
| Background similar to hand color | 87 |

```
def get_train_test_loaders(batch_size=32):
    trainset = SignLanguageMNIST('data/sign_mnist_train.csv')
    trainloader = torch.utils.data.DataLoader(trainset, batch_size=batch_size,
shuffle=True)

    testset = SignLanguageMNIST('data/sign_mnist_test.csv')
    testloader = torch.utils.data.DataLoader(testset, batch_size=batch_size,
shuffle=False)
    return trainloader, testloader
```

**Fig. 5** Initialization of the data set

Data incrementation specifically included randomly scaling down the scale of the image by changing values and locations via RandomResizedCrop, so that the scaling down would not affect the final sign class.

The testing was performed on a test set of images with different negative factors. The test results are presented in Table 2.

It should be noted that when using even a simple CNN neural network model an accuracy of 88% is achieved which is a good result.

Low resolution, palm rotation along the rotation axis and insufficient illumination were also influential factors in reducing the recognition quality.

Popular deep learning neural network models VGG16 and ResNet-50 [19, 20] were used as the third tested method.

**Table 2** Testing the convolutional neural network model

| Negative factor | Recognition quality (%) |
|---|---|
| No | 88 |
| Insufficient lighting | 76 |
| Excessive lighting | 85 |
| Low resolution | 77 |
| Palm orientation pointing more towards the camera | 82 |
| Palm rotated on the axis of rotation | 72 |
| Background similar to hand color | 80 |

VGG networks are known for their depth and use of small convolutional filters while ResNet is known for its residual connections that allow to create deeper networks without problems with vanishing gradients.

What makes VGG16 unique is that it uses architecture with small convolution filters of $3 \times 3$ with a step size of 1 and a MaxPooling filter layer of $2 \times 2$ with a step size of 2. The number 16 in VGG16 means that it has 16 convolution layers. The ResNet-50 structure includes 50 layers and uses residue-connected blocks consisting of convolution layers, batch-normalization and activation functions.

This architecture demonstrates high performance in image classification tasks and has become popular in the field of computer vision. The models were tested on a test set of images with different negative factors (Table 3).

The critical factors that reduced the quality of recognition were background similar to the color of the hand and changing the standard orientation of the palm.

Thus, as a result of the research, the main factors that negatively affect the recognition of sign language were identified and evaluated.

The following factors had the greatest impact: palm rotated along the axis of rotation; low image resolution and insufficient lighting. Knowledge of these factors and their contribution to the quality of recognition will help improve the performance of recognition systems.

**Table 3** Testing of VGG16 and ResNet-50 models

| Network | Negative factor | Recognition quality (%) |
|---|---|---|
| VGG16 | No | 94 |
| | Insufficient lighting | 87 |
| | Excessive lighting | 92 |
| | Low resolution | 92 |
| | Palm orientation pointing more towards the camera | 85 |
| | Palm rotated on the axis of rotation | 82 |
| | Background similar to hand color | 83 |
| ResNet-50 | No | 95 |
| | Insufficient lighting | 90 |
| | Excessive lighting | 93 |
| | Low resolution | 92 |
| | Palm orientation pointing more towards the camera | 88 |
| | Palm rotated on the axis of rotation | 82 |
| | Background similar to hand color | 81 |

**Table 4** Causes of poor recognition quality

| No | Problem | Recommendation |
|---|---|---|
| 1 | Some letters or gestures may be visually similar to each other, making it difficult to recognize them accurately | Additional learning of the model on a larger variety of data, including cases where gestures are close in appearance, can help improve recognition |
| 2 | The model may not have enough training data to fully cover a variety of scenarios | Supplementing the training dataset with new signs and situations can improve the generalization of the model |
| 3 | If the model is too complex or too simple, it can affect the model's ability to learn and recognize new data effectively | Selecting the optimal model architecture and tuning the training parameters can help achieve a balance between performance and accuracy |
| 4 | Changes in lighting conditions, background or foreshortening can affect recognition quality | Ensuring stable shooting conditions or improving the model's stability to changes in the environment can reduce the impact of this factor |

## 5 Consideration of the Influence of Negative Factors in Recognition Systems

During testing the main aspects that can be optimized to improve the performance and accuracy of the sign language recognition system are also identified. Recommendations for improving the recognition quality are given in Table 4.

If necessary, after corrections have been made, testing should be repeated to evaluate improvements and later on develop the sign language recognition system.

## 6 Conclusion

In the process of research, the main negative factors affecting the recognition of sign language are identified. For this purpose, we created our own data set and performed the integration and preparation of external data.

The obtained results will help to improve the efficiency and accuracy of sign recognition in different conditions as well as to form requirements for the use of sign language recognition system.

## References

1. Valdivieso, L., Vásconez, J.P., Barona, L., Benalcázar, M.E.: Recognition of hand gestures based on EMG signals with deep and double-deep Q-networks. Sensors **23**(8), 3905 (2023). https://doi.org/10.3390/s23083905

2. Ma'asum, F.F.M., Sulaiman, S., Saparon, A.: An overview of hand gestures recognition system techniques. In: IOP Conference Series: Materials Science and Engineering, vol. 99, p. 012012. IOP Publishing, Bristol (2015). https://doi.org/10.1088/1757-899X/99/1/012012
3. Elakkiya, R., Grif, M.G., Prikhodko, A.L., Bakaev, M.A.: Recognition of Russian and Indian sign languages used by the deaf people. Sci. Bull. Novosib. State Tech. Univ. **2–3**(79), 57–76 (2020). https://doi.org/10.17212/1814-1196-2020-2-3-57-76.
4. Agrawal, S.C., Jalal, A.S., Tripathi, R.K.: A survey on manual and non-manual sign language recognition for isolated and continuous sign. Int. J. Appl. Pattern Recognit. **3**(2), 99–134 (2016). https://doi.org/10.1504/ijapr.2016.079048
5. Yasen, M., Jusoh, S.S.: A systematic review on hand gesture recognition techniques, challenges and applications. PeerJ Comput. Sci. **5**, 218 (2019). https://doi.org/10.7717/peerj-cs.218
6. Cheok, M.J., Omar, Z., Jaward, M.H.: A review of hand gesture and sign language recognition techniques. Int. J. Mach. Learn. Cybern. **10**, 131–153 (2019). https://doi.org/10.1007/s13042-017-0705-5
7. Anju, S.R., Surendran, S.: A study on different hand gesture recognition techniques. Int. J. Eng. Res. Technol. (IJERT) **3**(4), 2275–2278 (2014)
8. Shovkoplias, G.F., Strokov, D.A., Kasantsev, D.V., et al.: Improving sign language processing via few-shot machine learning. Sci. Tech. J. Inf. Technol. Mech. Opt. **22**(3), 559–566 (2022). https://doi.org/10.17586/2226-1494-2022-22-3-559-566
9. Escudeiro, P., Escudeiro, N., Reis, R., Maciel-Barbosa, F., Bidarra, J., Gouveia, B.: Automatic sign language translator model. Adv. Sci. Lett. **20**, 531–533 (2014). https://doi.org/10.1166/asl.2014.5344
10. Zhang, H.: A review of convolutional neural network development in computer vision. EAI Endorsed Trans. Internet Things **7**(28), 1–11 (2022). https://doi.org/10.4108/eetiot.v7i28.445
11. Doroshchenko, I.A.: Analysis of the experimental flow shadowgraph images by computer vision methods. Numer. Methods Program. **24**(2), 231–242 (2023). https://doi.org/10.26089/NumMet.v24r217
12. Basarab, M.A., Buldakova, T.I., Smolyaninova, K.A., Sokolov, M.N.: User identification based on the vein pattern in biometric immobilizer. CEUR Work. Proc. **2603**, 1–5 (2019)
13. Fadlilah, U., Prasetyo, R.A.R., Mahamad, A.K., et al.: Modelling of basic Indonesian sign language translator based on raspberry Pi technology. Sci. Tech. J. Inf. Technol. Mech. Opt. **22**(3), 574–584 (2022). https://doi.org/10.17586/2226-1494-2022-22-3-574-584
14. Buldakova, T.I., Suyatinov, S.I.: Biological principles of integration information at big data processing. In: 2019 International Russian automation conference (RusAutoCon), p. 8867710. Sochi, Russia (2019). https://doi.org/10.1109/RUSAUTOCON.2019.8867710
15. Suyatinov, S.I., Buldakova, T.I., Vishnevskaya, Y.A.: Synergistic model of situational awareness of the human operator. In: Kravets, A.G., Bolshakov, A.A., Shcherbakov, M. (eds.) Society 5.0: Human-Centered Society Challenges and Solutions. Studies in Systems, Decision and Control, vol. 416, pp. 331–340. Springer, Cham (2022). https://doi.org/10.1007/978-3-030-95112-2_27
16. Camgoz, N.C., Hadfield, S., Koller, O., Ney, H., Bowden, R.: Neural sign language translation. In: IEEE Conference on Computer Vision and Pattern Recognition (CVPR) (2018). https://doi.org/10.1109/CVPR.2018.00812
17. Liu, W., et al.: SSD: single shot MultiBox detector. In: Leibe, B., Matas, J., Sebe, N., Welling, M. (eds.) Computer Vision—ECCV 2016. Lecture Notes in Computer Science, vol. 9905, pp. 21–37. Springer, Cham (2016). https://doi.org/10.1007/978-3-319-46448-0_2
18. Sufian, A., et al.: Fundamental concepts of convolutional neural network. In: Balas, V., Kumar, R., Srivastava, R. (eds.) Recent Trends and Advances in Artificial Intelligence and Internet of Things. Intelligent Systems Reference Library, vol. 172. Springer, Cham (2020). https://doi.org/10.1007/978-3-030-32644-9_36
19. Tammina, S.: Transfer learning using VGG-16 with deep convolutional neural network for classifying images. Int. J. Sci. Res. Publ. (IJSRP) **9**(10) (2019). https://doi.org/10.29322/IJSRP.9.10.2019.p9420

20. Theckedath, D., Sedamkar, R.R.: Detecting affect states using VGG16, ResNet50 and SE-ResNet50 networks. SN Comput. Sci. **1**(79) (2020). https://doi.org/10.1007/s42979-020-0114-9

# Computer Games and eXtended Reality

# New Level-Generation Method for a Single-Player 3D Game

**Alla G. Kravets, Anna V. Matokhina, Stanislav E. Dragunov, and N. A. Salnikova**

**Abstract** Creating an endless level for single-player 3D games is a difficult task. The work examines the problems of level generation and proposes a method aimed at speeding up and reducing the cost of the process of developing game levels, revealing the approaches and technologies used to solve the problems of level generation, implementing algorithms for generating rooms and paths between them. After analyzing existing level generation methods, the BSP-tree generation method was selected, which can create unique levels based on input parameters, allowing to reduce the development time for game levels. To achieve the effect of an endless level, it is proposed to use the "scrolling" technique. This creates the illusion of infinity and allows the player to continue exploring new areas of the level. As a result, a level generation module for a single-player 3D game was designed, consisting of submodules (functions), each of which is responsible for generating a separate part of the level, has its own set of parameters that determine its behavior and output result, level generation can take place in real time or with using a pre-generated file. The module is flexible, allows you to add new parameters for level generation and has scalable code. When developing the architecture of the software solution, we chose the approach of dividing the code into separate components, each of which is responsible for a certain part of the level generation process, which made it possible to easily configure and change generation parameters, as well as modify the BSP algorithm to meet the specific requirements of the game.

**Keywords** Artificial intelligence · Machine learning · Video games · Single-player video games · Video game level generation algorithms · Procedural

---

A. G. Kravets · A. V. Matokhina (✉) · S. E. Dragunov
Volgograd State Technical University, Volgograd, Russian Federation
e-mail: matokhina.a.v@gmail.com

A. G. Kravets
Dubna State University, Dubna, Russian Federation

N. A. Salnikova
Volgograd Institute of Management—Branch of the Russian Presidential Academy of National Economy and Public Administration, Volgograd, Russian Federation



A. G. Kravets and A. A. Bolshakov (eds.), *Human-Centric Cyber-Society*, Studies in Systems, Decision and Control 639, https://doi.org/10.1007/978-3-032-06787-6_17

generation · BSP-tree · Tunneling algorithm · Cellular automata 3D models · Simulation · Virtual reality

## 1 Introduction

Computer game development is a rapidly growing and exciting area of the programming and entertainment industry. Mobile games have made huge leaps in quality in recent years. Procedural level generation is a key feature of the rogue-like and rogue-lite genre of computer games. Generating 3D levels is a more complex and less explored task compared to 2D generation.

In procedurally generated games, the content (environment, quests, characters, etc.) is not prepared in advance but is created using algorithms. This makes the gameplay less predictable and increases replayability [1, 2].

Procedurally generated maps are a core feature of roguelikes. For a genre that's almost synonymous with randomness (and for good reason), randomized maps have become the easiest way to showcase its core element because they influence so many aspects of gameplay, from exploration strategy and tactical positioning to item and enemy placement [3, 4].

Therefore, randomized maps provide endless replayability, offering a different challenge each time. In addition, the player's progress depends on his own skill and not on trial and error. The design of each new card is 100% unknown, which adds tension to the process of learning it.

In this regard, the goal was to analyze procedural level generation methods for integration into a project in roguelike games, analyze existing solutions used in games, and select one of the existing level generation methods for implementing a single-player 3D game.

Creating an endless level may seem like a daunting task, but with some helpful tips and techniques, it becomes much easier. The first step to creating an infinite level is to create an empty object that will serve as the base for the level. Various environmental elements, such as blocks, plants, or obstacles, can then be added to create interesting play situations [5, 6].

To achieve the effect of an infinite level, you need to use the "scrolling" technique. This means that when the player moves in one direction, objects in the level will move in the opposite direction. This creates the illusion of infinity and allows the player to continue exploring new areas of the level.

## 2 Analysis of the Level Generation Process and Existing Methods

### *2.1 Analysis of Procedural Level Generation in 3D Games*

Procedural generation (PCG) is the automatic creation of game content using algorithms. In other words, PCG is software that can create game content on its own, or in collaboration with players or game designers. Content means the creation of game levels, a map of the game world, game rules, textures, plots, items, quests, music, weapons, vehicles, characters, etc. [4, 7].

Level generation modules for 3D games can be useful for game developers who need to generate a large number of levels for their games. This can be especially useful for open-ended games, where players can travel through randomly generated worlds.

A practical application of level generation modules in a 3D game could be to reduce the time it would take to create a large number of levels manually. Instead of creating each level manually, developers can use level generation modules to create levels automatically. This can reduce development time and increase the number of levels available in the game.

In addition, level-generation modules can make the gameplay more interesting and varied. Since each level will be generated randomly, each time the player completes the game, he will experience new challenges and new sensations. This can make the game more exciting and extend its life [8, 9]. Methods for generating levels in games have been around for a long time and continue to evolve. In the past, game developers used simple level-generation methods such as random object placement, tilesets, etc. However, with the development of technologies for generating random numbers and machine learning, methods for generating levels in games have become more complex and interesting [10, 11].

For example, some games use procedural generation algorithms that allow you to create levels based on certain parameters and rules [12, 13]. Such algorithms can create levels that will meet different conditions, for example, levels that will change depending on the player's actions or levels that will be generated based on analysis of past levels and player actions.

Additionally, with the advancement of technology and machine learning, some game developers have begun to use generation algorithms that can analyze players and create unique levels that will suit each player's abilities and preferences.

Thus, we can say that level-generation methods in games continue to evolve and become more complex and interesting.

## 2.2 *Review of Existing Level Generation Methods*

Methods for generating levels in games have been around for a long time and they continue to evolve. In the past, game developers used simple level-generation methods such as random object placement, tilesets, etc. However, with the development of technologies for generating random numbers and machine learning, methods for generating levels in games have become more complex and interesting.

### 2.2.1 BSP-Tree Method

Let's analyze existing methods for generating levels in 3D games. Let's consider the BSP-tree method for generating levels in games. It was developed back in the early 90s and is still one of the most popular ways to create game levels. BSP-tree allows you to generate levels with a variety of different textures, shapes, and sizes. This method ensures that every part of the level is visible, which improves game performance and allows for faster level loading speed. BSP-tree can be used to optimize the path selection process, which improves the gameplay quality of the game [14, 15]. However, processing large levels can take a long time and consume a lot of computer resources. A BSP tree can generate complex relationships between connected sections of a layer. The difficulty of the levels can be high compared to other methods.

The BSP-tree level generation method is widely used in various games and simulations. For example, the BSP-tree method was used in the classic game Doom. Its use allowed us to create levels with a variety of different textures and shapes, which improved the gameplay and attracted the attention of many players. Quake, another popular game, also used the BSP-tree method to generate levels. As a result of this, the levels have become more dynamic and interesting for players. BSP-tree can also be used to create terrains and missions in flight simulators. For example, Microsoft Flight Simulator and X-Plane use this level-generation method to create unique and interesting missions.

Overall, BSP-tree can be a good choice for many games and leaves a lot of room for scaling and innovation.

### 2.2.2 Tunneling Algorithm

The level generation method using the tunneling algorithm is one of the ways to create random levels in games. It works by creating a maze of tunnels that connect different rooms and areas on a level. The tunneling algorithm can create levels with a wide variety of shapes and sizes, which can make the game more interesting for players [16, 17]. This method can be easily extended by adding more sophisticated algorithms for generating rooms and zones. The algorithm can be configured to create specified parameters, such as the number of rooms, level difficulty, passability, etc.

The disadvantage of this method is the possible creation of inefficient routes in the level, which can reduce the quality of gameplay. Some players may prefer more predictable levels, which may make the game less enjoyable. The generated level can be expensive in terms of performance if the number of objects and tunnel details are high.

The level generation method using a tunneling algorithm is also used in various games. Minecraft uses a tunneling algorithm to generate most of the levels in the game, allowing for the creation of large and varied worlds that can be explored by players. Game Dwarf Fortress. The tunneling algorithm is also used in the game Dwarf Fortress to create maze and cave levels that can contain a wide variety of treasures and dangers for players. Spelunky game uses a tunneling method to generate levels that can be recreated every time the player starts a new game. This makes the game more interesting and varied for players.

The tunneling algorithm should be chosen for developing random levels for games, it has its advantages and disadvantages that developers should consider when choosing this level generation method.

### 2.2.3 Cellular Automata

The method of level generation using cellular automata is one of the most popular ways to create random levels in games. It works by creating artificial cellular structures that can be used to generate levels. Cellular automata can create levels with a variety of shapes and sizes, which increases the interest of players. This method can be easily extended using various algorithms that can control the shape, color, and arrangement of cells, allowing the level-creation process to be significantly accelerated [18, 19]. However, creating layouts can be resource-intensive since generating levels based on cellular automata usually requires a large amount of computing resources. This method may have limited control over the generated levels in the textures, players may be able to determine patterns in the level generation, which may reduce interest in the game.

The method of level generation based on cellular automata is widely used in various games and applications. No Man's Sky uses a cellular-automata level-generation method to create random galaxies and planets in the game. This allows for the creation of many unique and diverse worlds for players to explore. Diablo III uses cellular automata to generate random mazes and dungeons for players to explore. This makes the game more interesting and varied for players. The RimWorld game also uses a cellular automata-based level generation method to create random worlds for players. This allows for a variety of unique and interesting situations and challenges for players to solve. The level generation method using cellular automata can be used to create random levels in games.

**Table 1** Comparison of level generation methods

| Name of method | Advantages | Disadvantages |
|---|---|---|
| BSP trees | Generation of levels with different textures and shapes, high game performance, increased gameplay quality, algorithm scalability, customizability | Increasing levels requires a lot of resources, creates a complex relationship between connected sections of the level, and high complexity of the level |
| Tunneling algorithms | Variety of shapes and sizes, easy to expand, highly customizable | Ineffective connections, high unpredictability, resource consumption |
| Cellular automata | Variety of shapes and sizes, easy to expand, high speed | Resource consumption, limited control, patterns in the level structure |

### 2.2.4 Results of Comparison of Existing Methods

The results of a comparison of the considered methods for generating 3D game levels are shown in Table 1.

After conducting a comparative analysis of the considered level generation methods, the BSP-tree method was chosen to design a level generation module for a single-player 3D game since it is characterized by high performance, improved gameplay quality, has algorithm scalability and flexible customization, and allows the generation of levels with different textures and shapes.

## 3 Method for Constructing BSP Trees

Binary space partitioning is a method of recursively partitioning Euclidean space into convex sets and hyperplanes. As a result, objects are represented in the form of a data structure called a BSP tree [20, 21].

The input to the algorithm is the overall size of the level and the minimum size of the expected rooms. The entire area of the level is taken as one leaf of the BSP tree, which is randomly divided into two parts. These parts become leaves in the next level of the tree (Fig. 1). The algorithm is executed recursively for each new sheet until its size is below the specified minimum threshold.

The method of binary partitioning of space is simple and convenient to implement. It allows you to quickly divide an initially given figure into the required number of evenly distributed smaller figures, leaving no empty spaces between them and creating geometrically strict spaces.

The level obtained using such an algorithm is well suited where there is a need to depict an artificially created landscape. This can be an urban environment or a floor of a separate room (Fig. 2).

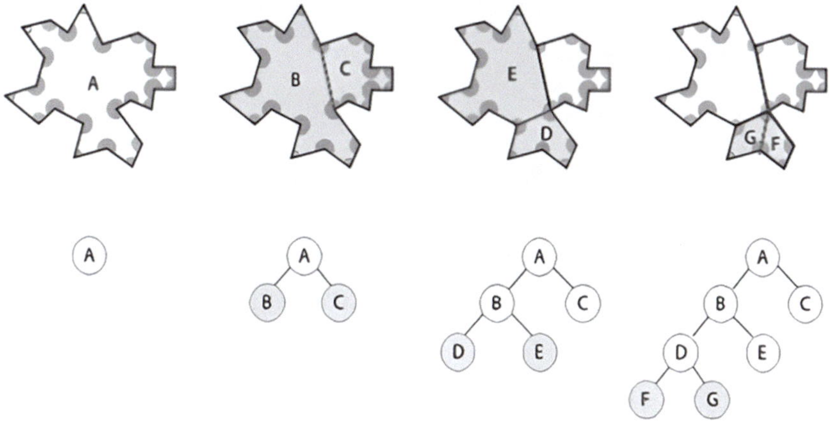


**Fig. 1** BSP algorithm execution diagram

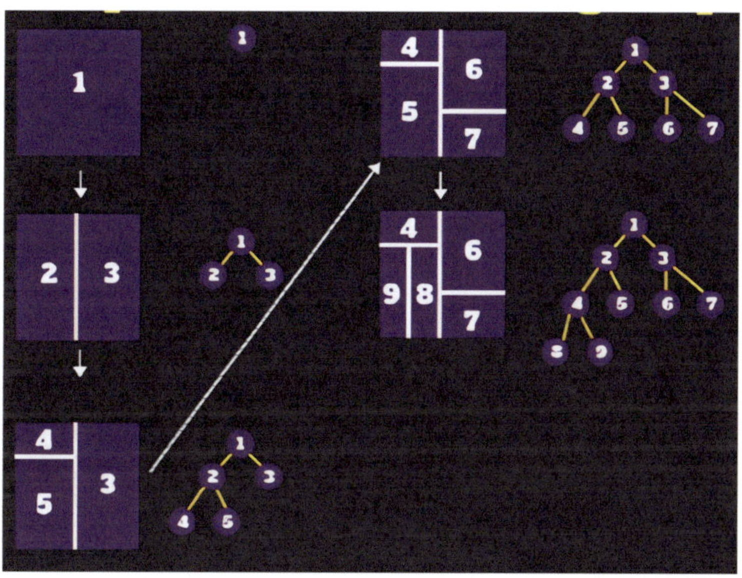


**Fig. 2** Space division scheme

## *3.1 Modification of the Algorithm for Constructing BSP Trees*

The basic algorithm does not involve the creation of corridors - transitions between rooms. This task can be performed manually, or the generation algorithm can be supplemented.

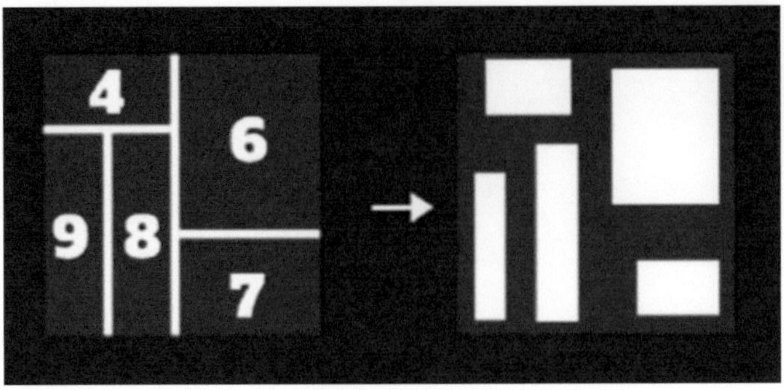


**Fig. 3** Scheme for creating rooms

To create passages between rooms, you need to find adjacent rooms that can be connected. To solve this problem, the leaves of the tree are represented in the form of a graph along which its vertices are traversed (Fig. 3).

Graph traversal is the process of visiting all the vertices of a graph or a specific part of a graph, which begins with some initial vertex and ends when all the vertices have been visited or a certain goal has been achieved.

To check, two vertices (of the same parent) are taken and their position relative to each other in space is checked. Based on determining their position relative to each other, a passage is added between them (Fig. 4).

The main stages of the new method's algorithm are as follows:

- first, you need to determine the space for partitioning:
  - set the initial rectangular space;
  - set the minimum room size;
- then divide the space:
  - check whether the space can be divided into two parts by a vertical or horizontal line;
  - if it is possible to divide, add new spaces to the tree;
  - check whether new spaces can be divided further;

**Fig. 4** Checking neighbors

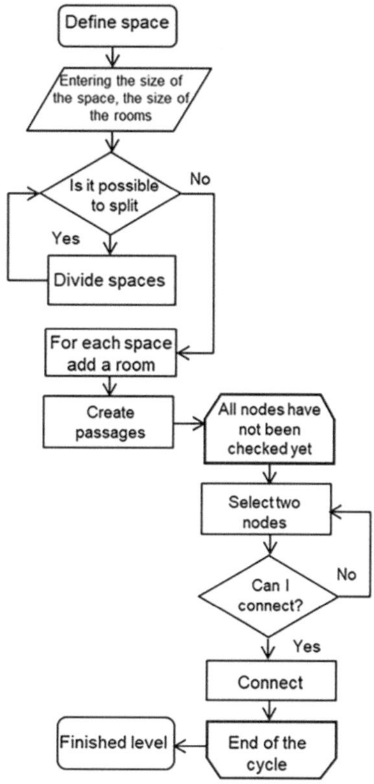


**Fig. 5** Block diagram of the algorithm

- if it is possible to divide, then add new spaces to the tree;

- at this stage, the rooms are created:
  - for each space, create a room by selecting random corner points within the space;
- then the creation of corridors is carried out;
  - starting from the youngest branches, draw corridors between the leaves of the same parent;
  - go up one layer and repeat the steps.

The flow diagram of the algorithm is shown in Fig. 5.

## 3.2 *Description of How the BinarySpacePartitioner Class Works*

Here is a description of how the BinarySpacePartitioner class works.

1. Create a root node rootNode.
2. The BinarySpacePartitioner constructor takes the width and length of the dungeon, dungeonWidth and dungeonLength, respectively. It creates a rootNode with coordinates (0, 0) and (dungeonWidth, dungeonLength).
3. The PrepareNodesCollection method takes the maximum number of iterations maxIterations, the minimum width and length of rooms roomWidthMin and roomLengthMin respectively. It creates a graph queue, adds a root node to the queue and a listToReturn list, and adds the root node to the list.
4. The method uses a while loop with the condition that the number of iterations is less than the maximum value and the queue is not empty. At each iteration, it fetches the first node from the graph queue, checks whether the current node can be split into two parts using the SplitTheSpace method, and adds new nodes to the listToReturn list and the graph queue.
5. The SplitTheSpace method takes the current node currentNode, a listToReturn, the minimum width and length of the rooms roomWidthMin and roomLengthMin, and a queue graph. It calls the GetLineDividingSpace method, which returns the space dividing line and divides the current node into two nodes node1 and node2. It then adds new nodes to the listToReturn list and graph queue using the AddNewNodeToCollections method.
6. The AddNewNodeToCollections method accepts a listToReturn, a graph queue, and a node. It adds the node to the listToReturn list and the graph queue.
7. The GetLineDividingSpace method takes the bottom left corner bottomLeftAreaCorner, the top right corner topRightAreaCorner, the minimum width and length of the rooms roomWidthMin and roomLengthMin. It determines the orientation of the space dividing line and returns a Line object containing the orientation and coordinates of the line.
8. The GetCoordinatesFororientation method takes the orientation of the line orientation, the bottom left corner bottomLeftAreaCorner, the top right corner topRightAreaCorner, the minimum width and length of the rooms roomWidthMin and roomLengthMin. It determines random coordinates for the space division line according to the orientation and returns them as a Vector2Int object.

## 4 Design of the Level-Generation Module Interface

When designing an interface for a level-generation module (LGM), it is also important to take into account the capabilities of the game engine and the platform on which the game will run. The interface must be adapted to different devices and screens and must be compatible with the technologies used in the Unity game engine.

It should also be taken into account that users have different levels of experience and skill, so the interface should be designed in such a way that it is clear and easy to use for both new and experienced users.

One of the key aspects of interface design for a LGM is usability.

The interface should contain level settings and the ability to create a new level, clear the scene, save the level, and load the level. An example of a ready-made interface integrated into Unity is shown in Fig. 6.

Key interface controls for the LGM include:

1. Settings panel: this is the main control element that allows you to configure level generation parameters. This panel may include controls such as sliders, radio

**Fig. 6** Settings panel

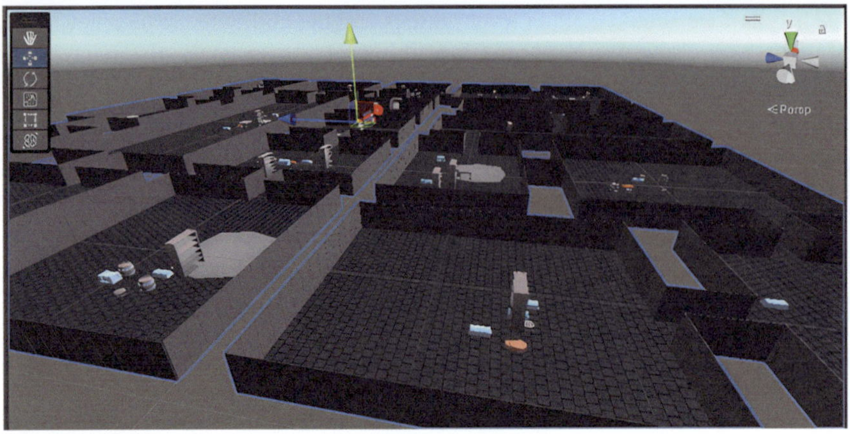

**Fig. 7** General view of the game level

buttons, drop-down lists, etc. that allow the user to adjust various parameters such as level size, number of rooms, difficulty, etc.
2. Generation button: This is a button that starts the level generation process with the current settings.
3. Scene Clear Button: This is a button that starts the process of clearing objects from the scene.
4. Save and Load Settings: These are controls that allow the player to save and load level generation settings. This is useful if the player wants to save settings for future games or share them with other players.
5. Level View: This is a control that allows the user to view the created level in a 3D view. This allows the user to evaluate the generation settings and make changes if necessary.

A well-designed interface can significantly improve the user experience and make the level-creation process more convenient and efficient.

Figure 7 shows a general view of the game level obtained as a result of the LGM.

## 5 Description of the Functionality of the LGM for a Single-Player 3D Game

The LGM for a single-player 3D game is designed to create random levels with different types of rooms, corridors, and objects according with the new method based on modified BSP algorithm.

The functionality of the LGM includes:

- creating a level with specified parameters, such as the size of the level, the number of rooms and corridors, the complexity of the labyrinth, the minimum and maximum size of the room, and the width of the corridor;
- placement of objects on the level, such as walls, doors, windows, furniture, etc.;
- checking the correct connection of rooms and corridors, the absence of intersections of walls and objects, the absence of closed paths in the maze (the ability to go from one point of the level to any other point);
- the ability to save the generated level to a file and load it for later use in the game;
- setting level generation parameters depending on the game requirements, such as play style.

The input data of the LGM must be the values for setting up the module, model, and texture from which the level will be assembled.

The output of the LGM should be the generated level. The level must be connected (the player character must reach any other place from any place). The level must stylistically match the game setting.

Fault tolerance testing consisted of reproducing the load on the level generation module. During the testing process, the functionality of the LGM and its fault tolerance were verified. The load was achieved by increasing the size of the map and the number of rooms. Incorrect use cases were also tested. During fault tolerance testing, it was verified that the system responds to non-standard situations without errors. Testing showed the correct operation of the LGM.

## 6 Conclusion

The developed level generation module for a single-player 3D game can be very useful for creating interesting and varied levels. Using the BSP-tree method allows you to effectively divide the level into different sectors, which can be filled with various objects and elements. This allows you to create levels with different characteristics and difficulties, which improves the gaming experience and makes the game more interesting for players. The BSP-tree algorithm can be easily integrated into the game engine, which simplifies the development process and allows you to quickly create new levels.

The BSP-tree method can be used to create levels with varying characteristics and difficulties, allowing for the creation of rich and engaging games. However, when using BSP-tree, potential problems must be considered, such as the need to optimize and balance the tree to avoid too many intersections and improve performance.

BSP-tree is a powerful level generation tool for 3D games that can greatly improve the quality of your gaming experience. However, its use requires careful optimization and balancing to achieve the best results.

The developed module allows you to generate new levels that differ in size, complexity and obstacles, while optimizing the process of creating levels, which significantly reduces the time spent on development. The level generation module is a promising direction in the field of game development. The module is capable of generating levels in real time and has the ability to add new parameters for level generation. The module is written using clear and easy to read code constructs, which makes the code easy to maintain and scalable. The module has a convenient and simple user interface, which provides the ability to create and configure new levels.

Testing the module's fault tolerance showed that the system handles non-standard situations without errors.

## References

1. Klimenko, R.G., Kravets, A.G., Strukova, I.V., Tatarova, N.I.: Analysis of people's fascinations for computer games. In: Information Innovative Technologies. International Scientific-Practical Conference, Moscow, pp. 48–57 (2022)
2. Kravets, A.G., Salnikova, N., Dmitrenko, K., Lempert, M.: Industrial cyber-physical systems: risks assessment and attacks modeling. In: Studies in Systems, Decision and Control, vol. 260, pp. 197–210 (2020)
3. Kravets, A.G., Salnikova, N.A., Mikhnev, I.P., Orudjev, N.Y., Poplavskaya, O.V.: Web portal for project management in electronics design software development. In: International Seminar on Electron Devices Design and Production (SED 2019), Proceedings, p. 8798472 (2019)
4. Matokhina, A.V., Shabalina, O.A., Kharlamova, N.V., Kulikov, E.A.: Game-based learning platform for integrating international IT-students into the Russian educational environment. In: Creativity in Intelligent Technologies and Data Science (CIT&DS 2017), Communications in Computer and Information Science, vol. 754, pp. 762–773. Springer (2017)
5. Shabalina, O.A., Vorobkalov, P.N., Kataev, A.V., Kravets, A.G.: Educational computer games development: methodology, techniques, implementation. In: Proceedings of the 2016 Conference on Information Technologies in Science, Management, Social Sphere and Medicine (ITSMSSM 2016), pp. 419–423 (2016)
6. Bolshakov, A.A., Klyuchikov, A.V.: Decision support system for selecting designs of autostereoscopic displays. In: Cyber-Physical Systems: Design and Application for Industry 4.0 (Studies in Systems, Decision and Control), vol. 342, pp.73–88 (2021)
7. Procedural generation in games. https://skillbox.ru/media/gamedev/protsedurnaya_generatsiya_v_igrakh/. Accessed 15 Mar 2024
8. Shcherbakov, M.V., Shcherbakova, N.L., Kamaev, V.A., Brebels, A., Gerget, O.M., Devyatykh, D.: Outlier detection and classification in sensor data streams for proactive decision support systems. J. Phys. Conf. Ser., 012143 (2017)
9. Chumak, A.A., Ukustov, S.S., Kravets, A.G.: Analysis of user profiles in social networks. In: Knowledge-Based Software Engineering (JCKBSE 2014), Communications in Computer and Information Science, vol. 466, pp. 70–76. Springer (2014)
10. Shcherbakov, M., Groumpos, P.P., Kravets, A.: A method and IR4I index indicating the readiness of business processes for data science solutions. In: Creativity in Intelligent Technologies and Data Science (CIT&DS 2017), Communications in Computer and Information Science, vol. 754, pp. 21–34. Springer (2017)
11. Sabbagh, A.A., Shcherbakov, M.V.: A hybrid clustering-based routing protocol for VANET using K-means and cuckoo search algorithm. In: Communications in Computer and Information Science, vol. 1552, pp. 48–61 (2022)
12. Matokhina, A.V., Kharlamova, N.V., Shabalina, O.A., Kulikov, E.A.: Using digital game-based technologies in a system of studying Russian as a foreign language in modern university. Open Educ. **21**(3), 39–47 (2017)

13. Fateenkov, D.V.: Procedural generation of game locations in Unity. Postulate **2**(76), 10–11 (2022)
14. Introduction to BSP-trees. https://gamedev.ru/code/articles/BSP/. Accessed 15 Mar 2024
15. Wang, T.: An object-oriented BSP-tree algorithm for hidden surface removal. Int. J. Image Graph. **2**(3), 395–411 (2002)
16. How the tunneling algorithm works to generate game levels. https://dtf.ru/gamedev/113303/. Accessed 15 Mar 2024
17. What is a VPN tunnel. https://selectel.ru/blog/vpn-tunnel/. Accessed 15 Mar 2024
18. Evolving cellular automata. https://habr.com/ru/articles/455958/. Accessed 15 Mar 2024
19. Matokhina, A.V., Kizim, A.V., Voronin, Y.F.: Automation of customer needs forecasting. World Appl. Sci. J. **24**(24), 126–132 (2013)
20. Using BSP-trees to create game maps. https://habr.com/ru/articles/332832/. Accessed 15 Mar 2024
21. DIY game engine #3: BSF-trees. https://xakep.ru/2001/12/24/14185/. Accessed 15 Mar 2024

# Scenarios Personification for Single-Player 3D Soulslike Games

**Alla G. Kravets, Anna V. Matokhina, N. A. Salnikova, and F. H. Khusenzoda**

**Abstract** The excessive complexity of games in the Action/RPG genre repels a large number of players, because not every gamer can master the complete pas-sage of a game of this genre. To solve this problem, it is necessary to have mechanisms that allow you to adjust the difficulty of the game for a specific player. This chapter analyzes existing methods for personifying game scenarios according to six criteria; it is concluded that the developers tried to work out the system of variability of game scenarios as best as possible, but do not disclose the exact methods for implementing personification. In this regard, the goal of this work is to develop a new video game with the ability to personalize scenarios by strengthening and weakening the player and opponents, as well as changing the environment. The chapter reveals the approaches and technologies used to solve the problem of developing a system for personalizing the gameplay, describes the stages of designing a new gaming application in the Soulslike style with the ability to personalize game scenarios, as well as the results of functional testing and game performance.

**Keywords** Video games · Gaming industry · Game design · Multiplayer battles · Scenario personification · Computer game architecture · Game engine · Scripts · Soulslike games · Performance · Functional testing · Programming · Game studies

A. G. Kravets · A. V. Matokhina (✉)
Volgograd State Technical University, Volgograd, Russian Federation
e-mail: matokhina.a.v@gmail.com

A. G. Kravets
e-mail: AllaGKravets@yandex.ru

A. G. Kravets
Dubna State University, Dubna, Russian Federation

N. A. Salnikova
Volgograd Institute of Management—Branch of the Russian Presidential Academy of National Economy and Public Administration, Volgograd, Russian Federation

F. H. Khusenzoda
Ministry of Internal Affairs of the Republic of Tajikistan for the Sughd Region, Khujand, Tajikistan



A. G. Kravets and A. A. Bolshakov (eds.), *Human-Centric Cyber-Society*, Studies in Systems, Decision and Control 639, https://doi.org/10.1007/978-3-032-06787-6_18

## 1 Introduction

Computer game development is a rapidly growing and exciting area of the programming and entertainment industry. There are many of news all over the world about how some game or studio has broken another record in the industry. Every year, video games are capturing an increasing share of the entertainment market and in this niche already they compete on an equal footing with movies [1, 2].

Video games are truly an art form that includes many elements, from graphic design and sound design to storylines and game mechanics. Video games not only influence the worldview of players but also raise important social issues that become the subject of discussion in society. They have a high experimental potential, and the ability to use unusual forms and themes to express ideas. Finally, video games have been criticized for promoting violence and lacking ethics, but have a strong scientific evidence base that they can develop players' technical, cognitive, and social skills.

The development of the gaming industry has contributed to the fact that many educational institutions have appeared on the market that offer evening courses in game development, in which students learn the basics of computer graphics, animation, game mechanics design, and many other disciplines that will be useful for artists, programmers, and game designers. A big plus of specialized courses is the systematic nature of the knowledge gained, a team of like-minded people, and the support of an experienced mentor [3].

The Soulslike genre is a conventionally distinguished subgenre of Action/RPG games. Souls-like computer games originate from the Souls series of games, such as Dark Souls, Demon Souls, Bloodborne, and Elden Ring, created by game designer Hidetaka Miyazaki and his studio FromSoftware.

The themes and mechanics of these games are directly carried over into many other similar games from third-party developers, hence the name, which can be deciphered as "similar to Dark Souls". Souls-like games are known for their unusually high difficulty levels, the presence of "bosses" (the boss is the strongest enemy in a level, usually the final enemy) at the end of each area, the loss of part of the player's progress upon death, and the emphasis on indirect storytelling. Such games are a bright representative of the dark fantasy genre with its inherent elements of horror and gothic.

The excessive complexity of such projects repels a large number of players because not every gamer can master the full passage of a game of this genre [4, 5]. In the vast majority of existing games, you can adjust the difficulty of the game to your liking. Depending on the parameters adopted, the passage of the game will vary from a walk with simple fights to a mortal battle that ends with the player's first mistake.

In the Soulslike genre, there is almost always no adjustment of the difficulty of the game for the player. In this case, many players will not be able to fully and with pleasure touch the huge cultural layer hidden in games of this genre [6, 7].

In this regard, the goal of this work is to develop a video game with the ability to personalize game scenarios. To achieve this goal, the following tasks were solved:

- approaches to the personification of game scenarios have been studied;

- an analysis of methods of personification in already released video games was carried out;
- Soulslike game "Dead Sands" was designed with the ability to personalize game scenarios;
- testing of the game has been completed.

## 2 Analysis of Methods for Personifying Game Scenarios

### *2.1 Scenario Personification*

Personalization of game scenarios is the process of a player creating personalized scenarios that take into account the player's preferences and playing style. Personalization is achieved by adapting the plot and gameplay of the game to the preferences of the player, for example, through building the plot by the player himself, which is called scenario personification, or gameplay by changing tactics or adjusting the difficulty of the game, which is called gameplay personification.

Scenario personification is a variation of the passage laid down at the level of the game's plot, where the player's actions will influence not so much the gameplay as the development of the game's plot. As a rule, subsequent events change either due to the choice of certain lines in dialogues, or due to certain actions of the character, for example, if the player stands still for too long when he needs to pursue the enemy, or if the player reaches the goal in an alternative way [8, 9]. Often, games with the greatest level of plot variability are games belonging to the RPG (CRPG) genre—Computer Role-Playing Game (Computer Role-Playing Game) or the "Interactive Cinema" genre.

### *2.2 Gameplay Personification*

Gameplay personification of the game is the ability of the player to choose different strategies to achieve game objectives. This may include the choice of different weapons and skills, as well as a number of different modifiers that can both simplify the passage of the game and complicate it [10, 11].

Gameplay variation is especially important in games where players are given freedom of action. Depending on their playstyle, players can discover new paths to complete both tasks and the game as a whole.

Creating such a version of personification will be no less difficult than its plot version because developers are required to not only create a wide variety of weapons, strategies, and settings but at the same time make them balanced and not violate the logic of the game world. In addition, gameplay personification may require the creation of a much larger amount of content, which will increase development costs [12, 13].

### 2.3 *Analysis of Games with Gameplay Personification of Scenarios*

To better understand and understand the personification of game scenarios, it is worth considering examples of games not only in the RPG genre but also in other genres, in which, one way or another, the topic of creating gameplay variability in the passage of the game was considered.

The selected analogs were compared according to 6 criteria:

- changing the characteristics of the player's character—strengthening or weakening the character by changing his parameters of health, stamina, mana, damage dealt, and others;
- changing the characteristics of NPCs—Non-Player Character—the complication of computer opponents is carried out only by increasing their basic characteristics, the behavior of opponents does not change;
- change in NPC behavior tactics—does the enemy change his tactics or set of skills when changing personification settings;
- changes to the in-game economy—as a rule, in-game resources become more valuable. It can be reflected both in the amount of game resources received and in the cost of using various abilities and spells;
- changing environmental parameters—damage values from traps placed at the level or the number of items collected at the level;
- entering global modifiers—time limit for completing a level, weather, and player assistance systems [14, 15].

The results of the comparison of all the above analogs are shown in Table 1.

Based on the analysis of existing games with personification of game scenarios, it was concluded that in all the above games, the developers tried to work out the system

**Table 1** Results of comparison of existing solutions

| Criterion | Hades | Frostpunk | Pathfinder: wrath of the righteous | NieR: automata |
|---|---|---|---|---|
| Changing player character characteristics | + | – | + | + |
| Changing NPC characteristics | + | + | + | + |
| Changing NPC behavior tactics | + | + | + | + |
| Changes to the in-game economy | + | + | – | – |
| Changing environment settings | + | + | + | – |
| Entering global modifiers | + | + | + | – |

of variability of game scenarios as best as possible [16, 17]. Through detailed personalization, these games can reach a large audience of players, making the game both friendly for beginners and challenging for experienced players, as well as creating repeat playability. The developers do not disclose the exact methods for implementing personification. It can be assumed that they used changing the parameters of characters, and objects in the location, and creating global game modifiers.

## 3 Review of Methods for Personifying Scenarios in Games

### *3.1 Random Appearance of Objects on Location*

The random appearance, or Spawn, of objects in a location is one of the methods to diversify the gameplay because each time the game will generate a modified location by changing the arrangement of objects. You can randomly place anything: elements of the environment, opponents, picked-up objects, and traps. For example, in games that use procedural level generation as core mechanics, not only individual objects but also entire sublevels are randomly placed [18, 19].

In this work, this type of personification is performed using the example of a random arrangement of static elements of the environment, opponents, and dangerous zones-objects that cause damage to the character upon contact.

### *3.2 Changing Player and Opponent Parameters*

The second option for creating personalization of game scenarios is to manually configure the parameters of the player characters and opponents. For example, in the game Hades, in this way you can adjust the damage dealt and received by the main character and his opponents, the speed of movement and attacks of enemies, the effectiveness of treatment, damage from traps, etc.

In this work, this type of personification is performed using the example of changing the number of health points and mana of the player character, the number of health points of ordinary opponents and the boss, the amount of damage they cause, and the effectiveness of the player's treatment.

### *3.3 Global Game Modifiers*

Global game modifiers are one of the options for personalizing game scenarios. For the most part, this can be a change in the weather conditions in the game, turning on or off the countdown timer for completing the level.

In this work, this type of personification is performed using the example of activating a countdown timer to complete a level, the value of which is randomly selected in a given interval.

## 4 Development of Dead Sands Game

### *4.1 The Architecture of the Game "Dead Sands"*

After a review and analysis of methods for personifying scenarios in games, options for personifying game scenarios were considered and identified, functional requirements for the game were created, as well as requirements for the game interface, the game architecture was designed and an analysis of use cases was carried out.

The architecture of the developed game is arranged as follows. The user enters commands from the keyboard, mouse, or gamepad. The commands he enters are sent to the Unity game engine. In turn, Unity stores all the necessary data for the normal functioning of the game system; in each frame of the game the engine processes in-game data and data from user-entered commands [20, 21].

The game engine processes data using scripts—components created by the game developer or initially integrated internal scripts of the engine. The DirectX API was chosen as the graphics API (Application Programming Interface) of the engine due to its performance and operability on the largest number of system configurations.

The visual engine receives the game data it needs to render, processes it, and then displays it. To calculate and implement game physics, Unity uses a physics engine to simulate a number of different phenomena, called PhysX.

The game personification system receives data entered by the user by pressing buttons, processes it, and changes the required values of static variables in the corresponding scripts [22]. The game architecture is shown in Fig. 1.

### *4.2 Game Loop of the Dead Sands Game*

At the beginning of the game, the player selects the difficulty settings for the passage, and the game applies them [19, 23]. Then the developed personification scripts, responsible for the random appearance of enemies, objects, and fire zones, distribute them in arrays of spawn points. The player appears at the fire in the game scene and goes to explore the location. While exploring, the player may come across an ordinary enemy, with whom he can either engage in combat or try to escape. Having entered into battle with an enemy, the player can either die or win. If defeated, a death animation plays, a death message is displayed on the screen, and the game begins again. In case of victory, the player continues to progress through either the battle with the boss or until the next detection of an ordinary enemy. If the player decides to

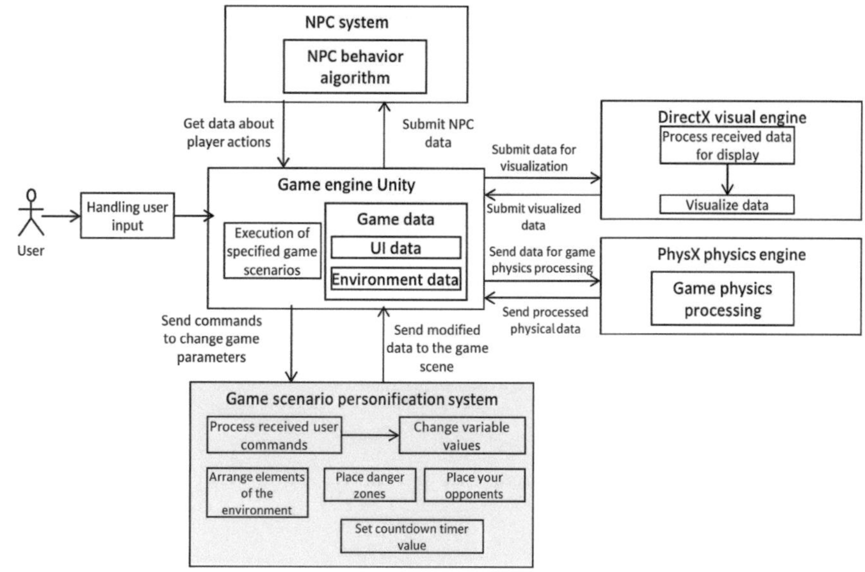


**Fig. 1** Game architecture

run away from the enemy, then either this will be successful and the enemy will lose sight of the player, or the enemy will catch up with the player. If the player reaches the boss arena, the location is considered successfully completed. In a boss fight, the player can either die, which will cause the level to load again, or win, which will lead to the end of the game, congratulations on the victory will appear on the screen, and the player will be transferred to the main menu of the game. The game cycle of the developed program is presented in Fig. 2.

## 4.3 Development of Settings for Personification of Game Scenarios

In this work, the development of game personification settings includes three stages:

- Development of personification of game scenarios through a series of small scripts that change the passage parameters.
- Introduction of a global timer modifier.
- Creating a module that randomizes the appearance of the desired objects on the map.

The development of the first stage involves the creation of the "Settings" menu, which is a separate game scene, and which can be accessed from the main menu of the game using the Play button. The menu contains several UI buttons with a

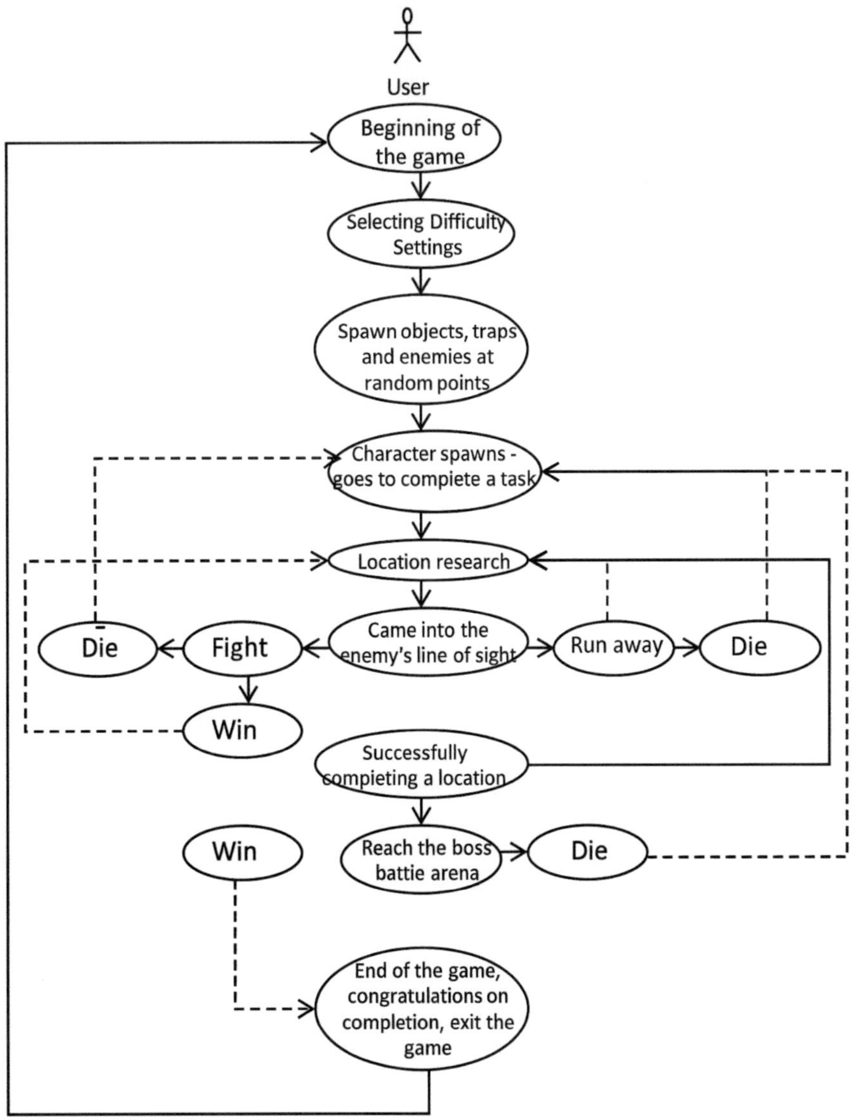


**Fig. 2** Game loop

description of what they do. Each button has an OnClick() method, and when clicked, the corresponding script is activated. There are only five such buttons, and they are responsible for strengthening the player's character, strengthening the boss of the location, increasing the number of health points of enemies, increasing the damage caused by enemies and reducing the effectiveness of the character's healing.

Let's consider the effect of this variant of personification using the example of the BuffedPlayer class, which strengthens the player character. In the excerpt below from the PlayerStats class, which is responsible for the character's characteristics, two static variables of type int are created, which are responsible for the number of health points and mana of the player. The variables are static because we need to use it in a different scene than where the object with the PlayerStats script is located (Fig. 3). In the BuffedPlayer class, a PlayerBuff method is created that changes the value of these variables to the ones we need. This method is hung on OnClick() of the desired UI button in the settings menu (Fig. 4).

A total of five options for changing character parameters were created; the class diagram is presented in Fig. 5.

Stage two development involves creating a random timer that determines the time required to complete the level. The timer ranges from 180 to 300 s. If the player does not complete the location within the allotted time, the game is considered lost, a corresponding message is displayed on the screen, and the level begins again. A variable levelTimer of type float is declared, in the Awake() method the value of this variable is set using Random.Range, which returns a random number between 180 and 300. Next, in the Update() method, levelTimer is bound to the Time.deltaTime function,

**Fig. 3** Player variables in the PlayerStats class method

```
public class PlayerStats : MonoBehaviour
{
    public GameObject DeadTest;
    public GameObject RanOutOfTime;
    public HealthBar healthbar;
    public ManaBar manabar;

    AnimatorHandler animatorHandler;

    public static int maxHealth = 100;
    public int currentHealth;
    public static int maxMana = 100;
    public int currentMana;
    public static bool isGameOver;
    public static int wasteMana = 10;
    public static int healplayer = 20;
    public float levelTimer;
```

**Fig. 4** Changed values by the PlayerBuff

```
public class BuffedPlayer : MonoBehaviour
{
    public void PlayerBuff()
    {
        PlayerStats.maxHealth = 150;
        PlayerStats.maxMana = 150;
    }
}
```

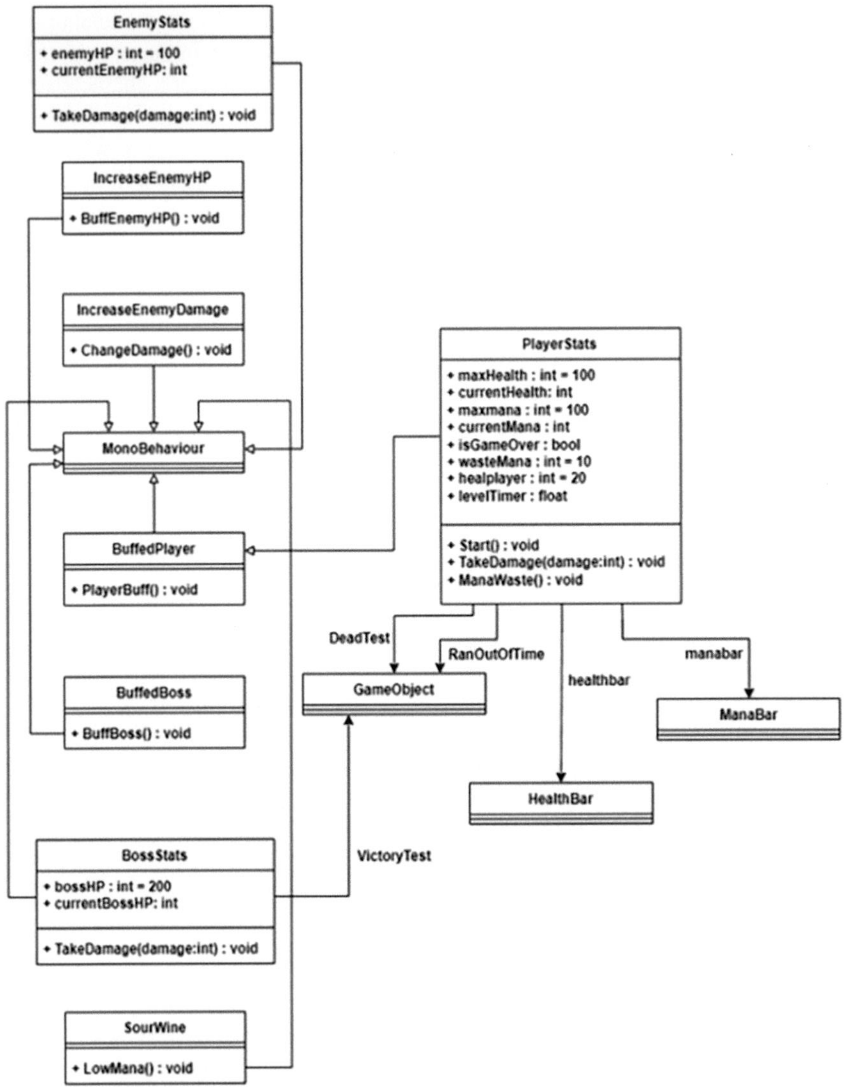


**Fig. 5** Personification system class diagram

which is responsible for the passage of time in the game, and if the levelTimer value is less than or equal to zero, then the TimeOut() and HandlePlayerDeath() methods are called with some delay, the first of which is responsible for displaying a message on the player's screen about exceeding the time limit, and the second is responsible for the death of the player (Fig. 6).

**Fig. 6** Countdown timer

```
void Update()
{

    levelTimer -= Time.deltaTime;
    if(levelTimer <= 0)
    {
        Invoke(nameof(TimeOut), 1f);
        Invoke(nameof(HandlePlayerDeath), 6f);
        RanOutOfTime.SetActive(false);
    }
```

To implement the third stage, the RandomSpawn, SpawnEnemy and SpawnWild-Fire classes were developed, which are responsible for random spawning at the location of obstacles, enemies and zones dangerous to the player.

Since the classes themselves are almost the same as each other, let's consider this option for personifying game scenarios using the example of the SpawnWildFire class.

To begin with, an empty object called SpawnWildFire is created in the scene, inside which the required number of small spheres are created that serve as places for objects to spawn. The spheres are placed at the required points throughout the location.

The SpawnWildFire script (Fig. 7) is added to the empty object, in which an array of spawn objects is created, a list of points at which objects can appear. Spawn points are randomly distributed and objects are placed in them.

Next, in the Inspector column in Unity, select the required number of objects that will appear in the scene, and select the points themselves.

The WildFire fire zone is a Particle System component object with a Capsule Collider, which is needed for collision with the player. The WildFire object (Fig. 8)

```
public class SpaunWildFire : NonoBabaviour
(
    public GaneObject() fireTraps;
    public ListcTrunsforn> fireTrapsPoints;

    void Start()
    (
        fireTrapsPoints - new ListcTransforн>(fireTrupsPoints);
        SpounTraps();
    )

    void SpaunTraps()
    (
        for (int I - 0; I < fireTraps.Length; I++)
        (
            ver spaun - Randon.Range(0, fireTrapsPoints.Count);
            Instantiate(fireTraps(i), fireTrapsPoints(spaun).transfore.position, Quaternion.identity);
            fireTrapsPoints.RenoveAt(spaun);
        )
    )
)
```

**Fig. 7** SpawnWildFire class code

has a script for causing damage to the player character DamagePlayer (Fig. 9), which is activated when the player comes into contact with the object's collider.

As part of the personalization of scenarios, objects for random appearance can be not only dangerous zones that cause damage to the player, but also ordinary objects such as houses, walls, and towers, as well as enemies, the appearance of which can also be adjusted randomly.

The following pictures show places where enemies, traps, and ordinary objects can appear (Fig. 10).

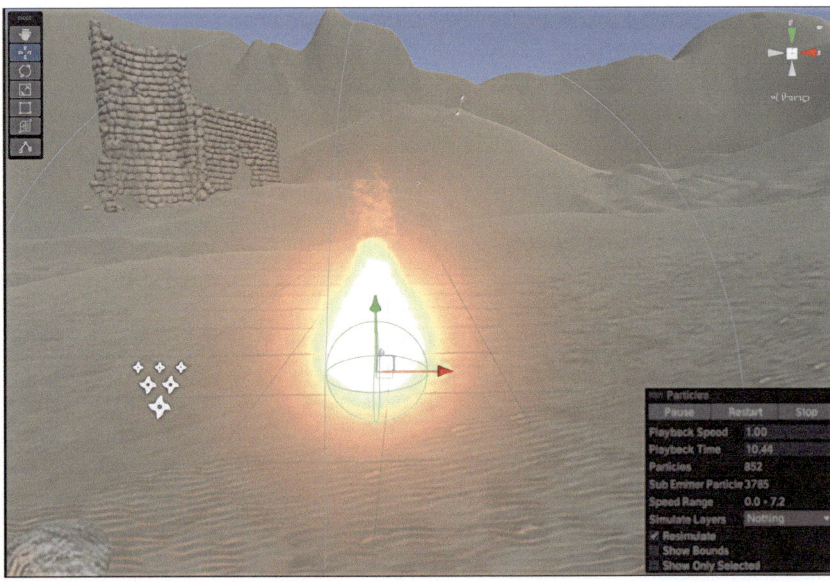


**Fig. 8** WildFire danger zone object

```
public class DanagePlayer : MonoBehaviour
{
    public int danage - 25;

    private void OnTriggerEnter(Collider other)
    {
        PlayerStats playerStats - other.GetConponent<PlayerStats>();

        if (playerStats !- null)
        {
            playerStats.TakeDanage(danage);
        }
    }
}
```

**Fig. 9** Damage player class code DamagePlayer

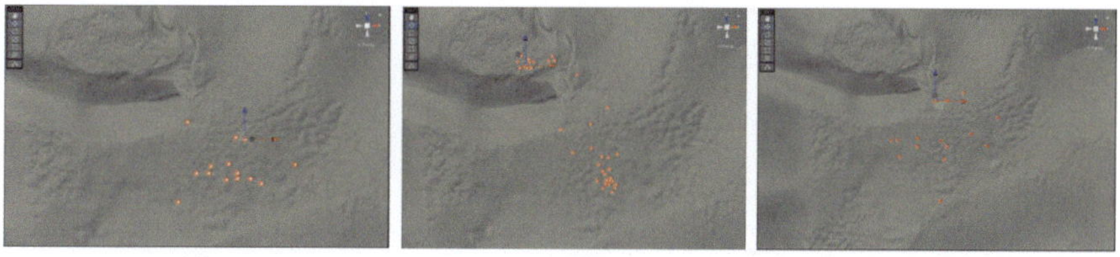

Fig. 10 Possible **a** enemy spawn locations, **b** locations of hazardous areas, **c** locations for other objects to appear

As can be seen in Fig. 10b and c, the spawn locations of dangerous zones and enemies overlap, which further forces the player to either be prepared for more serious battles or take a detour, which in turn wastes precious time.

The page for personifying game scenarios is represented by a separate scene. The background differs from the background of the main menu; it was similarly taken from open sources. This page displays UI buttons responsible for changing the game progression, as well as captions for them so that the player understands what he is choosing. By clicking the Onward button on the UI, the player is taken to the game itself, made as a separate scene in the engine. The player can also return to the main game menu using the UI Return button. All UI elements have been placed on the game scene and are standard Unity UI elements. After that, the Irish Grover font, previously imported into the engine, was superimposed on the text of the UI elements. All actions when pressing buttons were written in the game scripts and superimposed on the OnClick() method of the buttons.

The menu for personalizing game scenarios is presented in Fig. 11.

Indicators of the number of health points and mana of the player, images for these elements were also developed in the graphic editor Adobe Photoshop, after which

**Fig. 11** Menu for personalizing game scenarios

Fig. 12 Character stats interface

they were imported into Unity. These indicators are standard UI elements from Unity, with images superimposed on them. When the character's health indicators change, the health bar decreases by processing the image from Unity, namely by changing the Fill Amount indicator.

Indicators of the number of health points and mana of the player character are displayed in the upper left corner of the screen in Fig. 12.

## 5 Game Performance Testing

FPS (Frames Per Second) is the number of frames per second that the game draws. The larger this number, the smoother the animation in the game will look [24, 25].

FPS is the most important indicator for assessing the performance of graphics applications, as well as graphics systems (video card + driver).

However, this is only possible with vertical synchronization (VSync) turned off, since when vertical synchronization is turned on, the FPS indicator cannot be higher than the vertical scan frequency of the monitor (usually 60, 75, 120, 144 Hertz), so testing with vertical synchronization turned on is always incorrect.

When analyzing performance in video games, most attention is paid to the average and minimum FPS. The average indicates overall performance, while the minimum is used to determine the system's ability to handle the busiest parts of the game. Typically, in single-player games, a value of 30 frames per second is considered the minimum frame rate for comfortable gaming.

To test FPS in a Unity game, there is a built-in Profiler tool. This tool is very useful for getting a detailed understanding of how the game reacts to various changes in the editor. It displays categories such as CPU and GPU usage, rendering, physics, audio, and more.

Testing took place in three scenes of the game—in the main menu, in the settings menu for personalizing game scenarios, and in the main game scene.

In the menu, the average FPS was kept at 1000 frames per second. A decrease in frames per second was noticed when opening the game personalization settings page, at the moment the FPS indicator dropped to approximately 200 frames per second, this decrease in frames lasts approximately 0.2 s, and it is associated to a large extent with the loading of a new scene, which can be corrected by loading all scenes at the same time. The menu testing results are shown in Fig. 13.

In the main stage of the game, the average FPS remained at 200 frames per second. The FPS drop to 60 occurred in the first second after loading the game scene. This happens for two reasons:

- new scene and all models on it are loaded;
- when the game starts, many particles of the Particle System are loaded, which are responsible for displaying the effects of fire, sandstorms, and clouds. This can be corrected by reducing the number of rendered particles in these objects.

The test results on the main stage of the game are shown in Fig. 14.

Functional testing of the game, as well as testing of the game's performance in terms of frames per second, went satisfactorily.

## 6 Conclusion

As a result of the work, analog video games and available development tools were analyzed. A single-player 3D game in the Soulslike style with the ability to personalize game scenarios was designed, developed, and tested.

The scope of application of the developed game is for users to study various game scenarios while independently changing the parameters of the passage, as well as to attract players to study the variability in video games through entertainment in a playful way.

Possible areas for improving the game:

- Possibility of expanding enemy behavior states.
- Adding more items to the game, such as ranged weapons, different types of armor, and more spells.
- Creation of a class system and more characteristics of the player's character with the ability to level up the character in real-time.
- Improving the mechanics of the combat system and gameplay in general, will also have a better effect on users' perception of the game.

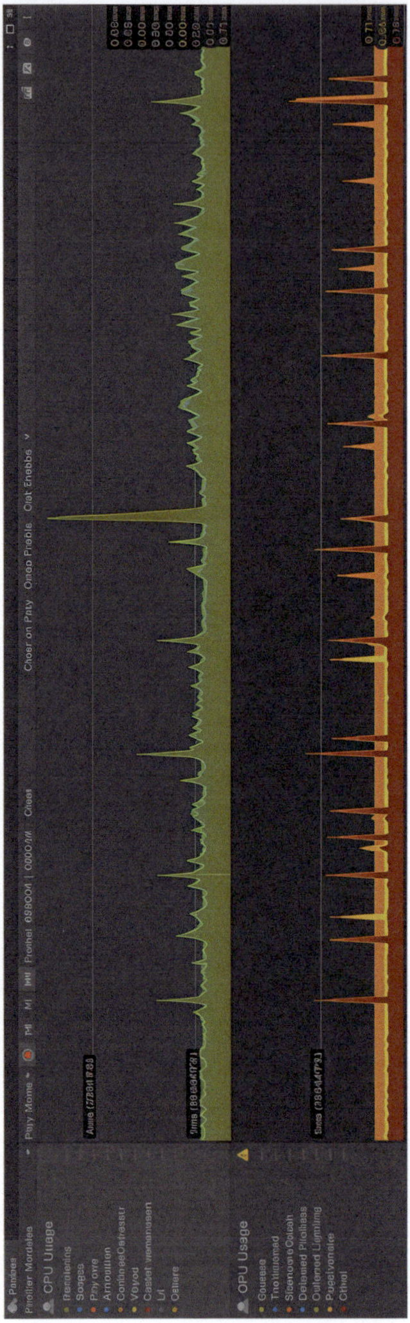

**Fig. 13** FPS test results in the menu

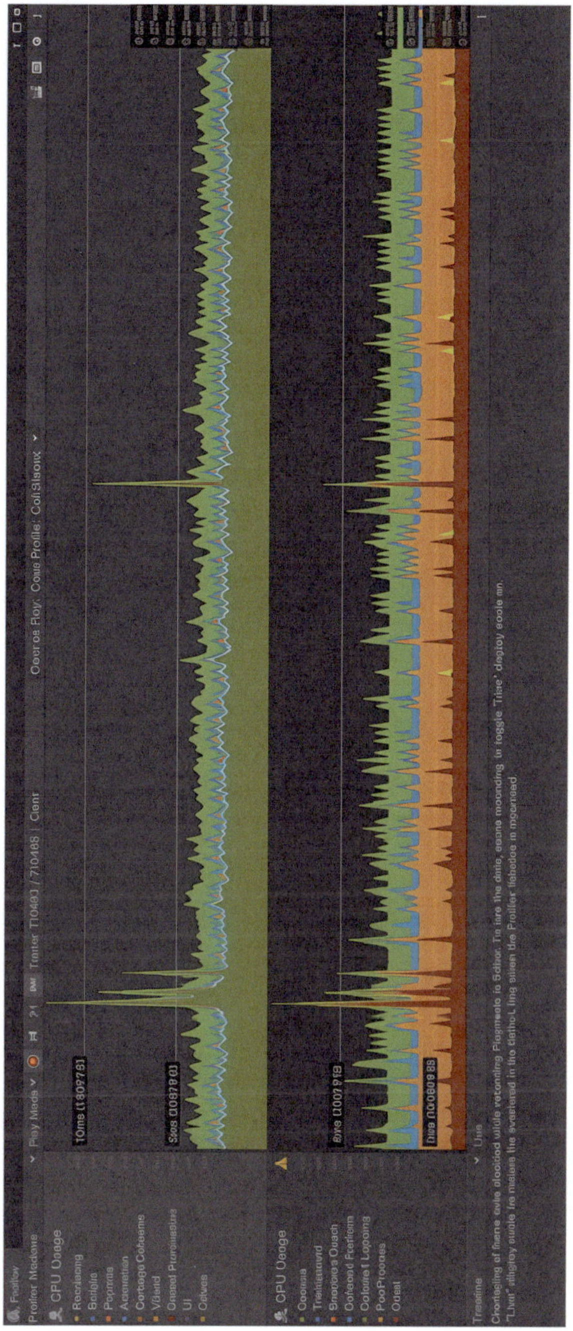

**Fig. 14** FPS test results in the main game

## References


1. Klimenko, R.G., Kravets, A.G., Strukova, I.V., Tatarova, N.I.: Analysis of people's fascinations for computer games. In: Information Innovative Technologies. International Scientific-Practical Conference, Moscow, pp. 48–57 (2022)
2. The game awards continues destroying the Oscars with record 103 million viewers in 2022. Forbes. https://www.forbes.com/sites/paultassi/2022/12/16/the-game-awards-continues-destroying-the-oscars-with-record-103-million-viewers-in-2022. Accessed 15 Mar 2024
3. Best video games with Hollywood actors. The gamer. 2019. https://www.thegamer.com/best-video-games-with-hollywood-actors/. Accessed 15 Mar 2024
4. Kravets, A.G., Salnikova, N., Mikhnev, I., Solovieva, N.: Relevant image search method when processing a patent array. In: Kravets, A.G., Shcherbakov, M., Parygin, D., Groumpos, P.P. (eds.) Creativity in Intelligent Technologies and Data Science. CIT&DS 2021. Communications in Computer and Information Science, vol. 1448. Springer, Cham (2021). https://doi.org/10.1007/978-3-030-87034-8_6
5. Khayrov, A., Shabalina, O., Sadovnikova, N., Kataev, A., Petrova, T.: Computer-aided development of adaptive learning games. In: Kabassi, K., Mylonas, P., Caro, J. (eds.) Novel & Intelligent Digital Systems: Proceedings of the 3rd International Conference (NiDS 2023). NiDS 2023. Lecture Notes in Networks and Systems, vol. 784. Springer, Cham (2023). https://doi.org/10.1007/978-3-031-44146-2_37
6. Bolshakov, A.A., Klyuchikov, A.V.: Decision support system for selecting designs of autostereoscopic displays. In: Cyber-Physical Systems: Design and Application for Industry 4.0 (Studies in Systems, Decision and Control), vol. 342, pp.73–88 (2021)
7. Kravets, A.G., Kolesnikov, S., Salnikova N., Lempert, M., Poplavskaya, O.: The study of neural networks effective architectures for patents images processing. In: Communications in Computer and Information Science, vol. 1084, pp. 27–41 (2019)
8. Matokhina, A.V., Shabalina, O.A., Kharlamova, N.V., Kulikov, E.A.: Game-based learning platform for integrating international IT-students into the Russian educational environment. In: Kravets, A., Shcherbakov, M., Kultsova, M., Groumpos, P. (eds.) Creativity in Intelligent Technologies and Data Science. CIT&DS 2017. Communications in Computer and Information Science, vol. 754. Springer, Cham (2017). https://doi.org/10.1007/978-3-319-65551-2_55
9. Kravets, A.G., Pavlenko, I.D., Egunov, V.A., Kravets, E.: Interaction with virtual objects in VR-applications. In: Kravets, A.G., Shcherbakov, M.V., Groumpos, P.P. (eds.) Creativity in Intelligent Technologies and Data Science. CIT&DS 2023. Communications in Computer and Information Science, vol. 1909. Springer, Cham (2023). https://doi.org/10.1007/978-3-031-44615-3_30
10. Marín-Lora, C., Chover, M.: GameScript: a simplified scripting language for video game development. Multimedia Syst. **31**, 70 (2025). https://doi.org/10.1007/s00530-024-01574-8
11. Shcherbakov, M., Groumpos, P.P., Kravets, A.: A method and IR4I index indicating the readiness of business processes for data science solutions. In: Kravets, A., Shcherbakov, M., Kultsova, M., Groumpos, P. (eds.) Creativity in Intelligent Technologies and Data Science. CIT&DS 2017. Communications in Computer and Information Science, vol. 754. Springer, Cham (2017). https://doi.org/10.1007/978-3-319-65551-2_2
12. Pact of punishment. Hades Wiki. https://hades.fandom.com/wiki/Pact_of_Punishment (2021). Accessed 15 Mar 2024
13. Dynamic game difficulty balancing. Википедия. https://en.wikipedia.org/wiki/Dynamic_game_difficulty_balancing (2023). Accessed 15 Mar 2024
14. Own game engines. A little research. Habr. https://habr.com/ru/articles/500320 (2020). Accessed 15 Mar 2024
15. Review of techniques for implementing gaming AI. Habr. https://habr.com/ru/articles/420219/ (2018). Accessed 15 Mar 2024
16. Role-playing video game. Britannica. https://www.britannica.com/topic/role-playing-video-game (2019). Accessed 15 Mar 2024

17. The most powerful real-time 3D creation tool. Unreal Engine. https://www.unrealengine.com/en-US (2023). Accessed 15 Mar 2024
18. Tips on designing non-linear single-player levels. 80LV. https://80.lv/articles/tips-on-designing-non-linear-single-player-levels/ (2021). Accessed 15 Mar 2024
19. GameMaker. Wikipedia. https://en.wikipedia.org/wiki/GameMaker (2023). Accessed 15 Mar 2024
20. Unity (game engine). Wikipedia. https://en.wikipedia.org/wiki/Unity_(game_engine) (2023). Accessed 15 Mar 2024
21. Kazakov, I.D., Shcherbakova, N.L., Rayushkin, E.S., Shcherbakov, M.V.: A method for change detection operating of power generating equipment. In: Cyber-Physical Systems: Design and Application for Industry 4.0, pp. 285–297. Springer Nature Switzerland (2021)
22. What is a gaming or game engine? Arm. https://gamedev.ru/code/terms/FPS (2023). Accessed 15 Mar 2024
23. Godot (game engine). Wikipedia. https://en.wikipedia.org/wiki/Godot_(game_engine) (2023). Accessed 15 Mar 2024
24. FPS: frames per second. GameDev.ru. https://www.arm.com/glossary/gaming-engines (2009). Accessed 15 Mar 2024
25. What FPS is considered optimal in games—what is it? What FPS in games is considered normal. Club-CS. https://club-cs.ru/news/kakoj-fps-v-igrah-schitaetsya-optimalnym-chto-eto-takoe-kakoj-fps-v-igrah-schitaetsya-normalnym-18316 (2022). Accessed 15 Mar 2024

# Depression Diagnosing Based on Analysis of Computer Game User Behavior

**Alla G. Kravets**

**Abstract** This study explores the potential of utilizing serious games as an innovative approach to diagnose depression by analyzing user behavior within a gaming environment. Traditional methods of depression diagnosis rely heavily on standardized questionnaires and clinical assessments, which can be subjective and prone to inaccuracies. Leveraging advances in information technology and machine learning, this research proposes a novel methodology where gamified environments provide controlled conditions for observing and evaluating user behaviors indicative of depressive symptoms. Key features of the study include the design of an interactive game scenario tailored to elicit behaviors relevant to depression diagnosis, the collection of objective indicators such as reaction times and number of actions, and the subsequent application of classification algorithms to distinguish between individuals with and without confirmed depression diagnoses. Preliminary findings suggest promising results, particularly with Support Vector Machines and Random Forests demonstrating higher accuracy in identifying depressive patterns. Future improvements could enhance the method's robustness by incorporating vocal and facial analyses, expanding the scope of the game narrative, and refining the underlying predictive models. Ultimately, this study highlights the potential of gamification techniques in enhancing early detection and intervention efforts for depression.

**Keywords** Depression diagnosis · Gaming environment · Machine learning · Behavioral data · Cognitive abilities · Classification models · Interactive scenario

A. G. Kravets (✉)
Volgograd State Technical University, Volgograd, Russia
e-mail: allagkravets@yandex.ru

Dubna State University, Dubna, Russia

 
A. G. Kravets and A. A. Bolshakov (eds.), *Human-Centric Cyber-Society*, Studies in Systems, Decision and Control 639, https://doi.org/10.1007/978-3-032-06787-6_19

# 1 Introduction

Depression is a common mental health disorder. According to the World Health Organization, 3.8% of the world's population suffers from depression, which is about 280 million people [1]. The main symptoms of depression include a depressed, anxious, and fearful mood; depression is accompanied by decreased concentration, a feeling of fatigue or loss of strength, an inadequate sense of guilt, and pessimism.

In all countries, regardless of income level, people suffering from depression often go undiagnosed [2], but with the development of digital technologies, it has become possible to track user behavior and, based on the collected information, recommend a specialist examination or even immediately diagnose the disease, thereby automating the process of diagnosing depression.

Typically, questionnaires or long-term specialized therapy sessions are used to diagnose depression [3], but with the development of information technology, it has become possible to use machine learning.

"Serious games" are proposed as an alternative way to collect data on user behavior games [4]. Video gaming provides a more active environment for the user, which can provide a more accurate picture of the emotional and psychological state, as it allows for controlled conditions for observation and analysis, which reduces the influence of external factors on the results. The gaming environment in virtual space allows for the collection of objective indicators, such as reaction time, number of actions, etc., which increases the accuracy of the analysis. Thus, the novelty of the method lies in the use of an interactive environment to collect data on user behavior.

## *1.1 Standardized Questionnaires and Scales*

The use of standardized questionnaires and scales allows for an objective assessment of the presence and degree of depression.

PHQ-9 (Patient Health Questionnaire-9) is a multipurpose instrument for screening, monitoring and measuring the severity of depression [5]. It is A self-assessment questionnaire consisting of 9 questions covering the main symptoms of depression. The scale helps to determine the severity of symptoms. The questions are based on the criteria for depression from the Diagnostic and Statistical Manual of Mental Disorders, 5th edition (DSM-5) [6]. Patients rate the frequency of symptoms over the past two weeks on a 4-point scale, where 0 is never and 3 is almost every day. A total score from 0 to 27 is used to determine the level of depression: 0–4—none, 5–9—mild, 10–14—moderate, 15–19—moderately severe, 20–27—severe. The questions focus on determining loss of interest or pleasure in activities, feeling depressed, sleep problems, feeling tired, loss of appetite or overeating, feeling excessively tired, difficulty concentrating, anxiety, and suicidal thoughts. This type of diagnosis depends on the honesty and self-awareness of the patient, and results may be inaccurate due to social stigma or missing atypical symptoms.

BDI (Beck Depression The Beck Depression Inventory is a self-assessment questionnaire developed by Aaron Beck. The Beck Depression Inventory contains 21 categories of symptoms and complaints from among the most common in patients with depression [7]. Each category contains 4–5 statements corresponding to specific signs of depression. The statements are distributed taking into account their contribution to the overall severity of depression. Each item in the category is assessed from 0 to 3 points, and the total score can range from 0 to 62. The severity of depression can be assessed by the cognitive-affective subscale and the subscale of somatic manifestations of depression: 0–9 no depressive symptoms, 10–15 mild depression (subdepression), 16–19 moderate depression, 20–29 severe depression (moderate), 30–63 severe depression. The questionnaire allows you to assess the intensity of depressive symptoms. The BDI questions cover feelings of sadness, pessimism, failure, guilt, dissatisfaction with oneself, self-criticism, irritability, indecisiveness and insomnia. This type of assessment can be distorted by misinterpretation of questions without professional guidance, so it may be difficult for patients with a low level of education.

HAM-D (Hamilton Depression Rating The Depression Inventory (DIS) is a clinical scale used by physicians to assess the severity of depression. Originally developed in the 1960s by Max Hamilton, it consists of 17–21 questions that are rated by a physician. The scale allows for a quantitative assessment of the dynamics of depressive manifestations and is an excellent tool for monitoring the effectiveness of treatment [8]. The scale questions concern the patient's condition over the past few days or the previous week. It is important not to allow the patient to deviate from the topic of the question and to give enough time for a detailed answer. The questions cover signs of low mood, guilt, suicidal tendencies, difficulty falling asleep, insomnia, and difficulties at work. The following gradations of results are considered: 0–6 points—normal, 7–16 points—mild depressive disorder, 17–24 points—moderate depressive disorder, more than 24 points—severe depressive disorder. This method requires a highly qualified specialist to perform and interpret, and the testing itself is a time-consuming procedure.

These questionnaires and scales are important components of a comprehensive approach to diagnosing depression, providing a structured and standardized assessment of symptoms.

## 1.2 Comparison of Information Methods of Diagnostics

We will evaluate the functionality, accessibility, effectiveness and limitations of existing depression diagnostic applications.

Thymia is an application for diagnosing depression and mental health. It uses gaming technologies and video analysis to identify behavioral and speech markers of depression [9]. The main data for analysis are video recordings of the user's face during gaming activity, audio recordings of the voice during gaming activity, and gaming behavior. The behavioral data includes reaction time, errors, and correct choices in mini-games. The program uses machine learning and behavior analysis

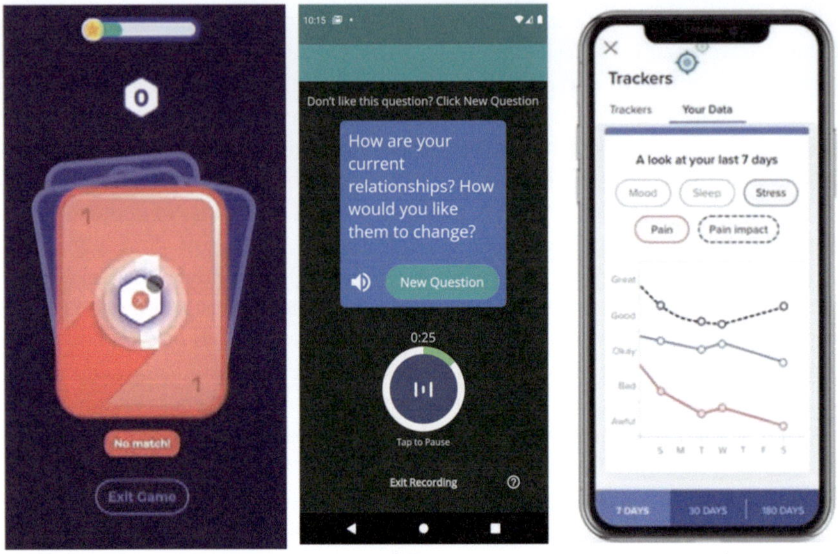


**Fig. 1** Analysis of the functionality of applications for the diagnosis of depression: **a** the gameplay of Thymia; **b** the speech analysis process in InSight; **c** the user analysis interface in Mindset

in the gaming environment. The program is distributed on a subscription basis and requires an internet connection. The program is available in a browser and on mobile devices. The Thymia gameplay is shown in Fig. 1a.

InSight is a program developed by Elipsis Health for diagnosing depression based on the analysis of the user's speech and voice [10]. The program uses speech recognition and natural language processing technologies. The program analyzes changes in the lexical and acoustic characteristics of speech. The process of speech analysis by the application is shown in Fig. 1b.

Mindset is a mobile application developed by NeuroFlow. It uses data on physical activity, sleep, and physiological indicators to monitor the user's mental health [11]. The application provides a comprehensive assessment of activity and uses data from all registered gadgets to classify depression in the patient. Figure 1c shows the user analysis interface based on data from trackers.

Youper is an application that uses a chatbot to regularly assess the user's mental state and diagnose depression [12]. The analysis uses the user's text responses to the chatbot's questions. Thus, the application offers the possibility of regularly assessing the emotional state. The application interface is shown in Fig. 2.

Every from these solutions demonstrates their unique Features and approach to diagnostics depression. From listed only Thymia has functionality diagnostics by behavior user in in-game surroundings. Comparison each from solutions presented in Table 1.

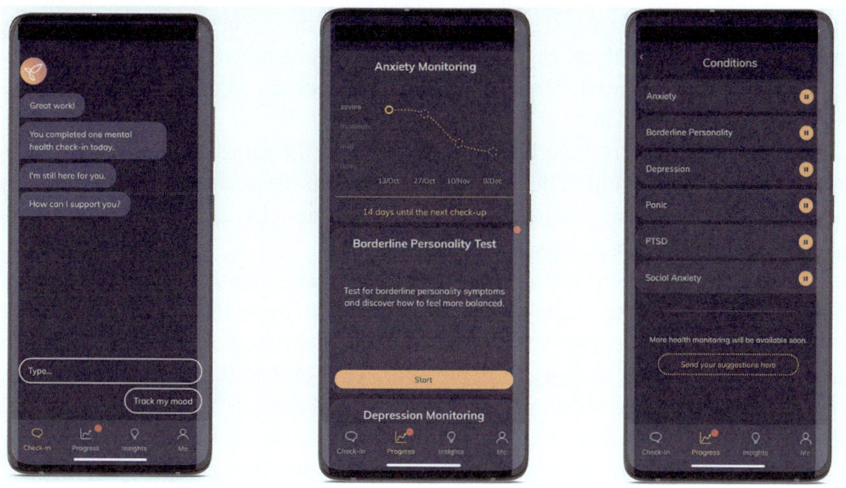


**Fig. 2** Youper application interface

**Table 1** Results of comparison of existing information methods of diagnostics

| Criterion/decision | Thymia | InSight | Mindset | Youper |
|---|---|---|---|---|
| Voice analysis | + | + | – | – |
| Facial analysis | + | – | – | – |
| Behavior analysis | – | – | + | + |
| Physical activity analysis | – | – | + | – |

## 2 Approaches to the Diagnosis of Depression in a Gaming Environment

To create a platform for collecting data on user behavior, it is necessary to analyze in detail possible scenarios of interaction between the actor and the environment and gameplay mechanics, according to which the analysis of depression will be implemented.

### *2.1 Interaction Scenario*

In the game approach to diagnosing depression, the interaction scenario plays a key role, as it must be carefully designed to meet the various criteria used in diagnosing a depressive episode.

The player assumes the role of a delivery man in a large city. This allows for a realistic and interactive environment to observe the user's behavior.

During the game, the player must perform various tasks, such as delivering orders and communicating with other characters. These interactions are designed to highlight symptoms of depression, such as low mood, decreased concentration, feelings of fatigue, and pessimism.

The player encounters disturbing situations during the game, which require appropriate reactions from him. The player's reactions to these situations can serve as indicators of a depressive state, such as decreased activity, slowness of action, or inadequate emotional reactions.

Each choice made by the player in dialogues and interactions has a certain "weight" in relation to specific diagnostic criteria from the DSM-5. Negative responses related to depressive symptoms will influence the final outcome. This allows us to assess the severity of depressive symptoms in the player. The criteria were extracted from the Diagnostic and Statistical Manual of Mental Disorders, 5th edition (DSM-5).

In order to prevent user fatigue, the game scenario assumes a 30-min playthrough.

The game belongs to the first-person genre of games, the character's appearance and voice are never shown so that the player can associate himself with his character.

It is important that the game does not provide the possibility of losing or being defeated. This allows for a less stressful environment for the player and avoids additional emotional reactions not related to depression.

The game is developed from a first-person perspective, where the character's appearance and voice are not shown, so that the player can fully associate himself with his character. This contributes to a deeper immersion in the gameplay and increases the reliability of the data collected about the player's behavior.

The game scenario is a fundamental basis for effective diagnosis of depression in a game environment.

## 2.2 Game Cycle

The gameplay cycle consists of completing mandatory story missions and balancing the character's health and fatigue indicators.

In addition to the mandatory plot tasks, the game environment is expected to contain secondary tasks. The tasks are minigames that test the user's short-term or long-term memory. After completing tasks, the player directly increases the character's statistics or receives currency for further increases.

The game environment is interactive—you can talk to characters, and the player's answers in the dialogue will affect the plot. For certain answers, the player can also receive in-game items to improve the character's statistics.

Interactive objects can be used to activate minigames or collect in-game items.

As the story progresses, the player must monitor their character's statistics—health, fatigue, and sanity. To keep the statistics within normal limits, the player must explore the environment, find hidden objects, or complete side quests.

The gameplay loop is designed to create an immersive [13] experience for the player, so that they become more invested in their character and the diagnosis becomes more accurate.

## 2.3 Dialogue System

The dialogue system is an important element in the development of video games, including in the context of diagnosing depression in the gaming environment provides interaction between player and others characters in the game, and plays key role in the collection behavioral data user.

The dialogue system is a mechanic that allows the player to interact with non-player characters (NPCs) through text dialogue. It gives the player the ability to choose from a variety of responses or lines that affect the development of the plot, relationships with characters, and other aspects of the game.

The dialogue system is usually implemented using a tree structure, where each player choice leads to certain branches of dialogue. Each branch may contain additional answer options, character animation, sound effects, and other elements that provide interactivity and player involvement.

In the gaming environment, this mechanic is used not only to collect information about the player's actions, but also to influence the surrounding world. After the end of the gaming session, the collected player data is used to create classifiers that can predict whether the player has a depressive disorder.

Each answer during the dialogue has a different "weight" relative to a certain diagnostic criterion from the DSM-5. For example, a positive answer that the player usually does not sleep well at night signals confirmation of the symptom of a sleep disorder. Showing sensitivity to characters and confident actions in difficult situations, on the contrary, changes the final score in favor of the absence of depression in the player.

Thus, the dialogue system provides interactive interaction with the player and allows collecting data related to diagnostic criteria.

## 2.4 Analysis of Cognitive Abilities

Analyzing cognitive abilities through a gaming environment provides a unique opportunity to identify signs of depression. Depression is often associated with cognitive deficits, including problems with memory, attention, and executive functions. Using minigames to test cognitive abilities allows you to monitor the player's condition.

Mini-games aimed at assessing short-term memory may include tasks involving memorization and reproduction of sequences or objects.

An example of a memory game is a sequence of cards from which you need to select repeating cards after a certain number of elements. With each level, the number of elements after which you need to remember repeating cards increases. The minigame tests the player's ability to memorize and reproduce information in short-term memory.

An example of a matching game is a series of closed cards that the player must turn over and look for pairs of identical images. The time to remember the positions of the cards is limited. This minigame tests visual short-term memory.

Mini-games that require remembering and using information throughout the entire gaming session can be used to assess long-term memory.

For example, story missions that require remembering previously provided details, such as character names, locations, or passwords, to complete.

Another example of a long-term memory minigame are navigational tasks, in which the player must remember and follow complex routes in a virtual world. Such tasks may involve repeated returns to previously visited locations, which requires good long-term memory.

# 3 Development of a Classification Model

## *3.1 Data Preparation*

Was 100 games recorded sessions from two groups: players who have was confirmed diagnosis depression (32%), and players who diagnosis Not was confirmed and with high probability absent (68%). For more accurate samples to the players from second groups was suggested pass PHQ-9 test, so that confirm absence depression and filter players with high probability collisions with this disorder. This was necessity For accurate markings tested users. It follows note that members second groups Not are considered "healthy", i.e. they can have others mental disorders.

The resulting data set is shown in Fig. 3. It consists of the user name, the final values of each of the nine DSM-5 depression characteristics based on the results of the dialogues, the value of the adequacy of the player's behavior in the virtual world (points are deducted for moving along the road and rude communication in dialogues), the value of the average player's health indicators, the number of dialogues conducted, the average number of actions per minute, and confirmation of depression in the user.

## *3.2 Model Training*

First, a visualization procedure was conducted using the principal component analysis to assess the game's ability to differentiate the behavior of players from the first and second groups.

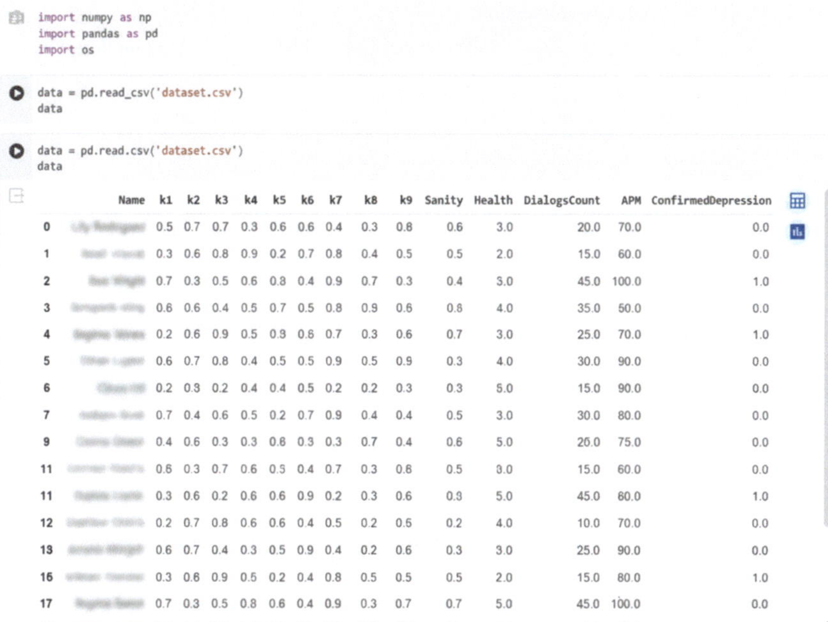

```
import numpy as np
import pandas as pd
import os
```

```
data = pd.read_csv('dataset.csv')
data
```

```
data = pd.read.csv('dataset.csv')
data
```

| | Name | k1 | k2 | k3 | k4 | k5 | k6 | k7 | k8 | k9 | Sanity | Health | DialogsCount | APM | ConfirmedDepression |
|---|---|---|---|---|---|---|---|---|---|---|---|---|---|---|---|
| 0 | [illegible] | 0.5 | 0.7 | 0.7 | 0.3 | 0.6 | 0.6 | 0.4 | 0.3 | 0.8 | 0.6 | 3.0 | 20.0 | 70.0 | 0.0 |
| 1 | [illegible] | 0.3 | 0.6 | 0.8 | 0.9 | 0.2 | 0.7 | 0.8 | 0.4 | 0.5 | 0.5 | 2.0 | 15.0 | 60.0 | 0.0 |
| 2 | [illegible] | 0.7 | 0.3 | 0.5 | 0.6 | 0.8 | 0.4 | 0.9 | 0.7 | 0.3 | 0.4 | 3.0 | 45.0 | 100.0 | 1.0 |
| 3 | [illegible] | 0.6 | 0.6 | 0.4 | 0.5 | 0.7 | 0.5 | 0.8 | 0.9 | 0.6 | 0.8 | 4.0 | 35.0 | 50.0 | 0.0 |
| 4 | [illegible] | 0.2 | 0.6 | 0.9 | 0.5 | 0.3 | 0.6 | 0.7 | 0.3 | 0.6 | 0.7 | 3.0 | 25.0 | 70.0 | 1.0 |
| 5 | [illegible] | 0.6 | 0.7 | 0.8 | 0.4 | 0.5 | 0.5 | 0.9 | 0.5 | 0.9 | 0.3 | 4.0 | 30.0 | 90.0 | 0.0 |
| 6 | [illegible] | 0.2 | 0.3 | 0.2 | 0.4 | 0.4 | 0.5 | 0.2 | 0.2 | 0.3 | 0.3 | 5.0 | 15.0 | 90.0 | 0.0 |
| 7 | [illegible] | 0.7 | 0.4 | 0.6 | 0.5 | 0.2 | 0.7 | 0.9 | 0.4 | 0.4 | 0.5 | 3.0 | 30.0 | 80.0 | 0.0 |
| 9 | [illegible] | 0.4 | 0.6 | 0.3 | 0.3 | 0.8 | 0.3 | 0.3 | 0.7 | 0.4 | 0.6 | 5.0 | 20.0 | 75.0 | 0.0 |
| 11 | [illegible] | 0.6 | 0.3 | 0.7 | 0.6 | 0.5 | 0.4 | 0.7 | 0.3 | 0.8 | 0.5 | 3.0 | 15.0 | 60.0 | 0.0 |
| 11 | [illegible] | 0.3 | 0.6 | 0.2 | 0.6 | 0.6 | 0.9 | 0.2 | 0.3 | 0.6 | 0.3 | 5.0 | 45.0 | 60.0 | 1.0 |
| 12 | [illegible] | 0.2 | 0.7 | 0.8 | 0.6 | 0.6 | 0.4 | 0.5 | 0.2 | 0.5 | 0.2 | 4.0 | 10.0 | 70.0 | 0.0 |
| 13 | [illegible] | 0.6 | 0.7 | 0.4 | 0.3 | 0.5 | 0.9 | 0.4 | 0.2 | 0.6 | 0.3 | 3.0 | 25.0 | 90.0 | 0.0 |
| 16 | [illegible] | 0.3 | 0.6 | 0.9 | 0.5 | 0.2 | 0.4 | 0.8 | 0.5 | 0.5 | 0.5 | 2.0 | 15.0 | 80.0 | 1.0 |
| 17 | [illegible] | 0.7 | 0.3 | 0.5 | 0.8 | 0.6 | 0.4 | 0.9 | 0.3 | 0.7 | 0.7 | 5.0 | 45.0 | 100.0 | 0.0 |

**Fig. 3** Dataset for training the model

Four classification methods were applied: decision tree, k-nearest neighbors, support vector machine with linear kernel, and random forest. The resulting model is able to predict and classify players with positive and negative depression diagnoses.

According to the resulting model, the optimal number of trees is 20 for the random forest method and 6 neighbors for the k-nearest neighbors method with Euclidean distance.

Different classification models were trained on the basis of gaming session data to diagnose depression in a user. Despite the resulting dataset being small and unbalanced, all models showed promising results after training and were able to distinguish gaming sessions between users with a confirmed diagnosis of severe depression and users with a negative diagnosis of severe depression.

The performance of each model was measured using the Accuracy, Precision, Recall, F1-score and F1-weighted metrics score, designed to use element-wise cross-validation (Table 2).

Hyperparameter tuning was performed using a grid search procedure. The best results were demonstrated by the SVM and Random classifiers. Forest. Graph of the resulting Decision model Tree is shown in Fig. 4.

**Table 2** Metrics for predicting depression

| | Decision Tree (%) | SVM (linear) (%) | kNN (%) | Random Forest (%) |
|---|---|---|---|---|
| Accuracy | 63.8 | 80.8 | 77 | 72.9 |
| Precision | 51.7 | 71.3 | 36.2 | 63.4 |
| Recall | 54.3 | 69 | 50 | 64.2 |
| F1 | 53.8 | 70.1 | 43.4 | 64.1 |
| F1-weighted | 67.1 | 78.4 | 62.2 | 73.8 |

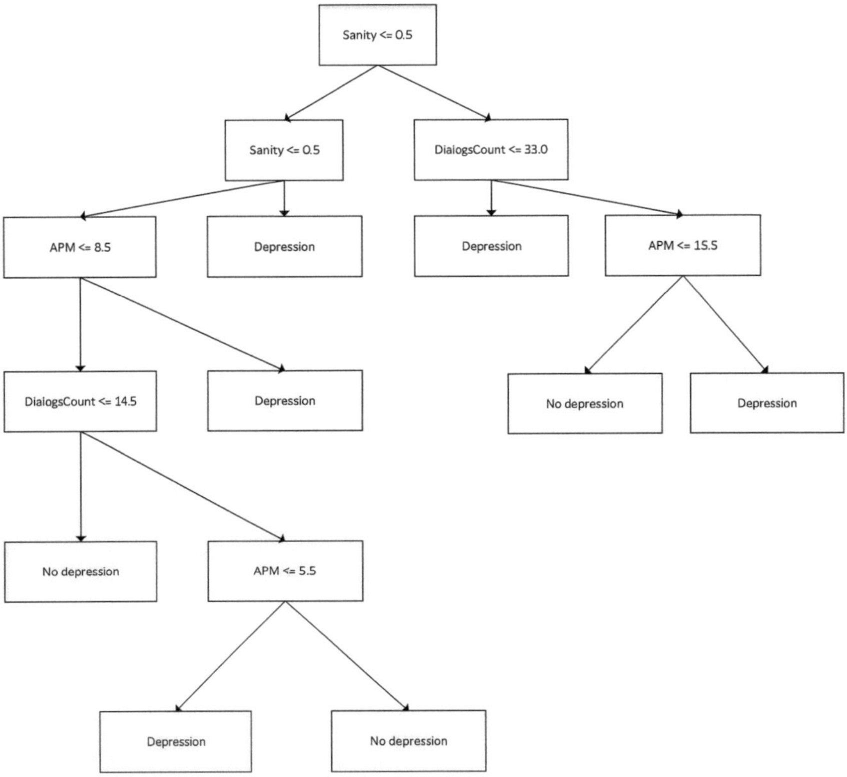


**Fig. 4** Schedule Decision Tree models

## 4 Game Development and Testing

Android operating system. The minimum supported API level is 22 (Android 5.1 “Lolipop”) [14] with support for the graphics API Vulkan, OpenGL ES 2.0 and above.

### 4.1 Game Menu

After launching the game, the user will see the main menu with the following items:

1. Start the game;
2. Settings;
3. Exit the game.

The exit point requires additional confirmation. In the settings, the user has the ability to change:

1. Sound effects;
2. Music;
3. Control scheme;
4. Language.

During gameplay, Head—Up is displayed on the screen Display (HUD) with elements:

1. Map of the game area;
2. Health;
3. Fatigue;
4. Currency;
5. Current task.

### 4.2 Built-In Mini-Games

Minigames interface is a window with an option to immediately complete the minigame without saving progress. Minigames are controlled by clicking on an area of the screen. Game logic blocks the possibility of using clicks when they can disrupt the behavior of the minigame. Minigames have animations for visual confirmation of the current stage of the game.

During the dialogue, an interface with a choice of dialogue options appears. The player is not shown the effects and results of choosing specific options. The dialogue system does not have the functionality of immediately closing the dialogue interface; exiting the dialogue is provided only after choosing one of the options. The game has implemented the functionality of saving dialogue options, as well as the functionality of branching dialogue options depending on previous answers.

The user is given a choice of 2 control schemes: with a locked area for dragging a finger across the screen and an unlocked area for dragging a finger across the screen.

The player is taught gameplay mechanics through in-game hints.

The game has 2 types of minigames:

- Memorizing a sequence of numbers;
- Memorizing pairs of numbers.

Minigame states are determined by a finite state machine.

Gameplay is not saved between game sessions. After completing the game, the game session data is saved. After completing the game, a window opens with a choice of specifying a pre-set diagnosis.

## 4.3 Development Technology Stack

The game engine is the cross-platform Unity 2021.3.21 f1 [15] (Fig. 5) development environment. It has a visual development environment and cross-platform support.

The game is being developed as a mobile application with data transfer via Firebase Realtime Database. Google packages were used to interact with the database from Unity LLCs that provide additional dependencies to allow integration of Firebase packages Realtime Database, Firebase Authentication and Firebase App (Core).

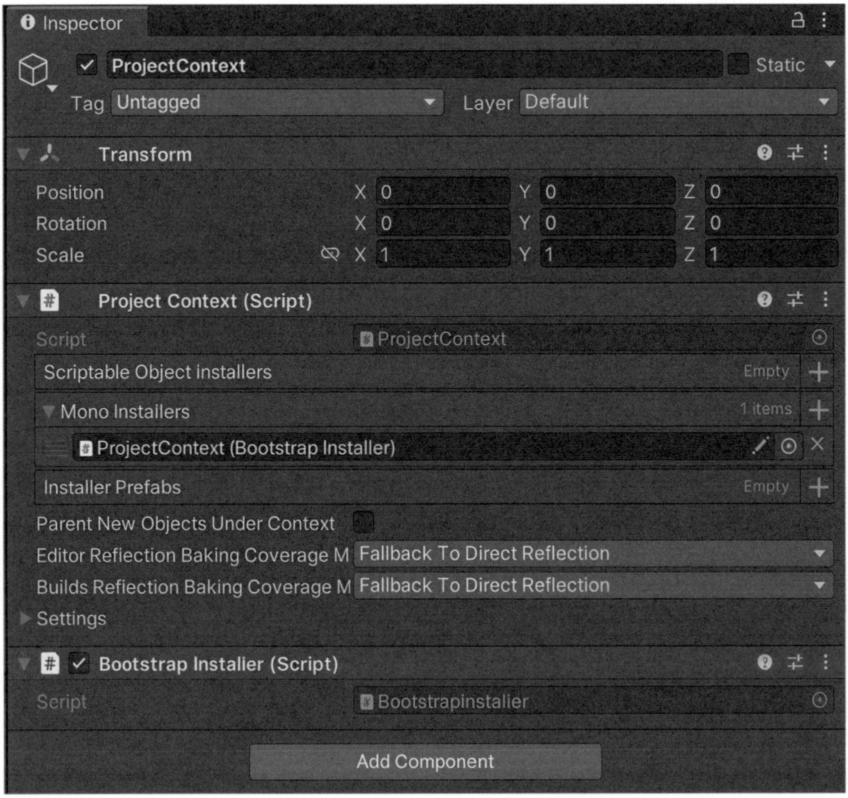


**Fig. 5** Project context

To transfer data, player model data is used, serialized using JsonUtility and transferred to the database instance. For testing, the game has implemented data display functionality, which contains the PlayerStatistics class.

To develop the game's graphics, we use the Universal Rendering Pipeline (URP, formerly LWRP)—a ready-made Unity rendering pipeline with scripting support that removes the post-processing limitations of the standard rendering pipeline and allows you to create optimized graphics for various platforms, which is especially useful when developing browser games, which place a huge emphasis on optimization.

Game scripts are written in C #, using standard Unity libraries and additional libraries from Asset Store.

The game uses Unity's physics model for the 3D environment. The game is developed with a top-down perspective view. 3D models are used for display. Rigidbody, Collider components are used for physics manipulation.

The following additional libraries were used to implement the game environment:

- 2 D Sprite is a library for working with two-dimensional images in the interface;
- ProBuilder—a library for fast creation of three-dimensional objects and prototyping;
- TextMeshPro—a library for working with text, replaces the standard components for working with text in the user interface;
- Zenject—a dependency injection framework (Fig. 6);
- Doozy is a framework for creating interfaces;
- Test Framework—testing framework;

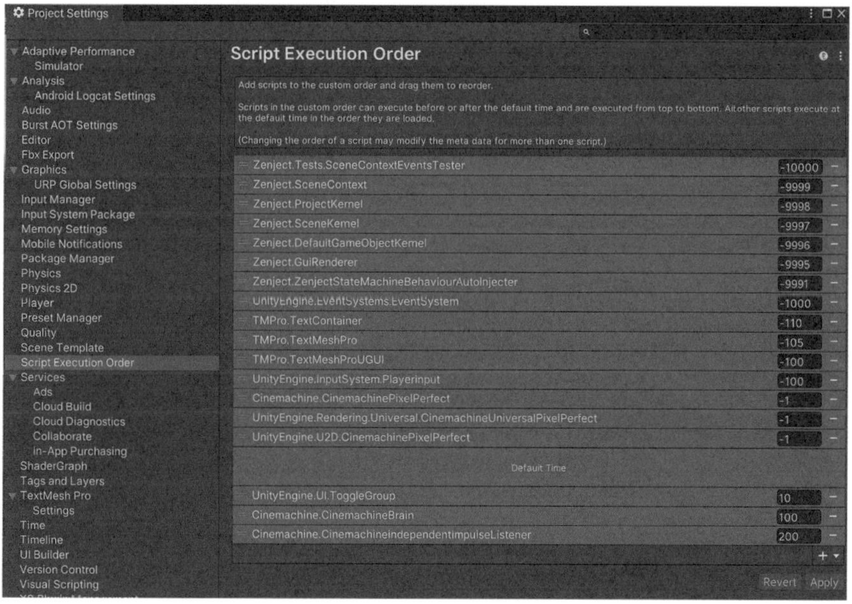


**Fig. 6** Changing the order of script initialization in Zenject

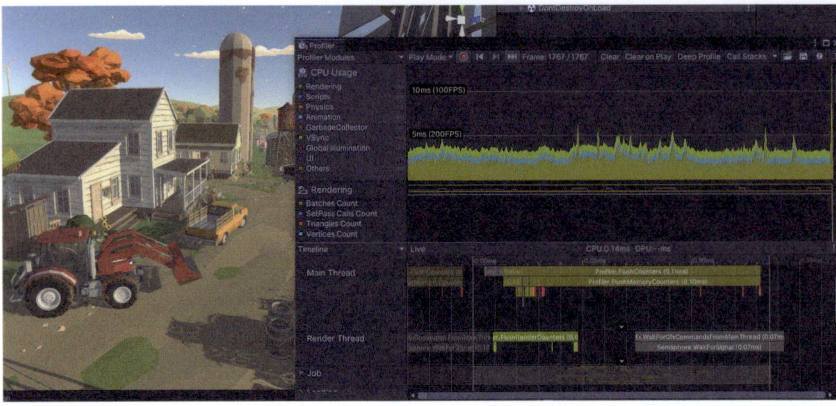


**Fig. 7** Job profiler

- GPU Instancer is a tool for monitoring GPU load.

### *4.4 Testing the Game*

One of the key technical aspects of the game is its smoothness and stability of frames per second (FPS). We strive to ensure that the game runs smoothly and without glitches at any FPS values. The minimum acceptable value for us is 30 FPS, and the recommended one is 60 FPS.

The main system load in the game is visual effects, particles and high-poly models. Lack of optimization in these areas can lead to decreased performance. For technical debugging, we use the Profiler tool in Unity, which allows you to track the game components that consume the most system resources.

Figure 7 shows an example of Profiler in action. Sharp spikes in resource usage are a sign of optimization issues that need to be addressed. The ideal situation is when the system operates stably, without spikes or fluctuations in frame rates.

## 5 Conclusion

During development, special attention was paid to the implementation of game mechanics that meet the diagnostic criteria of DSM-5. After conducting many game sessions and collecting data on the user's progress, a machine learning model was trained for diagnosis. The model showed promising efficiency in detecting depression based on user behavior. The most effective was the support vector model with a linear kernel.

In conclusion, it should be noted that this method of diagnosing depression based on user behavior analysis will help to remotely identify the possibility of the disease, but for a more accurate diagnosis, it is necessary to contact a specialist [16, 17]. As possible improvements to the method, it is proposed to use a microphone, record the user's voice, and use the AVEC 2016 model to diagnose depression and the player's mood. A larger volume of data on the gameplay, an expansion of the game scenario, and adaptation to identify other forms of depression will also have a positive impact on the development of this diagnostic method.

**Acknowledgements** I would like to express my gratitude to the graduate master student S. Khomutsov for the game technical design.

## References

1. Depressive disorder, World Health Organization. 2023. Mode access. https://www.who.int/news-room/fact-sheets/detail/depression
2. Depression (major depressive disorder), Mayo Clinic. Mode access. https://www.mayoclinic.org/diseases-conditions/depression/diagnosis-treatment/drc-20356013 (2022)
3. Carr, V.: Assessment of depression by questionnaire compared to DSM-III diagnosis. J. Affect Disorder **8**(2), 167–237 (1985)
4. Laamarti, F., Eid, M., El Saddik, A.: An overview of serious games. Int. J. Comput. Games Technol. **2014**(1), 1–15 (2014)
5. Kroenke, K., Spitzer, R.L., Williams, J.B.: The PHQ-9: validity of a brief depression severity measure. J. Gen. Intern. Med. **16**(9), 606–613 (2001)
6. DSM-5, American Psychiatric Association. Mode access. https://www.psychiatry.org/psychiatrists/practice/dsm (2022)
7. Beck Depression Inventory, American Psychological Association. Mode access. https://www.apa.org/pi/about/publications/caregivers/practice-settings/assessment/tools/beck-depression (2011)
8. Standardizing the Hamilton depression rating scale: past, present, and future. National Library of Medicine. Mode access. https://pubmed.ncbi.nlm.nih.gov/11824839/ (2001)
9. Cutting edge AI models uniquely combine three types of data, Thymia. Mode access. https://thymia.ai/#about-us (2023)
10. About elipsis health, CBInsights. Mode access. https://www.cbinsights.com/company/ellipsis-health (2024)
11. About NeuroFlow, NeuroFlow. Mode access. https://www.neuroflow.com/for-individuals/ (2024)
12. Safe and effective artificial intelligence for mental healthcare, Youper. Mode access. https://www.youper.ai/tech (2024)
13. Game Loop, Game Programming Patterns. Mode access. https://gameprogrammingpatterns.com/game-loop.html (2020)
14. Android requirements and compatibility, Unity Documentation. Mode access. https://docs.unity3d.com/Manual/android-requirements-and-compatibility.html#:~:text=Unity%20supports%20Android%205.1%20%E2%80%9CLollipop,API%20level%2022)%20and%20above (2022)
15. CrossPlatformConsiderations, Unity Documentation. Mode access. https://docs.unity3d.com/Manual/CrossPlatformConsiderations.html (2022)

16. Kravets, A., Poplavskaya, O., Lempert, L., Salnikova, N., Medintseva, I.: The development of medical diagnostics module for psychotherapeutic practice. In: Kravets, A., Shcherbakov, M., Kultsova, M., Groumpos, P. (eds.) Creativity in Intelligent Technologies and Data Science. CIT&DS 2017. Communications in Computer and Information Science, vol. 754. Springer, Cham (2017). https://doi.org/10.1007/978-3-319-65551-2_63
17. Lempert, L.B., Kravets, A.G., Lempert, B.A., Poplavskaya, O.V., Salnikova, N.A.: Development of the intellectual decision-making support method for medical diagnostics in psychiatric practice. In: 2018 9th International Conference on Information, Intelligence, Systems and Applications (IISA), pp. 1–5. Zakynthos, Greece (2018). https://doi.org/10.1109/IISA.2018.8633671

# VR Simulator for Scene Investigation Based on Real 3D Panoramas

**Anna V. Matokhina, Alla G. Kravets, Stanislav E. Dragunov, Mikhail Bugera, and Maksat A. Zhumashev**

**Abstract** Virtual reality (VR) is widely used in various fields, including training law enforcement officers. The paper discusses the directions and advantages of using VR for training law enforcement officers, provides a list and description of VR applications developed for training. This paper describes the development of a virtual reality simulator designed for crime scene inspection using 3D panoramic reconstruction. The functional requirements for the VR application are presented, the system architecture is described, including the VR application, RESTful Web API and Web client. A description and requirements for each component of the developed simulator are provided separately. Sequence diagrams and precedents for the cadet and educator on using the VR simulator and Web client are provided. A list of technical support for the VR simulator is provided, as well as tools for creating 360 photographs and a method for converting them into a 3D panorama. Methods for recognizing speech in a VR headset and creating inspection reports based on audio and photo comments are presented. The structure of the database is presented, which stores information about scenes, as well as data on the passage of cadets' task of inspecting a specific scene of an incident, including fragments of photographs taken using a VR headset and comments recorded to photographs, both voice and recognized and converted into text. The article contains the results of testing the simulator at the Volgograd Academy of the Ministry of Internal Affairs. As a result of the tests, the effectiveness of using the simulator was confirmed. At present, the Academy of the Ministry of Internal Affairs uses the developed simulator, including at the state final certification.

A. V. Matokhina · A. G. Kravets (✉) · S. E. Dragunov
Volgograd State Technical University, Volgograd, Russian Federation
e-mail: allagkravets@yandex.ru

A. G. Kravets
Dubna State University, Dubna, Russian Federation

M. Bugera
Volgograd Academy of the Ministry of Internal Affairs of Russia, Volgograd, Russian Federation

M. A. Zhumashev
Kostanay Academy of the Ministry of Internal Affairs of the Republic of Kazakhstan Named After Sh.Kabylbaev, Kostanay, Kazakhstan

 
A. G. Kravets and A. A. Bolshakov (eds.), *Human-Centric Cyber-Society*, Studies in Systems, Decision and Control 639, https://doi.org/10.1007/978-3-032-06787-6_20

**Keywords** VR simulator · Crime scene inspection · 3D panoramas · Virtual reality · Speech recognition · Personnel training · Educational technologies

## 1 Introduction

Virtual reality technologies are widely used in training, practicing skills in working with high-tech equipment and in medical practice, as well as in [1–3]. The use of VR technologies in the training of law enforcement officers has a few advantages and allows us to reduce the time of formation of professional competencies in comparison with classical approaches in training. VR technology enables near-realistic experiential learning, significantly enhancing cadet engagement and improving knowledge retention per training session [4].

The main advantages of VR simulators are [5]:

- Realism. VR tools allow you to create an environment as close to reality as possible, which significantly increases the effectiveness of training.
- Safety. Cadets can train in safe conditions, eliminating the risk of injury or other negative consequences.
- Individualization. Programs are easily adapted to the needs of a specific unit or employee, allowing you to consider the specifics of the tasks performed.
- Analysis of results. Modern VR applications provide the ability to track the progress of each cadet, identify weaknesses and adjust the training process.
- Saving resources. The use of virtual reality reduces the cost of conducting real exercises, since after implementation it does not require large material costs and the involvement of additional resources.

VR technologies in the education and training of law enforcement officers have already been successfully applied in several solutions from both Russian and foreign companies. The Russian company LikeVR provides many solutions for training specialists in various fields, including the military, using VR technologies. One example of the use of their software product in the Russian Ministry of Defense for training military personnel in parachute jumps, which eliminates fatalities during the training of the military. Also, according to the company itself, in this case, the use of VR technologies makes it possible to facilitate the determination and implementation of the flight path by 92% [6]. The company Streets Smarts provides solutions in the field of training military personnel and employees of the federal agency. Their software solution is designed to improve tactical skills, readiness to perform combat missions and decision-making in difficult conditions [7]. The company SIMX provides the US Air Force with the VALOR software solution for training military medical specialists. The training program includes a full range of injuries and diseases and covers the entire spectrum of medical care from the place of injury to fourth-role institutions [8].

Important tools for creating realistic and immersive virtual environments for VR simulators are 3D panoramas, or as they are also called 360° photographs [9]. These photographs allow you to reproduce detailed virtual spaces that accurately imitate real conditions, be it city streets, premises or natural landscapes. The use of 3D panoramas provides full immersion of the user in the environment, allowing you to view the space in all directions, which is especially valuable for training that requires high situational awareness. Thus, 3D panoramas significantly increase the effectiveness of training, allowing law enforcement officers to practice their skills in conditions as close as possible to real ones, which contributes to better preparation for the performance of official tasks.

Thus, the purpose of the study is to develop a VR simulator for training employees to conduct an inspection of the scene of the incident.

## 2 VR Simulator for Viewing Crime Scene Photographs

The software was created considering the following functional requirements:

- Identification upon entry. To analyze the inspection of the scene performed by the cadet, it is necessary to store all photo fragments and audio comments in one group, which requires identification data that allows you to link all the data into one protocol.
- Creation of audio comments. During the inspection of the scene in the VR simulator, the cadet must leave comments on what he saw for the protocol, for which audio files must be recorded from the microphone of the VR headset. The recorded audio files must be saved on the server.
- Creation of photo comments. Also, during the inspection of the scene in the VR simulator, the cadet must take photos of important details for inspection, for which screenshots of the 3D panorama area selected by the cadet must be created and saved on the server.
- Zooming in on the examined area of the 3D panorama. In some cases, at the scene of the incident, it is necessary to examine small objects that are difficult to see in the photograph, so the student must be able to zoom in on the area of the 3D panorama to obtain larger photo fragments.
- Examination of the photograph from the scene of the incident. To fully conduct an inspection of the scene of the incident, it is necessary to study the 3D panorama in all details, for which the VR simulator must provide the ability to freely move the camera in all planes.

## 3 Technical Specifications of the VR Application

The VR application has the following functional requirements:

- Identification upon entry. Since the software being developed will also be used to train cadets, it is necessary to ensure identification of cadets upon entry into the VR application. The data entered by the cadet will be used to save the audio and photo materials created by the cadet under his identification data.
- Creation of an audio commentary. During the inspection of the 3D panorama from the scene, the cadet will pronounce out loud all stages of the inspection, as well as say the details that he noticed during the inspection. These audio comments must be saved under the cadet's identification data for their subsequent transcription based on the received text of the inspection protocol of the scene of the incident.
- Creation of a photo commentary. Conducting an inspection of the scene of the incident also implies the creation of photo materials, which must be fully reflected in the application being developed. Photo comments created by the cadet will also be saved under the cadet's identification data, and a document containing the photo materials taken will be generated based on this data.
- Zooming in on parts of the crime scene photographs. In some cases, it is necessary to examine small objects at the crime scene that may be difficult to see in the photograph, so the student must be able to zoom in on parts of the crime scene photograph; inspection of the crime scene photograph. To conduct a full inspection, it is necessary to inspect the 3D panorama from the crime scene in all details, so inspection of the 3D panorama must allow the camera to be moved on the spot in all directions to carry out the inspection.

### *3.1 Photos from the Scene*

To train cadets to conduct the process of inspection of crime using the developed software, it is necessary to first make a 3D panorama from real crime scenes. For these purposes, the Insta360 Pro 2 camera was used [7]. This camera allows you to get pictures with a viewing angle of 360° and in 8 K resolution, which allows you to use them in a VR headset as a spatial environment in the virtual world, to simulate the inspection of the scene of the crime. Photos with a viewing angle of 360° are converted into a sphere around the user in the VR headset, as if stitching the left and right edges of the photo together and bringing the top and bottom edges of the photo together into two points, respectively. Figures 1 and 2 show examples of photos taken with this camera.

### *3.2 Speech Recognition in VR Headset*

As part of the inspection of the scene of the incident, the employee must keep the appropriate protocol. When using a VR headset, it is currently impossible to make a sufficiently accurate simulation of handwriting in the virtual space to implement this stage, so the recording of cadets' voice comments is used, with their subsequent

**Fig. 1** Example of 360° photograph from the scene of the theft

**Fig. 2** Example of 360° photograph from the scene of serious bodily harm

transcription into text and the formation of the required protocols. The cadet himself determines the start of recording an audio comment, no more than 10 min long, after which the generated audio file is sent to the server for processing. On the server, the audio file is transcribed by a neural network into text and saved under the identification data previously entered by the cadet. To transcribe audio into text, the vosk-model-ru-0.42 neural model is used, distributed under the Apache 2.0 license [10]. This model is specially trained to recognize audio in Russian. The version used is intended for transcribing audio into text on the server side, as it requires a lot of computing resources. This model was chosen because it is specifically focused on audio with Russian speech, a license that allows free use of the model in commercial

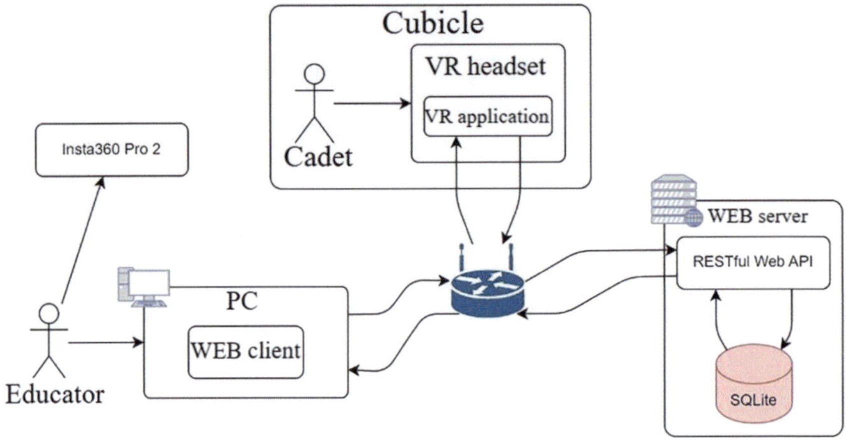


**Fig. 3** Architectural diagram of the VR simulator

software, convenient integration with many programming languages, and sufficient accuracy of audio-to-text transcription for the required purposes.

### 3.3 Architecture of VR Simulator

The VR simulator being developed consists of three software components: a VR application, a RESTful Web API application, and a Web client. The VR application runs on a VR headset and is used by students to inspect the scene of an incident [11]. The RESTful Web API application runs on a specialized Web server and accepts requests from both the VR application and the Web client. This component also interacts with a database that stores data on sessions, 3D panoramas, photos, and audio comments. The Web client runs on the teacher's personal computer and allows him to view students' sessions with inspections of the scene of an incident, as well as download protocols for inspection of the scene of an incident based on the sessions. Figure 3 shows the architectural diagram of the VR simulator.

### 3.4 Description of the VR Simulator Operating Sequences

Figure 4 shows a diagram of precedents for a student and a teacher for using the VR simulator and Web client, respectively [12, 13].

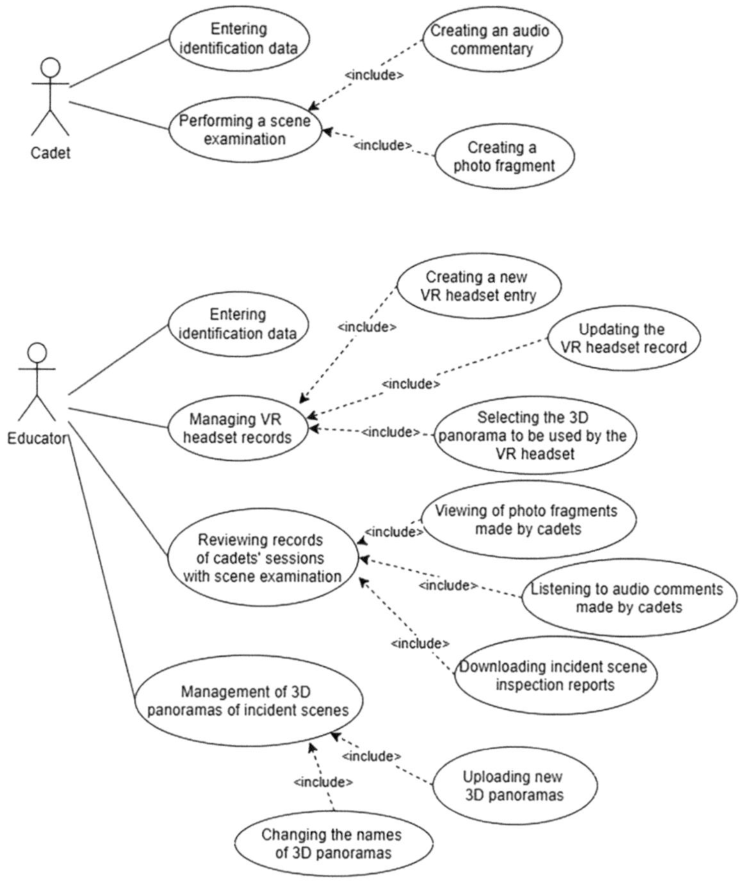


**Fig. 4** USE-CASE diagram for the VR simulator

## 4 RESTful Web API Application Functionality

RESTful Web API application is a link between the VR application and the Web client, so it must provide methods for saving, changing, creating and deleting all the necessary data, as well as storing data in the database, transcribing audio files into text and generating protocols. Let's describe the requirements in more detail:

- Methods for creating, updating and receiving VR headsets. Since the Web API will receive and send data to the VR headset, there must be corresponding methods for creating a record of the VR headset, updating data on the VR headset, and also receiving a complete list of all registered VR headsets.

- Registration and authorization for the educator. To verify the comments made by the cadet on the photographs from the scene, it is necessary to create an account for the teacher, as well as authorize and further provide access, only for authorized users, to the comments of the students about the scene.
- Creation, receiving a page-by-page list, receiving a separate file of 360 photographs from the scene. These methods will allow teachers to create records with new 360 photos from the scene, as well as assign a 360 photo for each VR headset separately for inspection of the scene. There should also be a method for obtaining a 3D panorama for the VR application, in accordance with the photo previously selected by the teacher for a specific VR headset; methods for creating, viewing and attaching photos and audio comments to the cadet's session. After identifying the cadet in the VR application, a record of a new session for the cadet should be created in the Web API application based on the transmitted identification data. Photos and audio comments created by the cadet will be attached to the created session. Also, methods should be provided for the authorized teacher to view all data on the cadet's final session and generate inspection protocols for the scene based on this data.

## *4.1 Web Client Functionality*

The web client should provide convenient access to the Web API application methods. Let's describe the requirements in more detail:

- registration and authorization form for teachers;
- page for viewing, creating and updating records about VR headsets;
- page for viewing cadet sessions and generating protocols based on them;
- page for creating and viewing 360 photos.

## *4.2 Database Structure*

Since the expected volume of stored data will be small, it is proposed to use the SQLite database. It is proposed to store cadets' audio and photo comments as files, and store paths to the corresponding files in the database tables. Figure 5 shows the ERD diagram of the database.

The diagram below provides a description for all the necessary tables: the examiners table for storing authorization data about teachers, the vr_headsets table for storing information about VR headsets, the scenes table for storing information about 360 photos from the scene, the sessions table for storing information about a student session with links to the vr_headsets and scenes tables, the comments table for storing information about audio comments with a link to the corresponding student session, and the photos table for storing information about photo comments with a link to the corresponding student session.

**Fig. 5** ERD diagram of the database

## 4.3 *Implementation of a VR Application*

Pico NEO 4 was used as the target VR headset, and Unity 2021.3.32 was used as the game engine for creating the VR application. The following libraries were used to use the capabilities of the VR headset: Unity.XR.Interaction.Toolkit 2.3.1 and Pico.Unity.Integration.SDK 2.5.0.

The right gamepad implements zooming in and out of the camera when the stick is deflected, starting to record an audio commentary when the trigger is pressed, and starting to record a photo commentary when pressing Grip.

The left gamepad implements moving the camera in space when the stick is deflected and exiting to the main menu when pressing the Primary button.

After identifying the cadet, a request is sent to the Web API to obtain a 3D panorama installed for this VR headset.

## 4.4 *Implementing a RESTful Web API Application*

The Web API application is implemented using C# 12,.NET 8 and ASP.NET Core 8 technologies. Entity Framework 8 is used to work with SQLite. The standard ASP.NET Core implementation of authorization using a JWT token is used to authorize teachers. The vosk-model-ru-0.42 neural network is used to transcribe audio files into text. Table 1 lists the methods implemented in the RESTful Web API application.

**Table 1** Methods of the RESTful Web API application

| Method | Endpoint method | Description |
|---|---|---|
| Post | /api/examiners/register | Registering a new account for a teacher |
| Post | /api/examiners/login | Authorization of the teacher account |
| Post | /api/scenes | Create a new entry about 3D panorama |
| Get | /api/scenes/page | Getting a list of 3D panoramas page by page |
| Get | /api/scenes | Getting a 3D panorama for a VR headset |
| Get | /api/scenes/{id}/preview | Getting a thumbnail image of a 3D panorama for preview |
| Post | /api/sessions | Create a new session record |
| Get | /api/sessions/page | Getting a list of sessions page by page |
| Post | /api/sessions/{id}/photos | Creating a photo commentary for a session |
| Get | /api/sessions/photos/{id}/screenshot | Getting a separate photo of a session |
| Get | /api/sessions/photos/{id}/audio | Getting a separate audio file created with a photo commentary |
| Post | /api/sessions/{id}/comments | Creating an audio commentary |
| Get | /api/sessions/photos/{id}/audio | Getting an audio file for an audio commentary |
| Post | /api/sessions/{id}/comments | Creating an audio commentary |
| Get | /api/sessions/photos/{id}/audio | Getting an audio file for an audio commentary |
| Post | /api/sessions/{id}/comments | Creating an audio commentary |
| Get | /api/sessions/comments/{id}/audio | Getting an audio file for an audio commentary |
| Get | /api/sessions/{id}/protocol | Getting a protocol for a selected session |
| Get | /api/sessions/{id} | Getting a session record |
| Delete | /api/sessions/{id} | Deleting a session record |
| Post | /api/vr-headsets | Creating a record about a VR headset |
| Put | /api/vr-headsets/{id} | Changing a record about a VR headset |
| Get | /api/vr-headsets | Getting a list of records about VR headsets |

## *4.5 Implementation of a Web Client*

The web client is also implemented using NET 8 and Blazor WebAssembly. The web client is launched together with the Web API application. To create screen forms, standard CSS, HTML and JavaScript tools are used together with WebAssembly technology, which allows writing most of the code in C# 12. Figure 6 shows the web client screen forms.

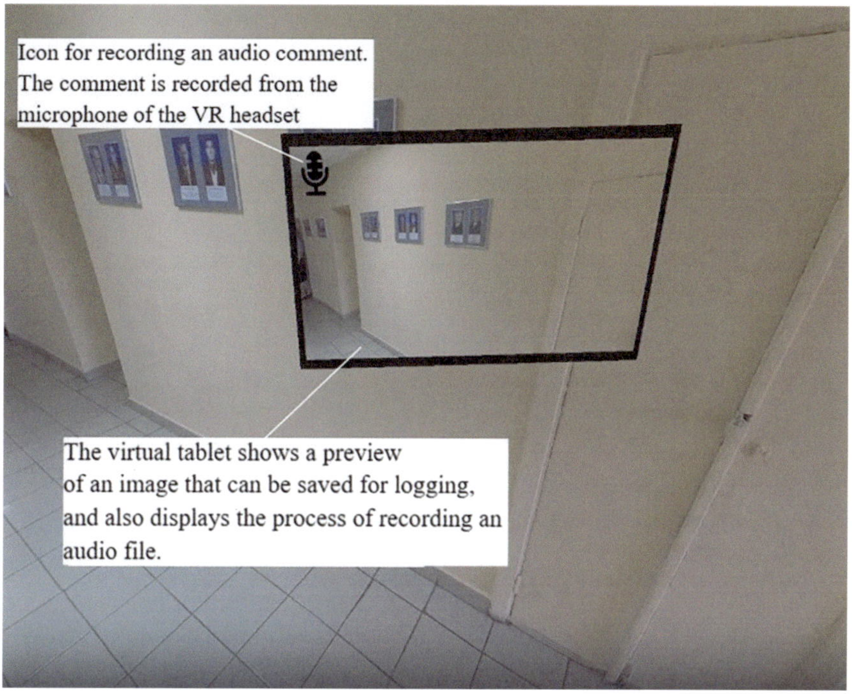


**Fig. 6** Screen form of the VR application

## 5 Results of Testing the Simulator in the Academy of the Ministry of Internal Affairs

The Volgograd Academy (VA) of the Ministry of Internal Affairs conducted a test run of the VR simulator in the implementation of the main programs of professional training in the academic disciplines taught at the Department of Preliminary Investigation of VA. To conduct the test, a training ground "Production of investigative actions using VR technologies" containing five individual workplaces was equipped.

To create 3D panoramas from the scenes of incidents and form a database of photo locations, an agreement was concluded between the VA of the Ministry of Internal Affairs and the Main Directorate of the Ministry of Internal Affairs of Russia for the Volgograd Region on the possibility of the department's employees traveling together with the investigative and operational group to shoot real scenes of incidents.

The capabilities of the VR simulator have been tested and are used in organizing and conducting practical classes, as well as in the course of independent training of cadets (students) in academic disciplines provided by the Department of Preliminary Investigation of the Educational and Scientific Complex for Preliminary Investigation in the Internal Affairs Bodies of the VA of the Ministry of Internal Affairs of Russia.

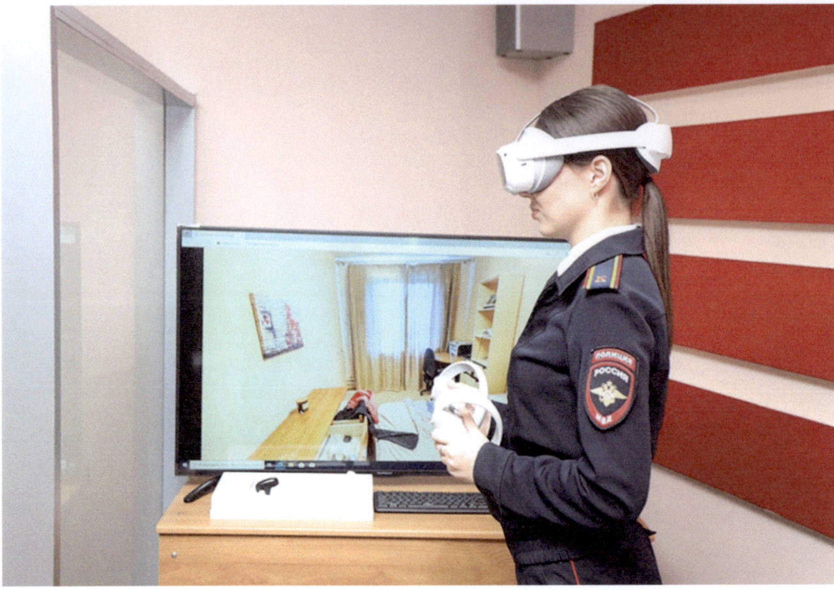

**Fig. 7** Cadet using VR simulator

In 2024, the VR simulator was first used at the state final certification of cadets in the specialty 40.05.01 Legal Support for National Security of the 2019 recruitment. Figures 7 and 8 show photographs from the final exams.

**Fig. 8** Conducting group training

## 6 Conclusion

The VR simulator for inspection of crime scenes has been developed in accordance with all functional requirements provided by the VA MIA. The simulator allows for realistic training in conditions as close to real as possible, which helps improve the quality of training of law enforcement officers. The system received positive feedback during testing at the Volgograd Academy of the Ministry of Internal Affairs and was used at the state final certification.

The application development prospects include expanding its functionality, for example: attaching several 3D photographs to one crime scene and adding the functionality of switching between 3D photographs of one crime scene. This will improve efficiency and detail, as well as realism. In addition, it is possible to add the functionality of joint inspection of the crime scene by several cadets, or to connect additional clarifying comments to the elements of the photograph [14].

## References


1. Application of VR technologies in the army and military training. VR APP. https://vr-app.ru/blog/primenenie-vr-texnologii-v-armii-i-voennoi-podgotovke/. Accessed 24 Nov 2024
2. Karagiannis, P., et al.: Operators training using simulation and VR technology. Procedia CIRP **96**(12), 290–294 (2021)

3. Matokhina, A.V., Dragunov, S.E., Volodina, D.A., Fokin, R.O.: VR simulators for teaching rapid prototyping and material processing techniques. In: Saxena, A.K., Parygin, D., Ater, D., Yadav, V. (eds.) 2019 8th International Conference System Modeling and Advancement in Research Trends (SMART), pp. 357–362, College of Computing Science and Information Technology, Tirthanker Mahavir University, Moradabad, India. IEEE (2019)
4. Spitsyn, P.A., Bavulya, A.A.: New level of training promising forms of training cadets using virtual reality tools. Bull. Mil. Train. **4**(31), 99–104 (2021)
5. Gryaznov, S.A.: Use of virtual reality technologies in the training of law enforcement officers. Bull. Samara Law Inst. **1**(42), 101–105 (2021)
6. VR for the army. LikeVR. https://likevr.ru/vr-industry/vr-dlya-armii/#project-3. Accessed 24 Nov 2024
7. Military Solution. Street Smarts. https://www.streetsmartsvr.com/military. Accessed 24 Nov 2024
8. Virtual reality operational military medical training. SIMX. https://www.simxvr.com/virtual-reality-medical-simulation-training-for-military/. Accessed 24 Nov 2024
9. Insta360 Pro 2—360 VR Camera | Professional 8K 360 Camera | 3D. Insta360. https://www.insta360.com/ru/product/insta360-pro2/. Accessed 24 Nov 2024
10. VOSK Models. Alpha Cephei. https://alphacephei.com/vosk/models. Accessed 24 Nov 2024
11. Kravets, A.G., Pavlenko, I.D., Egunov, V.A., Kravets, E.: Interaction with virtual objects in VR-applications. In: Kravets, A.G., Shcherbakov, M.V., Groumpos, P.P. (eds.) Creativity in Intelligent Technologies and Data Science: 5th International Conference CIT&DS 2023 (Volgograd, Russia, September 11–15, 2023): Proceedings, pp. 433–449. Springer Nature Switzerland, Cham (2023). https://doi.org/10.1007/978-3-031-44615-3_30 (Book ser.: Communications in Computer and Information Science (CCIS); vol. 1909)
12. Pavlenko, I.D., Egunov, V.A.: Study of the efficiency of methods for tracking the position of a helmet and virtual reality controllers. Informatiz. Commun. **3**, 110–119 (2023). https://doi.org/10.34219/2078-8320-2023-14-3-110-119
13. Pavlenko, I.D., Egunov, V.A.: Modeling virtual reality scenarios: efficiency assessment and practical results. Digit. Econ. **4**, 33–40 (2023)
14. Kravets, A.G., Matokhina, A.V., Plotnikov, I.N., Salnikova, N.A.: A new approach to personifying scenarios for a single-player 3D game in the Soulslike style. Softw. Eng. **15**(7), 372–384 (2024). https://doi.org/10.17587/prin.15.372-384

MIX
Papier aus verantwortungsvollen Quellen
Paper from responsible sources
FSC® C105338

If you have any concerns about our products,
you can contact us on
**ProductSafety@springernature.com**

In case Publisher is established outside the EU,
the EU authorized representative is:
**Springer Nature Customer Service Center GmbH**
**Europaplatz 3, 69115 Heidelberg, Germany**

Printed by Libri Plureos GmbH
in Hamburg, Germany